I0063629

Nanoelectronic Materials, Devices and Modeling

Nanoelectronic Materials, Devices and Modeling

Special Issue Editors

Qiliang Li
Hao Zhu

MDPI • Basel • Beijing • Wuhan • Barcelona • Belgrade

MDPI

Special Issue Editors

Qiliang Li
National Institute of Standards and
Technology, George Mason University,
Fairfax, VA, USA

Hao Zhu
School of Microelectronics,
Fudan University,
Shanghai, China

Editorial Office
MDPI
St. Alban-Anlage 66
4052 Basel, Switzerland

This is a reprint of articles from the Special Issue published online in the open access journal *Electronics* (ISSN 2079-9292) from 2018 to 2019 (available at: https://www.mdpi.com/journal/electronics/special_issues/nano_elec)

For citation purposes, cite each article independently as indicated on the article page online and as indicated below:

LastName, A.A.; LastName, B.B.; LastName, C.C. Article Title. *Journal Name* **Year**, *Article Number*, Page Range.

ISBN 978-3-03921-225-5 (Pbk)
ISBN 978-3-03921-226-2 (PDF)

© 2019 by the authors. Articles in this book are Open Access and distributed under the Creative Commons Attribution (CC BY) license, which allows users to download, copy and build upon published articles, as long as the author and publisher are properly credited, which ensures maximum dissemination and a wider impact of our publications.

The book as a whole is distributed by MDPI under the terms and conditions of the Creative Commons license CC BY-NC-ND.

Contents

About the Special Issue Editors

Qiliang Li (professor) received his B.S. in physics from Wuhan University, China, in 1996, and his M.S. in physics from Nanjing University, China, in 1999. He received his Ph.D. in electrical and computer engineering from North Carolina State University in 2004. His doctoral research was in the area of hybrid silicon/molecular field effect transistors and memories. In October 2004, he joined the Semiconductor Electronics Division of the National Institute of Standards and Technology (NIST), Gaithersburg, MD, as a research scientist, where he was involved in the fabrication, characterization, and simulation of advanced CMOS and nanoelectronics materials and devices. In August 2007, he joined the faculty of George Mason University as an assistant professor. He was promoted to tenured associate professor in May 2012 and professor in May 2017 in the Department of Electrical and Computer Engineering.

Hao Zhu (associate professor) received his B.S. and M.S. degrees in physics from Nanjing University, China, in 2007 and 2010, respectively, and his Ph.D. degree in electrical engineering from George Mason University in 2013. Afterwards, he worked as a research faculty professor in the Department of Electrical and Computer Engineering at George Mason University. From 2011 to 2015, he was a guest scientist at the National Institute of Standards and Technology (NIST), Gaithersburg, MD, USA. In October 2016, he joined the faculty of Fudan University, where he is currently an associate professor of microelectronics.

electronics

MDPI

Editorial

Nanoelectronic Materials, Devices and Modeling: Current Research Trends

Hao Zhu [1] and Qiliang Li [2,3,*]

1   State Key Laboratory of ASIC and System, School of Microelectronics, Fudan University, Shanghai 200433, China; hao_zhu@fudan.edu.cn
2   Quantum Materials Center, Department of Electrical and Computer Engineering, George Mason University, Fairfax, VA 22030, USA
3   Physical Measurement Laboratory, Nanoscale Device Characterization Division, National Institute of Standards and Technology, Gaithersburg, MD 20899, USA
*   Correspondence: qli6@gmu.edu

Received: 17 May 2019; Accepted: 21 May 2019; Published: 22 May 2019

1. Introduction

As CMOS scaling is approaching the fundamental physical limits, a wide range of new nanoelectronic materials and devices have been proposed and explored to extend and/or replace the current electronic devices and circuits so as to maintain progress in speed and integration density [1]. The major issues, including low carrier mobility, degraded subthreshold slope, and heat dissipation, have become worse as the size of the silicon-based metal oxide semiconductor field effect transistors (MOSFETs) decreased to nanometers while the device integration density increased. High electron mobility transistors (HEMTs) based on wide bandgap semiconductors, such as silicon carbides (SiC) and gallium nitrides (GaN) [1], are proposed to enhance the carrier mobility for high-speed logic devices. The HEMTs are also very attractive for high-power and high-frequency applications. While conventional semiconductors were studied to improve the current electronics, a new phase of materials is being explored and tested for new-concept devices. For example, topological insulators that have insulating bulk and gapless surfaces have exhibited unique properties for transistor applications [2].

2. The Current Research Trends

Each of the seventeen articles collected in this special issue proposes a solution to a specific problem related to the above-mentioned major challenges. The carrier mobility is a very good and convincing indicator in improving transistor performance. GaN vertical FETs with an additional back current blocking layer have been proposed and simulated for high-power electronics [3]. AlGaN/GaN metal–insulator–semiconductor HEMTs have been studied and exhibit a high breakdown voltage and an on–off current ratio [4]. The two-dimensional electron gas of an $In_xAl_{1-x}N/AlN/GaN$ HEMT has been studied and modeled by considering the polarization and quantum mechanical effects [5]. A steep subthreshold slope (SS) is another target for transistor performance improvement in switching speed. This issue collected two approaches to achieve a steep SS: (1) using the insulator–metal phase transition of VO_2 to achieve a decent SS of 42 mV/dec [6]; (2) using an L-shaped tunneling FET to improve the SS [7]. In addition, the drain-induced barrier lowering (DIBL) effect and leakage of a partial isolation FET for sub-0.1 μm have been studied [8].

New concepts of data storage and memory devices are another focus of this issue. A partial isolation type saddle-FinFET has been proposed for sub-30 nm DRAM applications [9]. A new method for neural networks based on resistive switches has been proposed for pattern storage and recognition [10]. A CMOS-compatible Ag/HfO_2-based synaptic was studied for application in an artificial neuromorphic system [11]. New analog memristive characteristics and conditioned reflex have been reported in Au/ZnO/ITO devices [12].

In addition to the research in transistors and memory devices, this issue has collected important research on solar cells based on ZnO/Si heterojunctions [13], Bi-doped and Bi-Er co-doped optical fibers [14], high-performance graphene electrolyte double-layer capacitors [15], quantum-dot and sample-grating semiconductor optical amplifiers [16], a transmission method to determine the complex conductivity of thin strips [17], and a high-efficiency CMOS power amplifier with a dual-switching transistor [18].

3. Future Trends

The future research in nanoelectronic materials and devices will continue to find the solutions to address the challenges of current electronics in switching speed, power consumption, and heat dissipation. New device concepts and new materials will be carried over to future nanoelectronics to enhance or replace the current devices. A new growing interest is the integration of nanomaterials and devices into smart systems for stand-alone applications. For example, a robotic vessel equipped with vision sensors [19] and a smart nanoelectronics sensor system governed by machine-learning intelligence will be of great interest to the academic society and industry.

Author Contributions: Q.L. and H.Z. worked together during the whole editorial process of the special issue entitled "Nanoelectronic Materials, Devices and Modeling" published in the MDPI journal Electronics. Z.H. and Q.L. drafted, reviewed, edited, and finalized this editorial summary.

Acknowledgments: We thank all the authors who submitted excellent research works to this special issue. We are very grateful to all reviewers for their evaluations of the merits and quality of the articles and valuable comments to improve the articles in this issue. We would also like to thank the editorial board and staff of MDPI journal Electronics for the opportunity to guest-edit this special issue.

Conflicts of Interest: The authors declare no conflict of interest.

References

1. Zeng, F.; An, J.X.; Zhou, G.; Li, W.; Wang, H.; Duan, T.; Jiang, L.; Yu, H. A Comprehensive Review of Recent Progress on GaN High Electron Mobility Transistors: Devices, Fabrication and Reliability. *Electronics* **2018**, *7*, 377. [CrossRef]
2. Yue, C.; Jiang, S.; Zhu, H.; Chen, L.; Sun, Q.; Zhang, D.W. Device Applications of Synthetic Topological Insulator Nanostructures. *Electronics* **2018**, *7*, 225. [CrossRef]
3. Huang, H.; Li, F.; Sun, Z.; Sun, N.; Zhang, F.; Cao, Y.; Zhang, H.; Tao, P. Gallium Nitride Normally-Off Vertical Field-Effect Transistor Featuring an Additional Back Current Blocking Layer Structure. *Electronics* **2019**, *8*, 241. [CrossRef]
4. Geng, K.; Chen, D.; Zhou, Q.; Wang, H. AlGaN/GaN MIS-HEMT with PECVD SiN$_x$, SiON, SiO$_2$ as Gate Dielectric and Passivation Layer. *Electronics* **2018**, *7*, 416. [CrossRef]
5. Qin, J.; Zhou, Q.; Liao, B.; Wang, H. Modeling of 2DEG characteristics of In$_x$Al$_{1-x}$N/AlN/GaN-Based HEMT Considering Polarization and Quantum Mechanical Effect. *Electronics* **2018**, *7*, 410. [CrossRef]
6. Tabib-Azar, M.; Likhite, R. Nano-Particle VO$_2$ Insulator-Metal Transition Field-Effect Switch with 42 mV/decade Sub-Threshold Slope. *Electronics* **2019**, *8*, 151. [CrossRef]
7. Najam, F.; Yu, Y.S. Optimization of Line-Tunneling Type L-Shaped Tunnel Field-Effect-Transistor for Steep Subthreshold Slope. *Electronics* **2018**, *7*, 275. [CrossRef]
8. Kim, Y.K.; Lee, J.S.; Kim, G.; Park, T.; Kim, H.; Cho, Y.P.; Park, Y.J.; Lee, M.J. Simulation Analysis in Sub-0.1 μm for Partial Isolation Field-Effect Transistors. *Electronics* **2018**, *7*, 227. [CrossRef]
9. Kim, Y.K.; Lee, J.S.; Kim, G.; Park, T.; Kim, H.J.; Cho, Y.P.; Park, Y.J.; Lee, M.J. Partial Isolation Type Saddle-FinFET(Pi-FinFET) for Sub-30 nm DRAM Cell Transistors. *Electronics* **2019**, *8*, 8. [CrossRef]
10. Velichko, A.; Belyaev, M.; Putrolaynen, V.; Boriskov, P. A New Method of the Pattern Storage and Recognition in Oscillatory Neural Networks Based on Resistive Switches. *Electronics* **2018**, *7*, 266. [CrossRef]
11. Chen, L.; He, Z.-Y.; Wang, T.-Y.; Dai, Y.-W.; Zhu, H.; Sun, Q.-Q.; Zhang, D.W. CMOS Compatible Bio-Realistic Implementation with Ag/HfO$_2$-Based Synaptic Nanoelectronics for Artificial Neuromorphic System. *Electronics* **2018**, *7*, 80. [CrossRef]

Electronics **2019**, *8*, 564

12. Cheng, T.; Rao, J.; Tang, X.; Yang, L.; Liu, N. Analog Memristive Characteristics and Conditioned Reflex Study Based on Au/ZnO/ITO Devices. *Electronics* **2018**, *7*, 141. [CrossRef]
13. Hussain, B.; Aslam, A.; Khan, T.M.; Creighton, M.; Zohuri, B. Electron Affinity and Bandgap Optimization of Zinc Oxide for Improved Performance of ZnO/Si Heterojunction Solar Cell Using PC1D Simulations. *Electronics* **2019**, *8*, 238. [CrossRef]
14. Uddin, R.; Wen, J.; He, T.; Pang, F.; Chen, Z.; Wang, T. Ultraviolet Irradiation Effects on luminescent Centres in Bismuth-Doped and Bismuth-Erbium Co-Doped Optical Fibers via Atomic Layer Deposition. *Electronics* **2018**, *7*, 259. [CrossRef]
15. Huffstutler, J.D.; Wasala, M.; Richie, J.; Barron, J.; Winchester, A.; Ghosh, S.; Yang, C.; Xu, W.; Song, L.; Kar, S.; Talapatra, S. High Performance Graphene-Based Electrochemical Double Layer Capacitors Using 1-Butyl-1-methylpyrrolidinium tris (pentafluoroethyl) trifluorophosphate Ionic Liquid as an Electrolyte. *Electronics* **2018**, *7*, 229. [CrossRef]
16. Qasaimeh, O. Multichannel and Multistate All-Optical Switch Using Quantum-Dot and Sample-Grating Semiconductor Optical Amplifier. *Electronics* **2018**, *7*, 166. [CrossRef]
17. Shahpari, M. Determination of Complex Conductivity of Thin Strips with a Transmission Method. *Electronics* **2019**, *8*, 21. [CrossRef]
18. Kurniawan, T.A.; Yoshimasu, T. A 2.5-GHz 1-V High Efficiency CMOS Power Amplifier IC with a Dual-Switching Transistor and Third Harmonic Tuning Technique. *Electronics* **2019**, *8*, 69. [CrossRef]
19. Yuan, H.; Xiao, C.; Xiu, S.; Zhan, W.; Ye, Z.; Zhang, F.; Zhou, C.; Wen, Y.; Li, Q. A Hierarchical Vision-Based UAV Localization for an Open Landing. *Electronics* **2018**, *7*, 68. [CrossRef]

© 2019 by the authors. Licensee MDPI, Basel, Switzerland. This article is an open access article distributed under the terms and conditions of the Creative Commons Attribution (CC BY) license (http://creativecommons.org/licenses/by/4.0/).

electronics

MDPI

Review

A Comprehensive Review of Recent Progress on GaN High Electron Mobility Transistors: Devices, Fabrication and Reliability

Fanming Zeng [1], Judy Xilin An [1,*], Guangnan Zhou [1], Wenmao Li [1], Hui Wang [1], Tianli Duan [2], Lingli Jiang [1] and Hongyu Yu [1,3,4,*]

[1]   Department of Electrical and Electronic Engineering, Southern University of Science and Technology, Shenzhen 518055, China; zengfm@sustc.edu.cn (F.Z.); 11761002@mail.sustc.edu.cn (G.Z.); 11410438@mail.sustc.edu.cn (W.L.); pabol2006@163.com (H.W.); jiangll@sustc.edu.cn (L.J.)
[2]   Materials Characterization and Preparation Center, Southern University of Science and Technology, Shenzhen 518055, China; duantl@sustc.edu.cn
[3]   Shenzhen Key Laboratory of the Third Generation Semi-conductor, Shenzhen 518055, China
[4]   GaN Device Engineering Technology Research Center of Guangdong, Shenzhen 518055, China
*   Correspondence: anjx@sustc.edu.cn (J.X.A.); yuhy@sustc.edu.cn (H.Y.); Tel.: +86-0755-88018508 (H.Y.)

Received: 31 October 2018; Accepted: 27 November 2018; Published: 3 December 2018

Abstract: GaN based high electron mobility transistors (HEMTs) have demonstrated extraordinary features in the applications of high power and high frequency devices. In this paper, we review recent progress in AlGaN/GaN HEMTs, including the following sections. First, challenges in device fabrication and optimizations will be discussed. Then, the latest progress in device fabrication technologies will be presented. Finally, some promising device structures from simulation studies will be discussed.

Keywords: AlGaN/GaN; high-electron mobility transistor (HEMTs); p-GaN; enhancement-mode

1. Introduction

Nitride-based wide band-gap semiconductors (WBS), such as GaN and related alloys, have been intriguing to high power and high frequency researches and applications over the past two decades, as reviewed in the literatures [1–10]. They are promising because of their intrinsic superior material properties, especially with their outstanding electrical performance exhibited in the GaN-based high-electron mobility transistors (HEMTs), also known as the heterojunction field-effect transistor (HFET). GaN-based HEMTs on silicon substrate are particularly appealing to the IC industry due to its compatibility with the industry-matured Si-CMOS IC technologies. Tremendous research and development work has been conducted and reported in the recent years with significant progresses on GaN-on-Si HEMTs covering the full scope of a new IC-chain, including the industrial acceptable-low defect quality of GaN-on-Si epitaxial materials, the optimized GaN-based HEMT devices and integration [11–17], the significantly enhanced reliability [18–24], the comprehensive circuit and device modeling [25–27] and the product designs [19,28–34].

The GaN-based HEMTs possess better performance than the silicon counterparts, such as they can be operated at higher frequency, higher power and higher temperature conditions. These advantages are benefited from III-nitrides' high electron saturation velocity, high breakdown electrical field, high electron mobility and high carrier density of the two-dimensional electron gas (2DEG) formed at the heterointerface between AlGaN and GaN layers in the GaN HEMTs.

GaN power transistors have been in volume production since 2010. AlGaN/GaN HEMT, however, is inherently normally-ON (depletion-mode, D-mode) as it was first reported in 1993 [35]. Thus their primary applications having been initially focused on the low-voltage and high-frequency applications.

For the power switching applications, on the other hand, normally-OFF (enhancement-mode, E-mode) transistors are required. It is not only for the concern of safety but also necessary to reduce the current leakages thus minimize the power loss, simplify the driving circuit and improve the device stability. Ever since the initial demonstration of the enhancement-mode AlGaN/GaN HEMT in 1996 [36] and furthermore in 2000 [37], several approaches have been used to make normally-OFF GaN-based HEMTs as reviewed previously [38] and more recently [8,39]. Besides the hybrid approach with GaN HV-HEMT and Si LV-MOSFET in a cascade configuration, AlGaN/GaN-on-Si HEMTs with direct GaN gate control [40,41] have attracted much attention commercially in the recent years due to its size thus cost. GaN-on-Si is acknowledged as a promising device platform for further exploration of commercial high-power modules with higher power density [4,9,33,34]. Moreover, owing to the availability of both depletion-mode and enhancement-mode of GaN HEMTs, they have invoked research interests and efforts in the IC industry for complementary logic applications [9,31].

This paper reviews the recent progress in the GaN-on-Si normally-OFF AlGaN/GaN HEMTs based on recent literature. The following aspects will be covered: devices in Section 2, device fabrication in Section 3 and some promising structures from simulation studies in Section 4.

2. AlGaN/GaN HEMT Device

For conventional semiconductor materials, the free charges come from the impurity ionization. In an AlGaN/GaN heterojunction, the polarization effects give rise to a high density ($>1 \times 10^{13}$ cm^{-2}) of electron gas at the interface between AlGaN and GaN, even without intentional doping. The AlGaN/GaN HEMTs can provide not only high electrical conductivity with low on-state resistance but also low input and output capacitance, thus inherently promising for high voltage and high frequency applications.

A conventional Schottky gate D-mode GaN HEMT device structure is schematically shown in Figure 1a. The source and drain metal stacks facilitate the Ohmic contacts. The gate metal stack provides a Schottky contact. As a voltage bias of V_{DS} is applied between the drain and source, a lateral electrical field is built and the 2DEG under the gate flows along the channel of AlGaN/GaN heterojunction as HEMT's current, I_{DS}. If the gate voltage is below the threshold voltage of HEMT while the drain is biased high, the device runs into the block region.

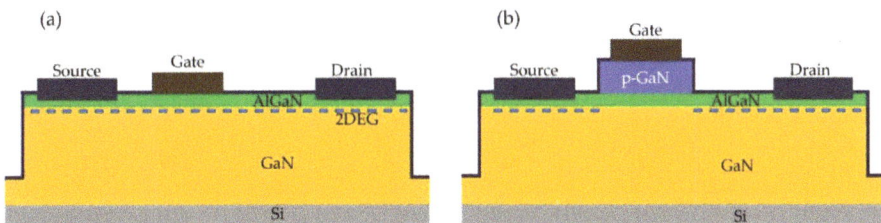

Figure 1. Structures of the GaN HEMTs, (**a**) the D-mode device and (**b**) the p-GaN E-mode device.

When the AlGaN/GaN HEMT in Figure 1a is inserted with a p-type GaN layer between the gate electrode and the AlGaN barrier layer, it forms a so-called p-GaN E-mode HEMT as shown in Figure 1b. With sufficiently high p-type doping (e.g., Mg) and an adequately thick p-GaN layer, the 2DEG below the gate would be depleted, which lead to a positive threshold voltage for the normally-OFF operation. Meanwhile, the low on-state resistance and high driving capability can be preserved with the 2DEG in the region between the gate edge and the drain. More advanced research work has been done in order to solve dynamic on-state resistance issues in GaN HEMTs. One of these successful examples has the current collapse suppressed up to 850 V [42]. There are also several varieties of p-GaN HEMTs as thoroughly reviewed in References [8,39], such as the gate injection transistor (GIT) [42].

Several advanced types of AlGaN/GaN FET devices have been explored recently. Among of them, the GaN MOS or MIS (metal-insulator-semiconductor) HEMTs has drawn most of the attention [9,14,28,43], so do the MISHEMTs with high-K gate dielectric [17,31,44]. Other new devices with a regrowth of AlGaN layer [41], or a regrowth of GaN drift channel layer [45] have also been reported. The MIS HEMT has the advantages over the Schottky gate GaN HEMT in terms of gate leakage and gate swing [46]. The gate dielectrics reliability, however, is one of the most important aspects to be well investigated and understood before it can be commercialized.

The performance of some recently published p-GaN HEMT devices [19,42,47,48] is compared and shown in Figure 2. It depicts the p-GaN HEMTs with excellent specific on-resistance $R_{on,sp}$ as low as a few m$\Omega \cdot$cm^2 while with the breakdown voltages larger than the theoretical limit of silicon.

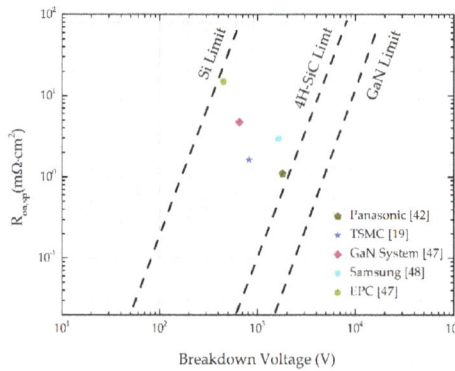

Figure 2. Specific on-resistance versus device breakdown voltage of the p-GaN HEMTs.

Figure 3 demonstrates the preliminary DC characteristics of the fabricated p-GaN E-mode AlGaN/GaN HEMT. The device gate width is about 500 μm. The metal stack for the source and drain Ohmic contacts was formed with Ti/TiN/Al/TiN and was annealed at 650 °C for 30 s. And the metal stack for Schottky contact for the p-GaN gate was formed with Ti/Al. The detailed process information will be discussed in a subsequent publication. The measurements were carried out with keithley 4200-SC. The transfer characteristics (I_{DS}-V_G) of the device is shown in Figure 3a and the output characteristics (I_{DS}-V_{DS}) of the device is illustrated in Figure 3b. The threshold voltage (V_{th}) of the fabricated device is about 1.5 V and the gate voltage swing is up to 10 V.

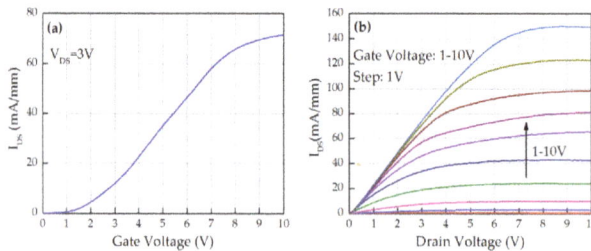

Figure 3. DC characteristics of a p-GaN HEMT, (**a**) the transfer characteristics, (**b**) the I-V output characteristics.

One of the challenges of making reliable p-GaN gate enhancement-mode HEMTs is to obtain controllable V_{th} with a reasonable variation across the wafer. It is found that V_{th} is sensitive to the p-GaN gate profiles, p-GaN layer residues, AlGaN layer thickness around the gate edges and also surface morphology of the AlGaN layer after the p-GaN etching. The threshold voltage instability can be explained by the charge control model due to the charge-transferring effect [49,50]. The enhancement of threshold voltage in E-mode AlGaN/GaN devices is mainly due to the space charge alteration under the gate, while defects can alter the charges via trapping and de-trapping processes during device operations. The fixed defects in the GaN and AlGaN layers can act like traps offering electron transferring paths between the barrier layer and the substrate, which can generate potential shift and energy band bowing, thus V_{th} drift. Therefore, the reduction of defects is mandatory for reducing the charge variations so as to control the threshold voltage instability and reliability. Achieving low defects in III-nitrides layers, however, requires careful process engineering and optimizations.

Regarding to the device reliability of GaN-on-Si HEMTs [51], it has been recently figured out that the epitaxial quality of materials is related to a key industrial reliability item: the high temperature reverse bias (HTRB) stress-induced on-state drain current degradation [18]. It was confirmed [18] that the optimization of epitaxial layers could significantly improve the device reliability of AlGaN/GaN HEMTs.

We have also studied the dielectric failure based on the time dependent dielectric breakdown (TDDB) measurement at different temperatures with statistical Weibull analysis [24]. The lifetime of the devices with 35-nm-thick LPCVD SiN_x gate dielectric was predicted to have a 10-year time-to-breakdown. This work also demonstrated the impacts of gate dielectric area and multi-fingers on the SiN_x TDDB characteristics.

3. AlGaN/GaN HEMT Device Fabrication

Device processing technologies play important roles to achieve the full-potential and remarkable features of AlGaN/GaN HEMTs. The process modules for the GaN HEMT device fabrication include device isolation, p-GaN gate formation, contacts for source and drain, contact for gate, surface passivation and so forth. In this section, we will focus on some of the key process technologies for the fabrication of p-GaN E-mode AlGaN/GaN HEMTs, discuss some of the challenges and review recent progresses.

3.1. Device Isolation

Device isolation is the fundamental process to separate the electrical connection of adjacent devices, so as to minimize the impact of current leakage from their neighbors. Mesa dry etching and ion implantation are two of the commonly used and effective isolation approaches as shown in Figure 4.

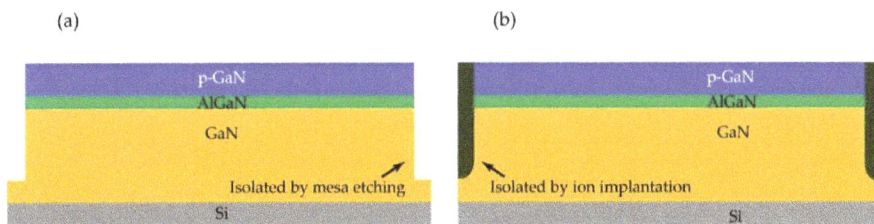

Figure 4. Schematic cross-section of (**a**) mesa etching and (**b**) ion implantation for device isolation.

3.1.1. Mesa Etching

The mesa etching step is to form an isolated area for each device and the material is removed physically around the devices by this process, as shown in Figure 4a. The etching depth is crucial for the isolation results or device leakage current. Under-etch of the mesa could result in an incomplete isolation, thus the devices performance would be impacted. The conventional wet etch techniques are not suitable for GaN device fabrication. The large bond energies and wide band-gaps make them highly resistant to acid and alkaline solutions at room temperature. Moreover, the wet etching of GaN shows isotropic profile and a slower etching rate than the dry etching techniques. Therefore, most of III-nitrides etching processes are conducted by dry etching [52].

Dry etching can be carried out by reactive ion etch (RIE), electron cyclotron resonance (ECR), inductively coupled plasma (ICP), electron cyclotron resonance (ECR), magnetron reactive ion etching (MIE) and so forth. [53]. RIE is one of the common methods for III-nitrides dry etching but it has slow etching rate, low degree of anisotropy and large surface damages due to the low plasma density and high operating pressure inherent in this technique [54]. Whereas ICP is expected to produce the etching results with low damages, high etching rates and high uniformity. Thus ICP etching has become the dominant dry etching technique for III-nitrides.

The dry etching rate can reach to the level of micron per minute handily [55,56]. However, many factors need to be considered to evaluate the dry etching quality of GaN and AlGaN, such as the surface morphology, sidewall profile, etching anisotropy, material selectivity, material damage, etching uniformity and so forth. These etching features can be affected by the etching conditions, including reactive chemistries and their fluxes, ICP antenna RF power, bias RF power, chamber pressure, chamber temperature, chamber configuration, etching technique and so forth.

The basic chemistries employed in the III-nitrides dry etching are chlorine based, such as Cl_2 and BCl_3 [57]. Moreover, Ar [52], N [54], H [58] and F [59] based chemistries can also be added to achieve better etching rate, selectivity and surface morphology [60]. Study [57] showed that the etching rate was found to increase with the ICP antenna RF power up to a certain level (e.g., 600 W). But further increased in antenna RF power could decrease the etching rate. It is noticed [54] that a high antenna RF power would cause the decrement of the DC bias voltage and thus resulting in a reduction of the ion-bombardment physical effect introduced by dry etching. High energy ion-bombardment would lead to surface roughness, material surface damages and low etching selectivity. Wakejima et al. [61] reported that the reasonably low bias RF power can be used to get high selectivity and low surface damage. Moreover, chamber pressure also has an impact on the etching rate. Increase chamber pressure (within the range of less than 10 mTorr) can help raise the etching rate. But further increase chamber pressure would not keep increasing the etching rate effectively. The mean free path of reacting molecules is short and the plasma density is low at high pressure. Re-deposition and polymer formation on the surfaces would also be favored at high pressure, which is undesirable [57].

3.1.2. Ion Implantation

Ion implantation is the other isolation process which is promising to product high performance GaN based HEMTs devices [15]. Ion implantation is an adding process, as shown in Figure 4b. The ion species can be used to isolate devices including H, He, N, P, Ar, O and so forth [62–67]. In the implantation process, atoms are added into the nitrides by means of energetic ion beam injection. Ion implantation has a better ability to control both of the dopant concentration and depth. The concentration can be adjusted by the ion beam current and the implantation duration, while the depth can be controlled by the ion energy.

The ion implantation is a room-temperature process; thus photoresist can be used as the mask. The ion species that implanted can be exactly selected by the mass analyzer, thus the contamination risk can be reduced. Moreover, the ion implantation is an anisotropic process, the side boundary of the implanted ions in the semiconductors can be straight. Thus the isolated area can be precisely defined. However, for the purpose of minimizing the channeling effect, the wafers are always placed with a tilted angle of about $7°$.

Comparing to the mesa etching process, the ion implantation creates the isolation boundary under the surface, which can protect the device sidewalls. Therefore, better device electrical performance and reliability are expected.

3.2. P-GaN Gate Formation

To achieve the E-mode GaN based HEMTs, the structure of p-GaN HEMTs has been widely used [51]. As shown in Figure 5, the basic structure of the p-GaN HEMT consists of the p-GaN cap layer, AlGaN barrier layer, GaN buffer layer, Si substrate and so forth. The energy bands of AlGaN are lifted up by the p-GaN cap layer, leading to the depletion of the 2DEG underneath. The p-GaN should be removed except for the gate area. The dry etching is widely used for this step. One of the key factors to achieve high performance device and high yield is the selectivity etching of GaN over AlGaN. The selectivity of Cl_2/BCl_3 plasma between GaN and AlGaN is unable to meet the requirement. Therefore, two approaches have been recommended to achieve this process of selective dry etching. One is the fluorine based chemistries etching technique and the other is oxygen based chemistries etching technique.

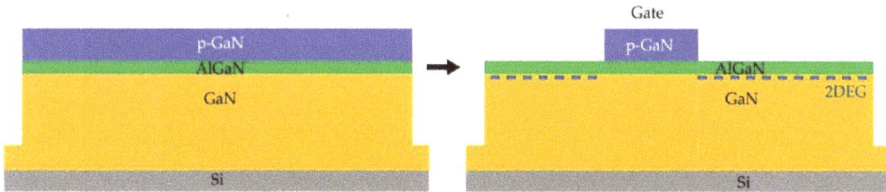

Figure 5. Gate area definition of a p-GaN HEMT.

The etching selectivity of compound semiconductor films containing Al can be optimized in Cl-based plasmas by adding F-containing gases. For example, the BCl_3/SF_6 selective dry etching on AlGaN/GaN, the decrease in AlGaN etching rate is due to the formation of non-volatile AlF_3 residues on AlGaN surface. This non-volatility of AlF_3 reduces the etching efficiency of chlorine, achieving selective etching of GaN over AlGaN.

The other option is to use the O_2 and chlorine based gases. Shawn et al. [68] demonstrated a cyclical two-step etching technique using separate O_2 and BCl_3 plasmas. The low power oxygen plasma creates an oxidized layer on the surface of AlGaN/GaN and then the low-power BCl_3 plasma is used to remove the surface oxide. The etching rate per cycle is about 0.7–2.5 nm, depending on the process parameters. Wong et al. [69] investigated highly selective dry etching with O_2 and chlorine gases. Their study showed that the O_2 flow has evident impacts on the etching rate and selectivity of AlGaN/GaN structure. Increasing O_2 flow (0–5 sccm) resulted in decrease etching rates for both GaN and AlGaN. The selectivity was increased at the O_2 flow range of 0–2 sccm and then descended when O_2 flow arrived at 3 sccm. Moreover, they reported that the surface roughness was improved with a higher O_2 flow.

To achieve high etching selectivity, some new structures have been developed. Chiu et al. [70] reported that they added an AlN etching stop layer between p-GaN and AlGaN, that is, the p-GaN/AlN/AlGaN/GaN structure, to improve the p-GaN etching selectivity. With the BCl_3 and CF_4 gas mixture, the etching rate can be slowed down remarkably on AlN layer due to the non-volatility of AlF_3. Then they obtained a better uniformity of V_{th} compared with the conventional structure. The standard deviation of V_{th} from the devices with AlN stop layer was about 0.06 V, whereas the data from the traditional ones was 0.2 V. Moreover, they further employed N_2O as the oxidant [71] with the same AlN stop layer and achieved further improved uniformity. The p-GaN was oxidized to Ga_2O_3 by N_2O plasma and then the Ga_2O_3 was removed using HCl solution in one cycle. After p-GaN was fully removed, the AlN etching stop layer was exposed and oxidized to Al_2O_3 by N_2O plasma. Since the etching rate of this Al_2O_3 in HCl solution was extremely low, the selective etching of AlGaN/GaN was achieved. They claimed that the uniformity of device on-state resistance (R_{on}) and the surface leakage current were improved both by this technique.

For the purpose of controlling the uniformity of threshold voltage, reducing the device leakage current and improving device reliability performance, more efforts have to be made to achieve smooth surface morphology, low material damages, high etching selectivity and outstanding etching uniformity.

3.3. Ohmic Contacts for Source and Drain

The Ohmic contacts provide the access points for the device to connect with external circuitry. Their resistance should be very low with respect to that of the channel drift region to reduce the device specific on-resistance. It is rather challenging to make good Ohmic contacts on GaN based materials, owing to the wide band-gap which naturally facilitates Schottky contacts. For the high performance p-GaN HEMT, the source/drain Ohmic contact resistance (R_c) should be minimized. Thus the work function metal layers, the thickness of metal layers, annealing temperature, semiconductor doping level, recess depth of AlGaN layer and so forth. are all important factors to be optimized. The structure of source and drain contacts are displayed in Figure 6.

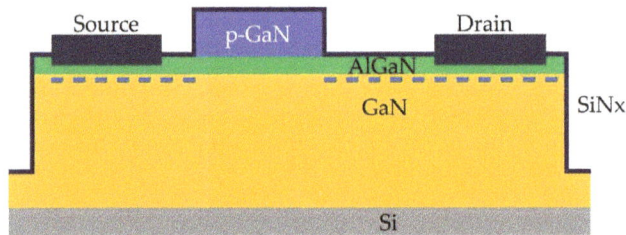

Figure 6. Ohmic contacts structure for source and drain of a p-GaN HEMT.

Single metal layer is not commonly used such as Ti or Al, due to their chemical activity with oxidant, especially for high power devices which often work at high temperatures. Therefore, different metal stacks with multiple layers have been proposed and explored. These metal stacks can be categorized into two groups, Au-based metal stacks and Au-free metal stacks. Au-based metal stacks have been studied for many years and become mature gradually. The most used stacks include Ti/x/x/Au structures [72], where x represents metals such as Al, Ni, Ti, Mo and so forth. On the other hand, the Au-free metal stacks are highly recommended recently for CMOS compatible process. The process compatibility with current Si wafer line is a fundamental factor of whether the GaN based devices can be fabricated in foundries, or even in the reused 6 or 8 inch Si wafer process lines. This means metals such as Au and Cu need to be excluded from the front-end-of-line.

Ti/Al/Ni/Au is widely used for source and drain Ohmic contacts. Studies [73,74] showed that the formation of Ohmic contacts for Ti based metal stacks is due to the reaction between Ti and GaN. TiN is formed at the interface by which N vacancies are created as donors to increase the carrier concentration resulting in a thinner barrier for electron tunneling [73]. In the Ti/Al/Ni/Au metal stack, thickness of each metal layer needs to be optimized to achieve low contact resistance and smooth surface. Ti/Al thickness ratio in the metal stack is one of the key factors for the formation of low contact resistance [75]. Research showed that for a given Ni/Au thickness ratio, the lowest contact resistance is obtained at Ti/Al ratio of 1/6 when annealing at 900 °C in N_2 ambient [76]. Moreover, the total thickness of Ti/Al also strongly influences the final contact resistance. Generally, the contact resistance below 0.6 $\Omega \cdot$mm can be commonly obtained in the Ti/x/x/Au metallization.

Au-free processes have been reported for many kinds of metal stacks. Beyond the consideration of compatibility with the Si-wafer-process-line, the research of Piazza et al. [77] demonstrated that long term thermal stress of Ti/Al/Ni/Au metal stack could led to a dramatic degradation of the structure. This could be a barrier for the high power applications of the GaN based devices. Lee et al. [11] reported Ti/Al/W metal stack for Ohmic contacts. The contact resistance was lower than 0.5 $\Omega \cdot$mm with the annealing temperature of 870 °C. The surface morphology of the Ti/Al/W metal stack was much smoother than that of the Ti/Al/Ni/Au metal stack.

Annealing temperature of metal stack is important for the formation of Ohmic contacts. Most Au-based and Au-free metal structures are annealed around 800 °C [72,78], while some work reported low temperature (lower than 700 °C) annealing to achieve Ohmic contacts [79,80].

Liu et al. [81] reported the Hf/Al/Ta metal stack on InAlN/GaN with the annealing temperature of 600 °C. The results from TEM and SIMS measurements revealed contacts with smooth metal-semiconductor interface and showed the formation of Hf-N and Hf-Al alloys near the interface. They achieved a specific contact resistance (ρ_c) of 6.7 $\pm$ 0.58 $\times$ 10^{-6} $\Omega\cdot cm^2$ at 600 °C annealing temperature. Lin et al. [82] presented a low Ohmic contacts resistance formed by sidewall contacts with Ohmic recess. They used a Ta/Al/Ta metal stack with the annealing temperature of lower than 600 °C. For the recess depth of 10 nm, they achieved contact resistance of lower than 0.25 $\Omega\cdot mm$. Ti/Al/TiN metal stack is also used for Ohmic contacts. Firrincieli et al. [80] employed the Ti/Al/TiN metal stack annealing at 550 °C in N_2 ambient and achieved a contact resistance of 0.62 $\Omega\cdot mm$. This metal stack was also shown to pass a high current of 0.6 A/mm when 10 V bias was applied.

Recently, Yoshida and Egawa [83] reported Ti/Al/W Ohmic contacts on AlGaN/GaN. The contact resistance was about 0.358 $\Omega\cdot mm$ with a low annealing temperature of 500 °C. They found that thinner Ti layer can lead to a lower annealing temperature, because Al needs to diffuse through the Ti layer in order to make Ohmic contact with AlGaN. Pozzovivo et al. [84] studied the effects of a special $SiCl_4$ plasma treatment prior to the Ohmic metallization. They achieved a sufficiently low contact resistance of 0.7 $\Omega\cdot mm$ at a reduced anneal temperature of 600 °C. The contact surfaces of AlGaN were directly treated by the $SiCl_4$ plasma for 15 s in RIE with self-induced bias of about 300 V. They suggested that the $SiCl_4$ plasma treatment of AlGaN surface would help to enhance the generation of N vacancies, which may be responsible for the high carrier concentration at the AlGaN surface to form a low contact resistance. The work from Graff et al. [75] exhibited consistent results of low contact resistance with the pre-treatment of $SiCl_4$.

The low annealing temperatures and the Au-free metal stacks would be beneficial to the device electrical and reliability performance [83] and are also compatible with foundry CMOS processes which is plausible for transferring to large scale production.

3.4. Contact for Gate

For the p-GaN E-mode HEMTs, the Schottky contacts on gate are generally adopted because it is expected to have lower gate leakage current and higher thermal stability. Figure 7 illustrates the gate structure of a typical p-GaN HEMT. The metal-semiconductor Schottky barrier height is depending on the difference between the work function of metal and the electron affinity of semiconductor. Metal with lower work function should give a higher barrier height for p-GaN contact. The Ti/Al [85], Ni/Au [86], W [87] and so forth. are used for gate metals. Greco et al. [85] studied the impacts of the thermal budget on the Schottky gate metals of Ti/Al. They found annealed with 800 °C of Ti/Al stack resulted in a higher gate leakage current than that of annealed at lower temperature. The decrease of the Schottky barrier height from 2.08 to 1.60 eV with annealing temperature increased to 800 °C was attribute to the structural modification occurring at the interface, which explained the increase of the leakage current.

Tapajna et al. [88] studied the Schottky gate reliability under forward bias stress of a p-GaN HEMT. They concluded that the generation-recombination centers were created during stress due to the defect percolation process. This stress-induced generation of defects formed the leakage path across the structure once sufficient defects are generated. Wu et al. [89] revealed that the gate breakdown phenomenon could be explained by the avalanche multiplication in the space charge region of the Schottky metal/p-GaN junction. Tallarico et al. [90] concluded that the time to failure is a function of the initial Schottky gate leakage and the breakdown mechanism is ascribed to the percolation path in the p-GaN layer. Stockman et al. [91] studied the Schottky gate leakage current under the reverse gate bias. Two efficient methods to suppress the perimeter-dependent leakage current were found. One is by improving Schottky/p-GaN interface quality to reduce the supply of carriers from the gate metal, the other is by making proper passivation of the p-GaN sidewalls to reduce the interface states and surface roughness.

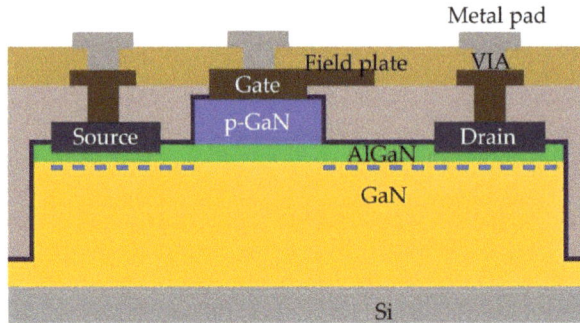

Figure 7. Schematic cross-section for gate structure of a p-GaN HEMT.

Yu et al. [92] also concluded that the gate leakage current is likely due to a highly doped and/or highly defective surface layer, resulting in carrier transport by tunneling across the Schottky barrier. Lu et al. [86] demonstrated the results of employing surface treatment and post-gate annealing to improve the Schottky gate leakage. The surface treatment reduced the lateral surface leakage by removal of the surface traps. The vertical gate tunneling current, on the other hand, was lowered through improved Schottky contact quality after thermal annealing. The off-state leakage current was reduced by about 7 orders after the implementation of the two-step treatment.

Besides the approaches of using the Schottky contacts, an Ohmic gate contact is suggested by Panasonic [42]. They reported that the hole-injection from the p-GaN gate into the GaN channel could have a positive impact on the p-GaN gate module.

As we can see, the gate metallization for p-GaN E-mode HEMTs has profound impacts on the device performance. For Schottky gate contacts, low gate leakage current and high thermal stability are required.

3.5. Surface Passivation

Although significant progress has been achieved to improve the performance of GaN devices, noticeable gaps still remain between demonstrated device performance in the real market applications and the theoretical expectations. The existence of large amount of surface states in AlGaN and GaN have been found to be the root cause of device current collapse and some reliability issues, which is one of the main challenges in fabricating high performance AlGaN/GaN HEMTs [93–97]. These surface states may come from dangling bonds of the surface atoms, defects at the surface as grown, plasma damages during the processes, the foreign contaminations and so forth. And they behave like carrier traps thus can seriously deteriorate the device performance. Furthermore, the charged surface states can act as a virtual gate and lead to the depletion of channel electrons, thus decreasing the channel 2DEG density [98]. The trapped charges at the surface can also contribute to the remote Coulomb scattering to the channel electrons. Therefore, surface passivation should be carefully optimized to prevent negative effects on device performance such as current collapse, frequency dispersion, large dynamic on-resistance and so forth.

The basic structure of surface passive layer is illustrated in the Figure 6. Many different insulation materials have been explored as the surface passivation layer for AlGaN/GaN HEMTs, including SiO_2 [93,99], Si_3N_4 [100], ZrO_2 [94,101], HfO_2 [95,96], Ga_2O_3 [43,102], AlN [103–106], Sc_2O_3 [107], TiO_2 [108], ZnO_2 [109], NiO [110], Ta_2O5 [111], Al_2O_3 [112–114], AlON [43] and so forth. In addition to the surface passivation, dielectric-free passivation technologies have also been proposed and demonstrated, for example, oxygen plasma oxidization [115,116], ozone oxidization [117], chemical oxidization [118], SiH_4 treatment [119] and so forth. AlN is found to be a good surface passivation layer which can create a sharp interface and suppress GaN surface oxidation effectively [120]. Recently, Panasonic developed a passivation technology with AlON layer [43]. The AlON layer was deposited by atomic layer deposition (ALD) possessing a better performance with less process damage, less extra fixed charges and less electron traps in comparison with Al_2O_3 layer. Moreover, O_2 annealing after the AlON deposition could further reduce the Al/Ga dangling bonds at the surface.

It is known that GaN can be oxidized in the air forming a layer of native gallium sub-oxide (GaO_x) on the surface. Defect states associated with the GaO_x further exacerbate gate leakage current and lead to drain-current collapse at high frequency [121–123]. Dong et al. [124] have proposed a method to reduce the surface states by changing the GaO_x morphology via an elevated-temperature. The ordering of native GaO_x layer can be improved towards the structure of bulk Ga_2O_3, therefore surface states can be reduced. Bae et al. [125] used a low-temperature treatment to enhance the quality of GaO_x formed at the SiO_2/GaN interface and thus reduce the interface defect density. Therrien et al. [126] demonstrated a microscopic mechanism that could reduce the defect density at the high-K/GaN interface through the formation of a high-quality Ga_2O_3 layer during remote-plasma-assisted oxidation. It should be noted that the high dielectric constant of the surface passivation layer can also increase the parasitic capacitance between the gate and source (C_{pgs})/drain (C_{pgd}) metals and thus decrease the device's speed [127]. In summary, surface passivation with dielectric layers should be carefully selected to improve device performance, eliminate the current collapse and reduce the dynamic on-resistance.

3.6. Field Plates

As we have mentioned above, GaN based HEMTs suffer the effect of current collapse [128], which can temporary increase the dynamic on-resistance due to the charge trapping effects. One reason is that the strong electric field at the gate edge on the drain side could enhance the charge trapping in the states at the interface between passivation and III-nitrides. Moreover, the hot-electron injection effect is considered to be the other reason for the current collapse [129]. Surface passivation [130] and field-plate [131] are two effective solutions for this issue.

The employment of a field plate on dielectric layer of the AlGaN/GaN HEMT has brought some of the most significant and exciting improvements on current collapse [129]. The function of the field plate is to redistribute the electric field profile and decrease the electric field peak value, hence reducing trapping effect and increasing breakdown voltage, as presented in Figure 7. Recently, Wong et al. [132] demonstrated a novel asymmetric field plate structure. The electric field was appropriately distributed by the slanted field plate. They obtained a uniform electric field distribution with the slant field plate, achieved a reduced dynamic on-resistance of 2.3 Ω·mm at the drain voltage of 50 V with a high breakdown voltage up to 138 V. Ma et al. [133] studied the effects of surface treatment and field plate structures on the current collapse of the AlGaN/GaN HEMTs. Their results showed that the gate field plate predominantly reduced the emission time constant of the trapped electron and O_2 plasma treatment decreased the density of traps. And they confirmed that the current collapse can be mitigated by the combination of these two methods.

4. AlGaN/GaN Device Simulation

The technology-computer-aided-design (TCAD) simulations are often applied to study the AlGaN/GaN HEMT. The focus is mainly on the E-mode mechanism as well as the methods of increasing the threshold voltage of the HEMTs [134,135]. And the simulations have also been used to study the gate leakage current [136,137], the device reliability [138], the short channel effects [139], current collapse [140] and so forth. For the simulation of the AlGaN/GaN HEMTs, the most important work is to solve the 2DEG concentration which is generated by spontaneous polarization (P_{sp}) and piezoelectric polarization (P_{pe}) [141], as shown in Figure 8. The two polarizations are influenced by many factors, such as the materials of substrate and epitaxial layers, crystallographic plane orientation of III-nitrides, AlN composition of AlGaN layer, thickness of AlGaN layer, relaxation of

AlGaN layer, doping concentration in AlGaN layer, thickness of GaN cap layer, fabrication process and so forth [142–149]. For the p-GaN HEMTs simulation, there are many valuable studies have been reported [39,142–145,150]. For example, Fujii et al. [151] presented the control of threshold voltage by the variation AlGaN barrier layer thickness and AlN molar fraction. Efthymiou et al. [152] investigated effects of the p-GaN doping and gate metal work function on the threshold voltage of devices. Bakeroot et al. [153] developed an analytical model for calculation of the threshold voltage for p-GaN HEMTs, they studied the contributions of p-type doping profile, AlN molar fraction of AlGaN layer and AlGaN layer thickness to the threshold voltage.

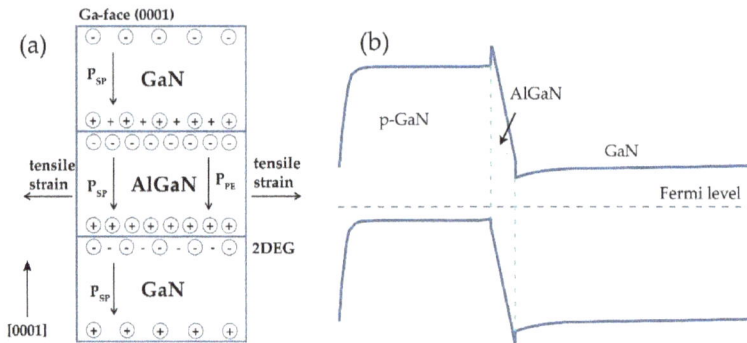

Figure 8. The schematics of (**a**) polarizations of AlGaN/GaN structure and (**b**) energy bands profile of the p-GaN HEMTs.

Moreover, in addition to the structures of p-GaN gate, recess gate and F-treatment, some novel E-mode HEMTs have been also proposed and studied through simulations. Huang et al. [154] simulated an E-mode HEMT with a floating gate in the gate dielectric. The results showed that combining with the gate recess, the V_{th} of the devices could be larger than 3 V with a low sheet density of ~10^{12} cm^{-2}. Wang et al. [155,156] simulated an E-mode HEMT with spilt floating gate. It is shown that such kind of device can improve the charge retention time comparing with the single floating gate HEMTs [157]. Duan et al. [158] simulated an E-mode HEMT with Groove-type channel and the normally-off function was based on the anisotropic polarization of GaN.

5. Summary

In this review, we have evaluated the recent progress of GaN based E-mode HEMTs. First, the typical device performance has been presented. And then we focused on some of the key technologies for fabricating p-GaN E-mode AlGaN/GaN HEMTs, including device isolation, p-GaN gate formation, Ohmic contacts for source and drain, contact for gate, surface passivation and field plate. For the material and device simulations, we reviewed some of the promising structures for achieving high performance E-mode devices. In summary, this paper provides the recent developments and existing challenges of p-GaN E-mode AlGaN/GaN HEMTs, we hope this could be useful for the study of E-mode AlGaN/GaN HEMTs.

Author Contributions: All the authors contributed equally. All the authors read and approved final manuscript.

Funding: This work was funded by Research of Low Cost Fabrication of GaN Power Devices and System Integration (grant number: JCYJ20160226192639004), Research of AlGaN HEMT MEMS Sensor for Work in Extreme Environment (grant number: JCYJ20170412153356899) and Research of the Reliability Mechanism and Circuit Simulation of GaN HEMT (grant number: 2017A050506002).

Conflicts of Interest: The authors declare no conflict of interest.

References

1. Treu, M.; Vecino, E.; Pippan, M.; Haberlen, O.; Curatola, G.; Deboy, G.; Kutschak, M.; Kirchner, U. The role of silicon, silicon carbide and gallium nitride in power electronics. In Proceedings of the 2012 International Electron Devices Meeting, San Francisco, CA, USA, 10–13 December 2012; pp. 7.1.1–7.1.4.
2. Ueda, D. Renovation of power devices by GaN-based materials. In Proceedings of the 2015 IEEE International Electron Devices Meeting (IEDM), Washington, DC, USA, 7–9 December 2015; pp. 16.4.1–16.4.4.
3. Piedra, D.; Lu, B.; Sun, M.; Zhang, Y.; Matioli, E.; Gao, F.; Chung, J.W.; Saadat, O.; Xia, L.; Azize, M.; et al. Advanced power electronic devices based on Gallium Nitride (GaN). In Proceedings of the 2015 IEEE International Electron Devices Meeting (IEDM), Washington, DC, USA, 7–9 December 2015; pp. 16.6.1–16.6.4.
4. Deboy, G.; Treu, M.; Haeberlen, O.; Neumayr, D. Si, SiC and GaN power devices: An unbiased view on key performance indicators. In Proceedings of the 2016 IEEE International Electron Devices Meeting (IEDM), San Francisco, CA, USA, 3–7 December 2016; pp. 20.2.1–20.2.4.
5. Yu, H.; Duan, T. *Gallium Nitride Power Devices*; Pan Stanford: Singapore, 2017.
6. Lidow, A.; Strydom, J.; de Rooij, M.; Reusch, D. *GaN Transistors for Efficient Power Conversion*; Willey: New York, NY, USA, 2015.
7. Meneghini, M.; Gaudenzio, M.; Zanoni, E. *Power GaN Devices—Materials, Applications and Reliability*; Springer: New York, NY, USA, 2017.
8. Chen, K.J.; Haberlen, O.; Lidow, A.; Tsai, C.L.; Ueda, T.; Uemoto, Y.; Wu, Y. GaN-on-Si Power Technology: Devices and Applications. *IEEE Trans. Electron Devices* **2017**, *64*, 779–795. [CrossRef]
9. Tsai, C.-L.; Wang, Y.-H.; Kwan, M.H.; Chen, P.C.; Yao, F.W.; Liu, S.C.; Yu, J.L.; Yeh, C.L.; Su, R.Y.; Wang, W.; et al. Smart GaN platform: Performance & challenges. In Proceedings of the 2017 IEEE International Electron Devices Meeting (IEDM), San Francisco, CA, USA, 2–6 December 2017; pp. 33.1.1–33.1.4.
10. Meneghesso, G.; Meneghini, M.; Zanoni, E. *Gallium Nitride-Enabled High Frequency and High Efficiency Power Conversion*; Springer: New York, NY, USA, 2018.
11. Hyung-Seok, L.; Dong Seup, L.; Palacios, T. AlGaN/GaN High-Electron-Mobility Transistors Fabricated through a Au-Free Technology. *IEEE Electron Device Lett.* **2011**, *32*, 623–625. [CrossRef]
12. Stoffels, S.; Zhao, M.; Venegas, R.; Kandaswamy, P.; You, S.; Novak, T.; Saripalli, Y.; Van Hove, M.; Decoutere, S. The physical mechanism of dispersion caused by AlGaN/GaN buffers on Si and optimization for low dispersion. In Proceedings of the 2015 IEEE International Electron Devices Meeting (IEDM), Washington, DC, USA, 7–9 December 2015; pp. 35.4.1–35.4.4.
13. Marcon, D.; Saripalli, Y.N.; Decoutere, S. 200mm GaN-on-Si epitaxy and e-mode device technology. In Proceedings of the 2015 IEEE International Electron Devices Meeting (IEDM), Washington, DC, USA, 7–9 December 2015; pp. 16.2.1–16.2.4.
14. Zhang, Z.; Li, B.; Tang, X.; Qian, Q.; Hua, M.; Huang, B.; Chen, K.J. Nitridation of GaN surface for power device application: A first-principles study. In Proceedings of the 2016 IEEE International Electron Devices Meeting (IEDM), San Francisco, CA, USA, 3–7 December 2016; pp. 36.2.1–36.2.4.
15. Wang, J.; Cao, L.; Xie, J.; Beam, E.; McCarthy, R.; Youtsey, C.; Fay, P. High voltage vertical p-n diodes with ion-implanted edge termination and sputtered SiNx passivation on GaN substrates. In Proceedings of the 2017 IEEE International Electron Devices Meeting (IEDM), San Francisco, CA, USA, 2–6 December 2017; pp. 9.6.1–9.6.4.
16. Hashizume, T.; Nishiguchi, K.; Kaneki, S.; Kuzmik, J.; Yatabe, Z. State of the art on gate insulation and surface passivation for GaN-based power HEMTs. *Mater. Sci. Semicond. Process.* **2018**, *78*, 85–95. [CrossRef]
17. Kim, Z.-S.; Lee, H.-S.; Na, J.; Bae, S.-B.; Nam, E.; Lim, J.-W. Ultra-low rate dry etching conditions for fabricating normally-off field effect transistors on AlGaN/GaN heterostructures. *Solid-State Electron.* **2018**, *140*, 12–17. [CrossRef]
18. Wong, K.-Y.; Lin, Y.S.; Hsiung, C.W.; Lansbergen, G.P.; Lin, M.C.; Yao, F.W.; Yu, C.J.; Chen, P.C.; Su, R.Y.; Yu, J.L.; et al. AlGaN/GaN MIS-HFET with improvement in high temperature gate bias stress-induced reliability. In Proceedings of the 2014 IEEE 26th International Symposium on Power Semiconductor Devices & IC's (ISPSD), Waikoloa, HI, USA, 15–19 June 2014; pp. 55–58.

19. Man Ho, K.; Wong, K.Y.; Lin, Y.S.; Yao, F.W.; Tsai, M.W.; Chang, Y.C.; Chen, P.C.; Su, R.Y.; Wu, C.H.; Yu, J.L.; et al. CMOS-compatible GaN-on-Si field-effect transistors for high voltage power applications. In Proceedings of the 2014 IEEE International Electron Devices Meeting (IEDM), San Francisco, CA, USA, 15–17 December 2014; pp. 17.6.1–17.6.4.

20. Moens, P.; Banerjee, A.; Uren, M.J.; Meneghini, M.; Karboyan, S.; Chatterjee, I.; Vanmeerbeek, P.; Casar, M.; Liu, C.; Salih, A.; et al. Impact of buffer leakage on intrinsic reliability of 650V AlGaN/GaN HEMTs. In Proceedings of the 2015 International Electron Devices Meeting, Washington, DC, USA, 7–9 December 2015; pp. 35.2.1–35.2.4.

21. Bahl, S.R.; Joh, J.; Fu, L.; Sasikumar, A.; Chatterjee, T.; Pendharkar, S. Application reliability validation of GaN power devices. In Proceedings of the International Electron Devices Meeting (IEDM), San Francisco, CA, USA, 3–7 December 2016; pp. 20.5.1–20.5.4.

22. Koller, C.; Pobegen, G.; Ostermaier, C.; Pogany, D. Evidence of defect band in carbon-doped GaN controlling leakage current and trapping dynamics. In Proceedings of the 2017 International Electron Devices Meeting (IEDM), San Francisco, CA, USA, 2–6 December 2017; pp. 33.4.1–33.4.4.

23. Meneghini, M.; Tajalli, A.; Moens, P.; Banerjee, A.; Stockman, A.; Tack, M.; Gerardin, S.; Bagatin, M.; Paccagnella, A.; Zanoni, E.; et al. Total suppression of dynamic-ron in AlGaN/GaN-HEMTs through proton irradiation. In Proceedings of the 2017 International Electron Devices Meeting (IEDM), San Francisco, CA, USA, 2–6 December 2017; pp. 33.5.1–33.5.4.

24. Qi, Y.; Zhu, Y.; Zhang, J.; Lin, X.; Cheng, K.; Jiang, L.; Yu, H. Evaluation of LPCVD SiN$_x$ Gate Dielectric Reliability by TDDB Measurement in Si-Substrate-Based AlGaN/GaN MIS-HEMT. *IEEE Trans. Electron Devices* **2018**, *65*, 1759–1764. [CrossRef]

25. Radhakrishna, U.; Lim, S.; Choi, P.; Palacios, T.; Antoniadis, D. GaNFET compact model for linking device physics, high voltage circuit design and technology optimization. In Proceedings of the 2015 International Electron Devices Meeting, Washington, DC, USA, 7–9 December 2015; pp. 9.6.1–9.6.4.

26. Cornigli, D.; Reggiani, S.; Gnani, E.; Gnudi, A.; Baccarani, G.; Moens, P.; Vanmeerbeek, P.; Banerjee, A.; Meneghesso, G. Numerical investigation of the lateral and vertical leakage currents and breakdown regimes in GaN-on-Silicon vertical structures. In Proceedings of the 2015 International Electron Devices Meeting, Washington, DC, USA, 7–9 December 2015; pp. 5.3.1–5.3.4.

27. Raciti, A.; Cristaldi, D.; Greco, G.; Vinci, G.; Bazzano, G. Integrated power electronics modules: Electro-thermal modeling flow and stress conditions overview. In Proceedings of the 2014 AEIT Annual Conference—From Research to Industry: The Need for a More Effective Technology Transfer (AEIT), Trieste, Italy, 18–19 September 2014; pp. 1–6.

28. Wong, K.Y.R.; Kwan, M.H.; Yao, F.W.; Tsai, M.W.; Lin, Y.S.; Chang, Y.C.; Chen, P.C.; Su, R.Y.; Yu, J.L.; Yang, F.J.; et al. A next generation CMOS-compatible GaN-on-Si transistors for high efficiency energy systems. In Proceedings of the 2015 International Electron Devices Meeting, Washington, DC, USA, 7–9 December 2015; pp. 9.5.1–9.5.4.

29. Di Cioccio, L.; Morvan, E.; Charles, M.; Perichon, P.; Torres, A.; Ayel, F.; Bergogne, D.; Baines, Y.; Fayolle, M.; Escoffier, R.; et al. From epitaxy to converters topologies what issues for 200 mm GaN/Si? In Proceedings of the 2015 International Electron Devices Meeting, Washington, DC, USA, 7–9 December 2015; pp. 16.5.1–16.5.4.

30. Hughes, B.; Chu, R.; Lazar, J.; Boutros, K. Increasing the switching frequency of GaN HFET converters. In Proceedings of the 2015 International Electron Devices Meeting, Washington, DC, USA, 7–9 December 2015; pp. 16.7.1–16.7.4.

31. Then, H.W.; Chow, L.A.; Dasgupta, S.; Gardner, S.; Radosavljevic, M.; Rao, V.R.; Sung, S.H.; Yang, G.; Fischer, P. High-K gate dielectric depletion-mode and enhancement-mode GaN MOS-HEMTs for improved OFF-state leakage and DIBL for power electronics and RF applications. In Proceedings of the 2015 International Electron Devices Meeting, Washington, DC, USA, 7–9 December 2015; pp. 16.3.1–16.3.4.

32. Coffa, S.; Saggio, M.; Patti, A. SiC- and GaN-based power devices: Technologies, products and applications. In Proceedings of the 2015 International Electron Devices Meeting, Washington, DC, USA, 7–9 December 2015; pp. 16.8.1–16.8.5.

33. Ishida, H.; Kajitani, R.; Kinoshita, Y.; Umeda, H.; Ujita, S.; Ogawa, M.; Tanaka, K.; Morita, T.; Tamura, S.; Ishida, M.; et al. GaN-based semiconductor devices for future power switching systems. In Proceedings of the International Electron Devices Meeting (IEDM), San Francisco, CA, USA, 3–7 December 2016; pp. 20.4.1–20.4.4.

34. Lidow, A.; Reusch, D.; Glaser, J. System level impact of GaN power devices in server architectures. In Proceedings of the International Electron Devices Meeting (IEDM), San Francisco, CA, USA, 3–7 December 2016; pp. 20.3.1–20.3.4.

35. Asif Khan, M.; Bhattarai, A.; Kuznia, J.N.; Olson, D.T. High electron mobility transistor based on a GaN-Al$_x$Ga$_{1-x}$N heterojunction. *Appl. Phys. Lett.* **1993**, *63*, 1214–1215. [CrossRef]

36. Khan, M.A.; Chen, Q.; Sun, C.J.; Yang, J.W.; Blasingame, M.; Shur, M.S.; Park, H. Enhancement and depletion mode GaN/AlGaN heterostructure field effect transistors. *Appl. Phys. Lett.* **1996**, *68*, 514–516. [CrossRef]

37. Hu, X.; Simin, G.; Yang, J.; Asif Khan, M.; Gaska, R.; Shur, M.S. Enhancement mode AlGaN/GaN HFET with selectively grown pn junction gate. *Electron. Lett.* **2000**, *36*, 753–754. [CrossRef]

38. Oka, T.; Nozawa, T. AlGaN/GaN Recessed MIS-Gate HFET With High-Threshold-Voltage Normally-Off Operation for Power Electronics Applications. *IEEE Electron Device Lett.* **2008**, *29*, 668–670. [CrossRef]

39. Greco, G.; Iucolano, F.; Roccaforte, F. Review of technology for normally-off HEMTs with p-GaN gate. *Mater. Sci. Semicond. Process.* **2018**, *78*, 96–106. [CrossRef]

40. Tanaka, K.; Morita, T.; Umeda, H.; Kaneko, S.; Kuroda, M.; Ikoshi, A.; Yamagiwa, H.; Okita, H.; Hikita, M.; Yanagihara, M.; et al. Suppression of current collapse by hole injection from drain in a normally-off GaN-based hybrid-drain-embedded gate injection transistor. *Appl. Phys. Lett.* **2015**, *107*, 163502. [CrossRef]

41. Kumar, A.; De Souza, M.M. Extending the bounds of performance in E-mode p-channel GaN MOSHFETs. In Proceedings of the International Electron Devices Meeting (IEDM), San Francisco, CA, USA, 3–7 December 2016; pp. 7.4.1–7.4.4.

42. Uemoto, Y.; Hikita, M.; Ueno, H.; Matsuo, H.; Ishida, H.; Yanagihara, M.; Ueda, T.; Tanaka, T.; Ueda, D. Gate Injection Transistor (GIT)—A Normally-Off AlGaN/GaN Power Transistor Using Conductivity Modulation. *IEEE Trans. Electron Devices* **2007**, *54*, 3393–3399. [CrossRef]

43. Nakazawa, S.; Shih, H.; Tsurumi, N.; Anda, Y.; Hatsuda, T.; Ueda, T.; Nozaki, M.; Yamada, T.; Hosoi, T.; Shimura, T.; et al. Fast switching performance by 20 A/730 V AlGaN/GaN MIS-HFET using AlON gate insulator. In Proceedings of the 2017 IEEE International Electron Devices Meeting (IEDM), San Francisco, CA, USA, 2–6 December 2017; pp. 25.1.1–25.1.4.

44. Son, D.-H.; Jo, Y.-W.; Won, C.-H.; Lee, J.-H.; Seo, J.H.; Lee, S.-H.; Lim, J.-W.; Kim, J.H.; Kang, I.M.; Cristoloveanu, S.; et al. Normally-off AlGaN/GaN-based MOS-HEMT with self-terminating TMAH wet recess etching. *Solid-State Electron.* **2018**, *141*, 7–12. [CrossRef]

45. Ji, D.; Gupta, C.; Chan, S.H.; Agarwal, A.; Li, W.; Keller, S.; Mishra, U.K.; Chowdhury, S. Demonstrating >1.4 kV OG-FET performance with a novel double field-plated geometry and the successful scaling of large-area devices. In Proceedings of the 2017 International Electron Devices Meeting (IEDM), San Francisco, CA, USA, 2–6 December 2017; pp. 9.4.1–9.4.4.

46. Zeng, C.; Wang, Y.-S.; Liao, X.-Y.; Li, R.-G.; Chen, Y.-Q.; Lai, P.; Huang, Y.; En, Y.-F. Reliability Investigations of AlGaN/GaN HEMTs Based on On-State Electroluminescence Characterization. *IEEE Trans. Device Mater. Reliab.* **2015**, *15*, 69–74. [CrossRef]

47. Amano, H.; Baines, Y.; Beam, E.; Borga, M.; Bouchet, T.; Chalker, P.R.; Charles, M.; Chen, K.J.; Chowdhury, N.; Chu, R.; et al. The 2018 GaN power electronics roadmap. *J. Phys. D Appl. Phys.* **2018**, *51*, 163001. [CrossRef]

48. Hwang, I.; Choi, H.; Lee, J.; Choi, H.S.; Kim, J.; Ha, J.; Um, C.-Y.; Hwang, S.-K.; Oh, J.; Kim, J.-Y.; et al. 1.6 kV, 2.9 mΩ cm^2 normally-off p-GaN HEMT device. In Proceedings of the 2012 24th International Symposium on Power Semiconductor Devices and ICs, Bruges, Belgium, 3–7 June 2012; pp. 41–44.

49. Wang, Z.; Zhang, B.; Chen, W.; Li, Z. A Closed-Form Charge Control Model for the Threshold Voltage of Depletion- and Enhancement-Mode AlGaN/GaN Devices. *IEEE Trans. Electron Devices* **2013**, *60*, 1607–1612. [CrossRef]

50. Ning, W.; Hui, W.; Xinpeng, L.; Yongle, Q.; Tianli, D.; Lingli, J.; Iervolino, E.; Kai, C.; Hongyu, Y. Investigation of AlGaN/GaN HEMTs degradation with gate pulse stressing at cryogenic temperature. *AIP Adv.* **2017**, *7*, 095317.

51. Meneghini, M.; Hilt, O.; Wuerfl, J.; Meneghesso, G. Technology and Reliability of Normally-Off GaN HEMTs with p-Type Gate. *Energies* **2017**, *10*, 153. [CrossRef]

52. Shul, R.J.; McClellan, G.B.; Casalnuovo, S.A.; Rieger, D.J.; Pearton, S.J.; Constantine, C.; Barratt, C.; Karlicek, R.F.; Tran, C.; Schurman, M. Inductively coupled plasma etching of GaN. *Appl. Phys. Lett.* **1996**, *69*, 1119–1121. [CrossRef]

53. Shul, R.J.; Vawter, G.A.; Willison, C.G.; Bridges, M.M.; Lee, J.W.; Pearton, S.J.; Abernathy, C.R. Comparison of plasma etch techniques for III–V nitrides. *Solid-State Electron.* **1998**, *42*, 2259–2267. [CrossRef]

54. Sheu, J.K.; Su, Y.K.; Chi, G.C.; Jou, M.J.; Liu, C.C.; Chang, C.M.; Hung, W.C. Inductively coupled plasma etching of GaN using Cl_2/Ar and Cl_2/N_2 gases. *J. Appl. Phys.* **1999**, *85*, 1970–1974. [CrossRef]

55. Pearton, S.J.; Shul, R.J.; Ren, F. A Review of Dry Etching of GaN and Related Materials. *MRS Internet J. Nitride Semicond. Res.* **2014**, *5*, 11. [CrossRef]

56. Kodera, M.; Yoshioka, A.; Sugiyama, T.; Ohguro, T.; Hamamoto, T.; Kawamoto, T.; Yamanaka, T.; Xinyu, Z.; Lester, S.; Miyashita, N. Impact of Plasma-Damaged-Layer Removal on GaN HEMT Devices. *Phys. Status Solidi* **2018**, *215*, 1700633. [CrossRef]

57. Tripathy, S.; Ramam, A.; Chua, S.J.; Pan, J.S.; Huan, A. Characterization of inductively coupled plasma etched surface of GaN using Cl_2/BCl_3 chemistry. *J. Vac. Sci. Technol. A Vac. Surf. Films* **2001**, *19*, 2522–2532. [CrossRef]

58. Pearton, S.J.; Abernathy, C.R.; Ren, F. Dry patterning of InGaN and InAlN. *Appl. Phys. Lett.* **1994**, *64*, 3643–3645. [CrossRef]

59. Zhe, X.; Jinyan, W.; Jingqian, L.; Chunyan, J.; Yong, C.; Zhenchuan, Y.; Maojun, W.; Min, Y.; Bing, X.; Wengang, W.; et al. Demonstration of Normally-Off Recess-Gated AlGaN/GaN MOSFET Using GaN Cap Layer as Recess Mask. *IEEE Electron Device Lett.* **2014**, *35*, 1197–1199. [CrossRef]

60. Shul, R.J.; Willison, C.G.; Bridges, M.M.; Han, J.; Lee, J.W.; Pearton, S.J.; Abernathy, C.R.; MacKenzie, J.D.; Donovan, S.M.; Zhang, L.; et al. Selective inductively coupled plasma etching of group-III nitrides in Cl_2- and BCl_3-based plasmas. *J. Vac. Sci. Technol. A Vac. Surf. Films* **1998**, *16*, 1621–1626. [CrossRef]

61. Wakejima, A.; Ando, A.; Watanabe, A.; Inoue, K.; Kubo, T.; Osada, Y.; Kamimura, R.; Egawa, T. Normally off AlGaN/GaN HEMT on Si substrate with selectively dry-etched recessed gate and polarization-charge-compensation δ-doped GaN cap layer. *Appl. Phys. Express* **2015**, *8*, 026502. [CrossRef]

62. Nanjo, T.; Miura, N.; Oishi, T.; Suita, M.; Abe, Y.; Ozeki, T.; Nakatsuka, S.; Inoue, A.; Ishikawa, T.; Matsuda, Y.; et al. Improvement of DC and RF Characteristics of AlGaN/GaN High Electron Mobility Transistors by Thermally Annealed Ni/Pt/Au Schottky Gate. *Jpn. J. Appl. Phys.* **2004**, *43*, 1925–1929. [CrossRef]

63. Umeda, H.; Takizawa, T.; Anda, Y.; Ueda, T.; Tanaka, T. High-Voltage Isolation Technique Using Fe Ion Implantation for Monolithic Integration of AlGaN/GaN Transistors. *IEEE Trans. Electron Devices* **2013**, *60*, 771–775. [CrossRef]

64. Ducatteau, D.; Minko, A.; Hoel, V.; Morvan, E.; Delos, E.; Grimbert, B.; Lahreche, H.; Bove, P.; Gaquiere, C.; De Jaeger, J.C.; et al. Output power density of 5.1/mm at 18 GHz with an AlGaN/GaN HEMT on Si substrate. *IEEE Electron Device Lett.* **2006**, *27*, 7–9. [CrossRef]

65. Sun, M.; Lee, H.-S.; Lu, B.; Piedra, D.; Palacios, T. Comparative Breakdown Study of Mesa- and Ion-Implantation-Isolated AlGaN/GaN High-Electron-Mobility Transistors on Si Substrate. *Appl. Phys. Express* **2012**, *5*, 074202. [CrossRef]

66. Shiu, J.-Y.; Lu, C.-Y.; Su, T.-Y.; Huang, R.-T.; Zirath, H.; Rorsman, N.; Chang, E.Y. Electrical Characterization and Transmission Electron Microscopy Assessment of Isolation of AlGaN/GaN High Electron Mobility Transistors with Oxygen Ion Implantation. *Jpn. J. Appl. Phys.* **2010**, *49*, 021001. [CrossRef]

67. Shiu, J.-Y.; Huang, J.-C.; Desmaris, V.; Chang, C.-T.; Lu, C.-Y.; Kumakura, K.; Makimoto, T.; Zirath, H.; Rorsman, N.; Chang, E.Y. Oxygen Ion Implantation Isolation Planar Process for AlGaN/GaN HEMTs. *IEEE Electron Device Lett.* **2007**, *28*, 476–478. [CrossRef]

68. Burnham, S.D.; Boutros, K.; Hashimoto, P.; Butler, C.; Wong, D.W.S.; Hu, M.; Micovic, M. Gate-recessed normally-off GaN-on- Si HEMT using a new O_2-BCl_3 digital etching technique. *Phys. Status Solidi* **2010**, *7*, 2010–2012. [CrossRef]

69. Wong, J.C.; Micovic, M.; Brown, D.F.; Khalaf, I.; Williams, A.; Corrion, A. Selective anisotropic etching of GaN over AlGaN for very thin films. *J. Vac. Sci. Technol. A Vac. Surf. Films* **2018**, *36*, 030603. [CrossRef]

70. Chiu, H.-C.; Chang, Y.-S.; Li, B.-H.; Wang, H.-C.; Kao, H.-L.; Hu, C.-W.; Xuan, R. High-Performance Normally off p-GaN Gate HEMT with Composite $AlN/Al_{0.17}Ga_{0.83}N/Al_{0.3}Ga_{0.7}N$ Barrier Layers Design. *IEEE J. Electron Devices Soc.* **2018**, *6*, 201–206. [CrossRef]

71. Chiu, H.-C.; Chang, Y.-S.; Li, B.-H.; Wang, H.-C.; Kao, H.-L.; Chien, F.-T.; Hu, C.-W.; Xuan, R. High Uniformity Normally-off p-GaN Gate HEMT Using Self-Terminated Digital Etching Technique. *IEEE Trans. Electron Devices* **2018**, *65*, 4820–4825. [CrossRef]

72. Greco, G.; Iucolano, F.; Roccaforte, F. Ohmic contacts to Gallium Nitride materials. *Appl. Surf. Sci.* **2016**, *383*, 324–345. [CrossRef]

73. Lin, M.E.; Ma, Z.; Huang, F.Y.; Fan, Z.F.; Allen, L.H.; Morkoç, H. Low resistance ohmic contacts on wide band-gap GaN. *Appl. Phys. Lett.* **1994**, *64*, 1003–1005. [CrossRef]

74. Luther, B.P.; Mohney, S.E.; Jackson, T.N.; Asif Khan, M.; Chen, Q.; Yang, J.W. Investigation of the mechanism for Ohmic contact formation in Al and Ti/Al contacts ton-type GaN. *Appl. Phys. Lett.* **1997**, *70*, 57–59. [CrossRef]

75. Graff, A.; Simon-Najasek, M.; Altmann, F.; Kuzmik, J.; Gregušova, D.; Haščík, Š.; Jung, H.; Baur, T.; Grünenpütt, J.; Blanck, H. High resolution physical analysis of ohmic contact formation at GaN-HEMT devices. *Microelectron. Reliab.* **2017**, *76–77*, 338–343. [CrossRef]

76. Jacobs, B.; Kramer, M.C.J.C.M.; Geluk, E.J.; Karouta, F. Optimisation of the Ti/Al/Ni/Au ohmic contact on AlGaN/GaN FET structures. *J. Cryst. Growth* **2002**, *241*, 15–18. [CrossRef]

77. Piazza, M.; Dua, C.; Oualli, M.; Morvan, E.; Carisetti, D.; Wyczisk, F. Degradation of TiAlNiAu as ohmic contact metal for GaN HEMTs. *Microelectron. Reliab.* **2009**, *49*, 1222–1225. [CrossRef]

78. Motayed, A.; Bathe, R.; Wood, M.C.; Diouf, O.S.; Vispute, R.D.; Mohammad, S.N. Electrical, thermal, and microstructural characteristics of Ti/Al/Ti/Au multilayer Ohmic contacts to n-type GaN. *J. Appl. Phys.* **2003**, *93*, 1087–1094. [CrossRef]

79. France, R.; Xu, T.; Chen, P.; Chandrasekaran, R.; Moustakas, T.D. Vanadium-based Ohmic contacts to n-AlGaN in the entire alloy composition. *Appl. Phys. Lett.* **2007**, *90*, 062115. [CrossRef]

80. Firrincieli, A.; De Jaeger, B.; You, S.; Wellekens, D.; Van Hove, M.; Decoutere, S. Au-free low temperature ohmic contacts for AlGaN/GaN power devices on 200 mm Si substrates. *Jpn. J. Appl. Phys.* **2014**, *53*, 04EF01. [CrossRef]

81. Liu, Z.; Sun, M.; Lee, H.-S.; Heuken, M.; Palacios, T. AlGaN/AlN/GaN High-Electron-Mobility Transistors Fabricated with Au-Free Technology. *Appl. Phys. Express* **2013**, *6*, 096502. [CrossRef]

82. Lin, Y.-K.; Bergsten, J.; Leong, H.; Malmros, A.; Chen, J.-T.; Chen, D.-Y.; Kordina, O.; Zirath, H.; Chang, E.Y.; Rorsman, N. A versatile low-resistance ohmic contact process with ohmic recess and low-temperature annealing for GaN HEMTs. *Semicond. Sci. Technol.* **2018**, *33*, 095019. [CrossRef]

83. Yoshida, T.; Egawa, T. Improvement of Au-Free, Ti/Al/W Ohmic Contact on AlGaN/GaN Heterostructure Featuring a Thin-Ti Layer and Low Temperature Annealing. *Phys. Status Solidi* **2018**, *215*, 1700825. [CrossRef]

84. Pozzovivo, G.; Kuzmik, J.; Giesen, C.; Heuken, M.; Liday, J.; Strasser, G.; Pogany, D. Low resistance ohmic contacts annealed at 600 °C on a InAlN/GaN heterostructure with SiCl4-reactive ion etching surface treatment. *Phys. Status Solidi* **2009**, *6*, S999–S1002. [CrossRef]

85. Greco, G.; Iucolano, F.; Di Franco, S.; Bongiorno, C.; Patti, A.; Roccaforte, F. Effects of Annealing Treatments on the Properties of Al/Ti/p-GaN Interfaces for Normally OFF p-GaN HEMTs. *IEEE Trans. Electron Devices* **2016**, *63*, 2735–2741. [CrossRef]

86. Lu, X.; Jiang, H.; Liu, C.; Zou, X.; Lau, K.M. Off-state leakage current reduction in AlGaN/GaN high electron mobility transistors by combining surface treatment and post-gate annealing. *Semicond. Sci. Technol.* **2016**, *31*, 055019. [CrossRef]

87. Hwang, I.; Kim, J.; Choi, H.S.; Choi, H.; Lee, J.; Kim, K.Y.; Park, J.-B.; Lee, J.C.; Ha, J.; Oh, J.; et al. p-GaN Gate HEMTs With Tungsten Gate Metal for High Threshold Voltage and Low Gate Current. *IEEE Electron Device Lett.* **2013**, *34*, 202–204. [CrossRef]

88. Ťapajna, M.; Hilt, O.; Bahat-Treidel, E.; Wurfl, J.; Kuzmik, J. Gate Reliability Investigation in Normally-Off p-Type-GaN Cap/AlGaN/GaN HEMTs Under Forward Bias Stress. *IEEE Electron Device Lett.* **2016**, *37*, 385–388. [CrossRef]

89. Wu, T.-L.; Marcon, D.; You, S.; Posthuma, N.; Bakeroot, B.; Stoffels, S.; Van Hove, M.; Groeseneken, G.; Decoutere, S. Forward Bias Gate Breakdown Mechanism in Enhancement-Mode p-GaN Gate AlGaN/GaN High-Electron Mobility Transistors. *IEEE Electron Device Lett.* **2015**, *36*, 1001–1003. [CrossRef]

90. Tallarico, A.N.; Stoffels, S.; Magnone, P.; Posthuma, N.; Sangiorgi, E.; Decoutere, S.; Fiegna, C. Investigation of the p-GaN Gate Breakdown in Forward-Biased GaN-Based Power HEMTs. *IEEE Electron Device Lett.* **2017**, *38*, 99–102. [CrossRef]

91. Stockman, A.; Canato, E.; Tajalli, A.; Meneghini, M.; Meneghesso, G.; Zanoni, E.; Moens, P.; Bakeroot, B. On the origin of the leakage current in p-gate AlGaN/GaN HEMTs. In Proceedings of the 2018 IEEE International Reliability Physics Symposium (IRPS), Burlingame, CA, USA, 11–15 March 2018; pp. 4B.5-1–4B.5-4.

92. Yu, L.S.; Jia, L.; Qiao, D.; Lau, S.S.; Li, J.; Lin, J.Y.; Jiang, H.X. The origins of leaky characteristics of schottky diodes on p-GaN. *IEEE Trans. Electron Devices* **2003**, *50*, 292–296. [CrossRef]

93. Zhu, G.C.; Wang, Y.M.; Xin, Q.; Xu, M.S.; Chen, X.F.; Xu, X.G.; Feng, X.J.; Song, A.M. GaN metal-oxide-semiconductor high-electron-mobility transistors using thermally evaporated SiO as the gate dielectric. *Semicond. Sci. Technol.* **2018**, *33*, 095023. [CrossRef]

94. Dora, Y.; Han, S.; Klenov, D.; Hansen, P.J.; No, K.-s.; Mishra, U.K.; Stemmer, S.; Speck, J.S. ZrO$_2$ gate dielectrics produced by ultraviolet ozone oxidation for GaN and AlGaN/GaN transistors. *J. Vac. Sci. Technol. B* **2006**, *24*, 575. [CrossRef]

95. Cook, T.E.; Fulton, C.C.; Mecouch, W.J.; Davis, R.F.; Lucovsky, G.; Nemanich, R.J. Band offset measurements of the GaN (0001)/HfO$_2$ interface. *J. Appl. Phys.* **2003**, *94*, 7155–7158. [CrossRef]

96. Gao, Z.; Romero, M.F.; Calle, F. Thermal and Electrical Stability Assessment of AlGaN/GaN Metal-Oxide-Semiconductor High-Electron Mobility Transistors (MOS-HEMTs) With HfO$_2$ Gate Dielectric. *IEEE Trans. Electron Devices* **2018**, *65*, 3142–3148. [CrossRef]

97. Roccaforte, F.; Fiorenza, P.; Greco, G.; Vivona, M.; Lo Nigro, R.; Giannazzo, F.; Patti, A.; Saggio, M. Recent advances on dielectrics technology for SiC and GaN power devices. *Appl. Surf. Sci.* **2014**, *301*, 9–18. [CrossRef]

98. Vetury, R.; Zhang, N.Q.; Keller, S.; Mishra, U.K. The impact of surface states on the DC and RF characteristics of AlGaN/GaN HFETs. *IEEE Trans. Electron Devices* **2001**, *48*, 560–566. [CrossRef]

99. Khan, M.A.; Hu, X.; Tarakji, A.; Simin, G.; Yang, J.; Gaska, R.; Shur, M.S. AlGaN/GaN metal-oxide-semiconductor heterostructure field-effect transistors on SiC substrates. *Appl. Phys. Lett.* **2000**, *77*, 1339–1341. [CrossRef]

100. Hua, M.; Liu, C.; Yang, S.; Liu, S.; Fu, K.; Dong, Z.; Cai, Y.; Zhang, B.; Chen, K.J. Characterization of Leakage and Reliability of SiNx Gate Dielectric by Low-Pressure Chemical Vapor Deposition for GaN-based MIS-HEMTs. *IEEE Trans. Electron Devices* **2015**, *62*, 3215–3222. [CrossRef]

101. Hatano, M.; Taniguchi, Y.; Kodama, S.; Tokuda, H.; Kuzuhara, M. Reduced gate leakage and high thermal stability of AlGaN/GaN MIS-HEMTs using ZrO$_2$/Al$_2$O$_3$gate dielectric stack. *Appl. Phys. Express* **2014**, *7*, 044101. [CrossRef]

102. Ueoka, Y.; Deki, M.; Honda, Y.; Amano, H. Improvement of breakdown voltage of vertical GaN p-n junction diode with Ga$_2$O$_3$ passivated by sputtering. *Jpn. J. Appl. Phys.* **2018**, *57*, 070302. [CrossRef]

103. Zhang, D.L.; Cheng, X.H.; Zheng, L.; Shen, L.Y.; Wang, Q.; Gu, Z.Y.; Qian, R.; Wu, D.P.; Zhou, W.; Cao, D.; et al. Effects of polycrystalline AlN filmon the dynamic performance of AlGaN/GaN high electron mobility transistors. *Mater. Des.* **2018**, *148*, 1–7. [CrossRef]

104. Zhang, L.Q.; Wang, P.F. AlN/GaN metal-insulator-semiconductor high-electron-mobility transistor with thermal atomic layer deposition AlN gate dielectric. *Jpn. J. Appl. Phys.* **2018**, *57*, 096502. [CrossRef]

105. Sen, H.; Qimeng, J.; Shu, Y.; Chunhua, Z.; Chen, K.J. Effective Passivation of AlGaN/GaN HEMTs by ALD-Grown AlN Thin Film. *IEEE Electron Device Lett.* **2012**, *33*, 516–518.

106. Koehler, A.D.; Nepal, N.; Anderson, T.J.; Tadjer, M.J.; Hobart, K.D.; Eddy, C.R.; Kub, F.J. Atomic Layer Epitaxy AlN for Enhanced AlGaN/GaN HEMT Passivation. *IEEE Electron Device Lett.* **2013**, *34*, 1115–1117. [CrossRef]

107. Luo, B.; Mehandru, R.; Kim, J.; Ren, F.; Gila, B.P.; Onstine, A.H.; Abernathy, C.R.; Pearton, S.J.; Gotthold, D.; Birkhahn, R.; et al. High three-terminal breakdown voltage and output power of Sc$_2$O$_3$ passivated AlGaN/GaN high electron mobility transistors. *Electron. Lett.* **2003**, *39*, 809–810. [CrossRef]

108. Tsu-Yi, W.; Lin, S.-K.; Po-Wen, S.; Jian-Jiun, H.; Wei-Chi, C.; Chih-Chun, H.; Ming-Ji, T.; Yeong-Her, W. AlGaN/GaN MOSHEMTs With Liquid-Phase-Deposited TiO$_2$ as Gate Dielectric. *IEEE Trans. Electron Devices* **2009**, *56*, 2911–2916.

109. Lee, C.-T.; Ya-Lan, C.; Chi-Sen, L. AlGaN/GaN MOS-HEMTs With Gate ZnO Dielectric Layer. *IEEE Electron Device Lett.* **2010**, *31*, 1220–1223. [CrossRef]

110. Oh, C.S.; Youn, C.J.; Yang, G.M.; Lim, K.Y.; Yang, J.W. AlGaN/GaN metal-oxide-semiconductor heterostructure field-effect transistor with oxidized Ni as a gate insulator. *Appl. Phys. Lett.* **2004**, *85*, 4214–4216. [CrossRef]

111. Kanamura, M.; Ohki, T.; Imanishi, K.; Makiyama, K.; Okamoto, N.; Kikkawa, T.; Hara, N.; Joshin, K. Joshin High power and high gain AlGaN/GaN MIS-HEMTs with high-k dielectric layer. *Phys. Stat. Sol.* **2008**, *5*, 2037–2040.

112. Hao, Y.; Yang, L.; Ma, X.H.; Ma, J.g.; Cao, M.y.; Pan, C.y.; Wang, C.; Zhang, J.c. High-performance microwave gate-recessed AlGaN/AlN/GaN MOS-HEMT with 73% power-added efficiency. *IEEE Electron Dev. Lett.* **2011**, *32*, 626–628. [CrossRef]

113. Liu, Z.H.; Ng, G.I.; Arulkumaran, S.; Maung, Y.K.T.; Teo, K.L.; Foo, S.C.; Sahmuganathan, V. Improved two-dimensional electron gas transport characteristics in AlGaN/GaN metal-insulator-semiconductor high electron mobility transistor with atomic layer-deposited Al$_2$O$_3$ as gate insulator. *Appl. Phys. Lett.* **2009**, *95*, 223501. [CrossRef]

114. Qin, X.; Wallace, R.M. In situ plasma enhanced atomic layer deposition half cycle study of Al$_2$O$_3$ on AlGaN/GaN high electron mobility transistors. *Appl. Phys. Lett.* **2015**, *107*, 081608. [CrossRef]

115. Dong Seup, L.; Chung, J.W.; Wang, H.; Xiang, G.; Shiping, G.; Fay, P.; Palacios, T. 245-GHz InAlN/GaN HEMTs with Oxygen Plasma Treatment. *IEEE Electron Device Lett.* **2011**, *32*, 755–757.

116. Ronghua, W.; Guowang, L.; Laboutin, O.; Cao, Y.; Johnson, W.; Snider, G.; Fay, P.; Jena, D.; Huili, X. 210-GHz InAlN/GaN HEMTs With Dielectric-Free Passivation. *IEEE Electron Device Lett.* **2011**, *32*, 892–894.

117. Liu, H.-Y.; Lee, C.-S.; Hsu, W.-C.; Tseng, L.-Y.; Chou, B.-Y.; Ho, C.-S.; Wu, C.-L. Investigations of AlGaN/AlN/GaN MOS-HEMTs on Si Substrate by Ozone Water Oxidation Method. *IEEE Trans. Electron Devices* **2013**, *60*, 2231–2237. [CrossRef]

118. Liu, H.-Y.; Chou, B.-Y.; Hsu, W.-C.; Lee, C.-S.; Sheu, J.-K.; Ho, C.-S. Enhanced AlGaN/GaN MOS-HEMT performance by using hydrogen peroxide oxidation technique. *IEEE Trans. Electron Devices* **2013**, *60*, 213–220. [CrossRef]

119. Liu, X.; Low, E.K.F.; Pan, J.; Liu, W.; Teo, K.L. Impact of In situ vacuum anneal and SiH$_4$ treatment on electrical characteristics of AlGaN/GaN metal-oxide-semiconductor high-electron mobility transistors. *Appl. Phys. Lett.* **2011**, *99*, 093504. [CrossRef]

120. Liu, S.; Yang, S.; Tang, Z.; Jiang, Q.; Liu, C.; Wang, M.; Shen, B.; Chen, K.J. Interface/border trap characterization of Al$_2$O$_3$/AlN/GaN metal-oxide-semiconductor structures with an AlN interfacial layer. *Appl. Phys. Lett.* **2015**, *106*, 051605. [CrossRef]

121. Miao, M.S.; Weber, J.R.; Van de Walle, C.G. Oxidation and the origin of the two-dimensional electron gas in AlGaN/GaN heterostructures. *J. Appl. Phys.* **2010**, *107*, 123713. [CrossRef]

122. Coan, M.R.; Woo, J.H.; Johnson, D.; Gatabi, I.R.; Harris, H.R. Band offset measurements of the GaN/dielectric interfaces. *J. Appl. Phys.* **2012**, *112*, 024508. [CrossRef]

123. Oyama, S.; Hashizume, T.; Hasegawa, H. Mechanism of current leakage through metal/n-GaN interfaces. *Appl. Surf. Sci.* **2002**, *190*, 322–325. [CrossRef]

124. Dong, Y.; Feenstra, R.M.; Northrup, J.E. Electronic states of oxidized GaN(0001) surfaces. *Appl. Phys. Lett.* **2006**, *89*, 171920. [CrossRef]

125. Bae, C.; Lucovsky, G. Low-temperature preparation of GaN-SiO$_2$ interfaces with low defect density. I. Two-step remote plasma-assisted oxidation-deposition process. *J. Vac. Sci. Technol. A* **2004**, *22*, 2402–2410. [CrossRef]

126. Therrien, R.; Lucovsky, G.; Davis, R. Charge redistribution at GaN–Ga$_2$O$_3$ interfaces: A microscopic mechanism for low defect density interfaces in remote-plasma-processed MOS devices prepared on polar GaN faces. *Appl. Surf. Sci.* **2000**, *166*, 513–519. [CrossRef]

127. Lu, W.; Kumar, V.; Schwindt, R.; Piner, E.; Adesida, I. Adesida A comparative study of surface passivation on AlGaN/GaN HEMTs. *Solid-State Electron.* **2002**, *46*, 1441–1444. [CrossRef]

128. Hasegawa, H.; Inagaki, T.; Ootomo, S.; Hashizume, T. Mechanisms of current collapse and gate leakage currents in AlGaN/GaN heterostructure field effect transistors. *J. Vac. Sci. Technol. B* **2003**, *21*, 1844–1855. [CrossRef]

129. Jones, E.A.; Wang, F.F.; Costinett, D. Review of Commercial GaN Power Devices and GaN-Based Converter Design Challenges. *IEEE J. Emerg. Sel. Top. Power Electron.* **2016**, *4*, 707–719. [CrossRef]

130. Green, B.M.; Chu, K.K.; Chumbes, E.M.; Smart, J.A.; Shealy, J.R.; Eastman, L.F. The effect of surface passivation on the microwave characteristics of undoped AlGaN/GaN HEMTs. *IEEE Electron Device Lett.* **2000**, *21*, 268–270. [CrossRef]

131. Chini, A.; Buttari, D.; Coffie, R.; Heikman, S.; Keller, S.; Mishra, U.K. 12 W/mm power density AlGaN/GaN HEMTs on sapphire substrate. *Electron. Lett.* **2004**, *40*, 73–74. [CrossRef]

132. Wong, J.; Shinohara, K.; Corrion, A.L.; Brown, D.F.; Carlos, Z.; Williams, A.; Tang, Y.; Robinson, J.F.; Khalaf, I.; Fung, H.; et al. Novel Asymmetric Slant Field Plate Technology for High-Speed Low-Dynamic Ron E/D-mode GaN HEMTs. *IEEE Electron Device Lett.* **2017**, *38*, 95–98. [CrossRef]

133. Ma, J.; Zanuz, D.C.; Matioli, E. Field Plate Design for Low Leakage Current in Lateral GaN Power Schottky Diodes: Role of the Pinch-off Voltage. *IEEE Electron Device Lett.* **2017**, *38*, 1298–1301. [CrossRef]

134. Bajaj, S.; Akyol, F.; Krishnamoorthy, S.; Hung, T.-H.; Rajan, S. Simulation of Enhancement Mode GaN HEMTs with Threshold> 5 V using P-type Buffer. *arXiv*, 2015; arXiv:1511.04438.

135. Gao, Z.; Hou, B.; Liu, Y.; Ma, X. Impact of fluorine plasma treatment on AlGaN/GaN high electronic mobility transistors by simulated and experimental results. *Microelectron. Eng.* **2016**, *154*, 22–25. [CrossRef]

136. Wang, A.; Zeng, L.; Wang, W. Simulation of Gate Leakage Current of AlGaN/GaN HEMTs: Effects of the Gate Edges and Self-Heating. *ECS J. Solid State Sci. Technol.* **2017**, *6*, S3025–S3029. [CrossRef]

137. Mukherjee, K.; Darracq, F.; Curutchet, A.; Malbert, N.; Labat, N. TCAD simulation capabilities towards gate leakage current analysis of advanced AlGaN/GaN HEMT devices. *Microelectron. Reliab.* **2017**, *76–77*, 350–356. [CrossRef]

138. Wong, H.Y.; Braga, N.; Mickevicius, R.; Gao, F.; Palacios, T. Study of AlGaN/GaN HEMT degradation through TCAD simulations. In Proceedings of the 2014 IEEE International Conference on Simulation of Semiconductor Processes and Devices (SISPAD), Yokohama, Japan, 9–11 September 2014; pp. 97–100.

139. Park, P.S.; Rajan, S. Simulation of Short-Channel Effects in N- and Ga-Polar AlGaN/GaN HEMTs. *IEEE Trans. Electron Devices* **2011**, *58*, 704–708. [CrossRef]

140. Jia, Y.; Xu, Y.; Lu, K.; Wen, Z.; Huang, A.-D.; Guo, Y.-X. Characterization of Buffer-Related Current Collapse by Buffer Potential Simulation in AlGaN/GaN HEMTs. *IEEE Trans. Electron Devices* **2018**, *65*, 3169–3175. [CrossRef]

141. Romanov, A.E.; Baker, T.J.; Nakamura, S.; Speck, J.S. Strain-induced polarization in wurtzite III-nitride semipolar layers. *J. Appl. Phys.* **2006**, *100*, 023522. [CrossRef]

142. Ambacher, O.; Smart, J.; Shealy, J.R.; Weimann, N.G.; Chu, K.; Murphy, M.; Schaff, W.J.; Eastman, L.F.; Dimitrov, R.; Wittmer, L.; et al. Two-dimensional electron gases induced by spontaneous and piezoelectric polarization charges in N- and Ga-face AlGaN/GaN heterostructures. *J. Appl. Phys.* **1999**, *85*, 3222–3233. [CrossRef]

143. Heikman, S.; Keller, S.; Wu, Y.; Speck, J.S.; DenBaars, S.P.; Mishra, U.K. Polarization effects in AlGaN/GaN and GaN/AlGaN/GaN heterostructures. *J. Appl. Phys.* **2003**, *93*, 10114–10118. [CrossRef]

144. Ibbetson, J.P.; Fini, P.T.; Ness, K.D.; DenBaars, S.P.; Speck, J.S.; Mishra, U.K. Polarization effects, surface states, and the source of electrons in AlGaN/GaN heterostructure field effect transistors. *Appl. Phys. Lett.* **2000**, *77*, 250–252. [CrossRef]

145. Smorchkova, I.P.; Elsass, C.R.; Ibbetson, J.P.; Vetury, R.; Heying, B.; Fini, P.; Haus, E.; DenBaars, S.P.; Speck, J.S.; Mishra, U.K. Polarization-induced charge and electron mobility in AlGaN/GaN heterostructures grown by plasma-assisted molecular-beam epitaxy. *J. Appl. Phys.* **1999**, *86*, 4520–4526. [CrossRef]

146. Fischer, A.; Kuhne, H.; Richter, H. New approach in equilibrium theory for strained layer relaxation. *Phys. Rev. Lett.* **1994**, *73*, 2712–2715. [CrossRef]

147. Bykhovski, A.D.; Gaska, R.; Shur, M.S. Piezoelectric doping and elastic strain relaxation in AlGaN–GaN heterostructure field effect transistors. *Appl. Phys. Lett.* **1998**, *73*, 3577–3579. [CrossRef]

148. Rashmi; Kranti, A.; Haldar, S.; Gupta, R.S. An accurate charge control model for spontaneous and piezoelectric polarization dependent two-dimensional electron gas sheet charge density of lattice-mismatched AlGaN/GaN HEMTs. *Solid-State Electron.* **2002**, *46*, 621–630.

149. Frayssinet, E.; Knap, W.; Lorenzini, P.; Grandjean, N.; Massies, J.; Skierbiszewski, C.; Suski, T.; Grzegory, I.; Porowski, S.; Simin, G.; et al. High electron mobility in AlGaN/GaN heterostructures grown on bulk GaN substrates. *Appl. Phys. Lett.* **2000**, *77*, 2551–2553. [CrossRef]

150. Ambacher, O.; Foutz, B.; Smart, J.; Shealy, J.R.; Weimann, N.G.; Chu, K.; Murphy, M.; Sierakowski, A.J.; Schaff, W.J.; Eastman, L.F.; et al. Two dimensional electron gases induced by spontaneous and piezoelectric polarization in undoped and doped AlGaN/GaN heterostructures. *J. Appl. Phys.* **2000**, *87*, 334–344. [CrossRef]

151. Fujii, T.; Tsuyukuchi, N.; Hirose, Y.; Iwaya, M.; Kamiyama, S.; Amano, H.; Akasaki, I. Control of Threshold Voltage of Enhancement-Mode AlxGa1-xN/GaN Junction Heterostructure Field-Effect Transistors Using p-GaN Gate Contact. *Jpn. J. Appl. Phys.* **2007**, *46*, 115–118. [CrossRef]

152. Efthymiou, L.; Longobardi, G.; Camuso, G.; Chien, T.; Chen, M.; Udrea, F. On the physical operation and optimization of the p-GaN gate in normally-off GaN HEMT devices. *Appl. Phys. Lett.* **2017**, *110*, 123502. [CrossRef]

153. Bakeroot, B.; Stockman, A.; Posthuma, N.; Stoffels, S.; Decoutere, S. Analytical Model for the Threshold Voltage of ${p}$ -(Al)GaN High-Electron-Mobility Transistors. *IEEE Trans. Electron Devices* **2018**, *65*, 79–86. [CrossRef]

154. Huang, H.; Liang, Y.C.; Samudra, G.S.; Huang, C.-F. Design of novel normally-off AlGaN/GaN HEMTs with combined gate recess and floating charge structures. In Proceedings of the IEEE 10th International Conference on Power Electronics and Drive Systems (PEDS), Kitakyushu, Japan, 22–25 April 2013; pp. 555–558.

155. Wang, H.; Wang, N.; Jiang, L.-L.; Zhao, H.-Y.; Lin, X.-P.; Yu, H.-Y. Study of the enhancement-mode AlGaN/GaN high electron mobility transistor with split floating gates. *Solid-State Electron.* **2017**, *137*, 52–57. [CrossRef]

156. Wang, H.; Wang, N.; Jiang, L.-L.; Lin, X.-P.; Zhao, H.-Y.; Yu, H.-Y. A novel enhancement mode AlGaN/GaN high electron mobility transistor with split floating gates. *Chin. Phys. B* **2017**, *26*, 047305. [CrossRef]

157. Kirkpatrick, C.; Lee, B.; Choi, Y.; Huang, A.; Misra, V. Threshold voltage stability comparison in AlGaN/GaN FLASH MOS-HFETs utilizing charge trap or floating gate charge storage. *Phys. Status Solidi* **2012**, *9*, 864–867. [CrossRef]

158. Duan, X.-L.; Zhang, J.-C.; Xiao, M.; Zhao, Y.; Ning, J.; Hao, Y. Groove-type channel enhancement-mode AlGaN/GaN MIS HEMT with combined polar and nonpolar AlGaN/GaN heterostructures. *Chin. Phys. B* **2016**, *25*, 087304. [CrossRef]

© 2018 by the authors. Licensee MDPI, Basel, Switzerland. This article is an open access article distributed under the terms and conditions of the Creative Commons Attribution (CC BY) license (http://creativecommons.org/licenses/by/4.0/).

electronics

MDPI

Review

Device Applications of Synthetic Topological Insulator Nanostructures

Chenxi Yue, Shuye Jiang, Hao Zhu *, Lin Chen, Qingqing Sun and David Wei Zhang

State Key Laboratory of ASIC and System, School of Microelectronics, Fudan University,
Shanghai 200433, China; 17212020056@fudan.edu.cn (C.Y.); 16210720065@fudan.edu.cn (S.J.);
linchen@fudan.edu.cn (L.C.); qqsun@fudan.edu.cn (Q.S.); dwzhang@fudan.edu.cn (D.W.Z.)
* Correspondence: hao_zhu@fudan.edu.cn; Tel.: +86-21-6564-7395

Received: 31 August 2018; Accepted: 26 September 2018; Published: 1 October 2018

Abstract: This review briefly describes the development of synthetic topological insulator materials in the application of advanced electronic devices. As a new class of quantum matter, topological insulators with insulating bulk and conducting surface states have attracted attention in more and more research fields other than condensed matter physics due to their intrinsic physical properties, which provides an excellent basis for novel nanoelectronic, optoelectronic, and spintronic device applications. In comparison to the mechanically exfoliated samples, the newly emerging topological insulator nanostructures prepared with various synthetical approaches are more intriguing because the conduction contribution of the surface states can be significantly enhanced due to the larger surface-to-volume ratio, better manifesting the unique properties of the gapless surface states. So far, these synthetic topological insulator nanostructures have been implemented in different electrically accessible device platforms via electrical, magnetic and optical characterizations for material investigations and device applications, which will be introduced in this review.

Keywords: topological insulator; field-effect transistor; nanostructure synthesis; optoelectronic devices; topological magnetoelectric effect

---

1. Introduction

In the past century, fundamental scientists and physicists have never stopped searching for new elementary particles. Instead of dealing with the atoms and electrons that were found centuries ago, there has been a growing interest in condensed matter physics in the formation of a new state of matter by putting the fundamental elements together. Other than the conventional states of matter such as conductors, insulators, semiconductors, superconductors and magnets, microcosmic mechanisms and quantum states of emerging states of matter are becoming more attractive with the boost of observation techniques and characterization capabilities. Since the first discovery of the quantum Hall (QH) state by Klitzing et al. in 1980 [1] which describes the electric current flow along the edges of a two-dimensional (2D) sample with insulating bulk, great effort has been made to seek the origin and manipulate such quantum state that is topologically different from all known states of matter.

The topological insulator is a newly discovered electronic phase which has insulating bulk and conducting surface [2–4]. Topological insulator is different from superconductors and magnets in that it has a topological order which is protected by time-reversal symmetry. These properties make it very attractive to realize high-mobility and non-dissipative electrical transmission. More interestingly, unlike the QH state, which requires the presence of a strong magnetic field, topological insulators can be observed without magnetic field. Instead, the spin-orbit coupling with heavy elements such as Hg and Bi induces magnetic field during the electron movement in 2D topological insulators. This is known as the quantum spin Hall (QSH) state, which was first experimentally observed in HgTe quantum wells in 2006 [5,6]. The three-dimensional (3D) topological insulator was predicted in

Bi_xSb_{1-x} alloys in 2007, and shortly, binary chalcogenide compounds with simpler structure such as Bi_2Se_3, Bi_2Te_3, and Sb_2Te_3 were predicted as 3D topological insulators, which were further confirmed by angle-resolved photoemission spectroscopy (ARPES) and scanning tunneling microscopic (STM) measurements [4,7].

So far, 3D topological insulators have been widely explored and studied. However, the experimental realization of the prediction and applications of 3D topological insulators still remain elusive largely due to the interference from the conduction contribution from the bulk in small bandgap semiconductor. In recent years, 3D topological insulator nanostructures with thickness shrinking down to nanoscale while retaining the 3D topological insulator characteristics and stoichiometry have aroused more and more attention because of the large surface-to-volume ratio highlighting the surface conduction. These topological insulator nanostructures have provided excellent platforms for creative and innovative research, and have great potential for future device applications. Benefiting from quintuple layer structure of typical topological insulators such as Bi_2Te_3 and Bi_2Se_3, mechanical exfoliation similar to that used on graphene was first tested to fabricate simple yet low-cost topological insulator devices [8]. Significant progress has been achieved in the understanding and applications of topological insulators. For example, a magnetotransport measurement on the exfoliated Bi_2Te_3 device suggested that the 2D conduction channels originate from the surface states in the 3D topological insulators [9], and the coupling between the top and bottom surface states can result in an energy barrier close to the bulk bandgap demonstrated by the insulating behavior in exfoliated Bi_2Se_3 FET device [10]. In addition, optoelectronic properties of the exfoliated topological insulators have also been characterized. The spin direction of the electrons on the surface states of the exfoliated Bi_2Se_3 flake have been confirmed to be perpendicular to the electron movement, and it can be manipulated by shedding circular polarized light [11].

Although mechanical exfoliation has made the preparation of topological insulator samples easier with simple process and low cost, the challenge for realizing state-of-the-art device and manipulating the surface states of the exfoliated flakes still lies in minimizing the bulk conduction, which is basically originated from the chalcogen vacancies or anti-site defects. Furthermore, the low yield and random nature of the exfoliated flakes also make this method not suitable for large-scale fabrication and device integration. Up to now, various synthetic approaches to grow topological insulator nanostructures have been studied and developed, enabling more delicate devices while maintaining high surface-to-volume ratio. In this paper, we review the development of synthetic approaches for topological insulators nanostructures and their applications in novel nanoelectronic, spintronic and optoelectronic devices.

2. Synthesis Approaches

Although most current experimental research based on topological insulator still focuses on the surface states of thin films prepared by mechanical exfoliation from bulk material, various physical and chemical approaches to synthesize topological insulator nanostructures have been widely investigated. Molecular beam epitaxy (MBE) is one of the earliest synthetic methods to grow high-quality topological insulator thin films. Nominally stoichiometric single crystal thin film can be easily prepared by MBE, and more significantly, MBE deposition of topological insulator is based on a growth unit of one quintuple layer, with both physical and chemical reaction involved in the process. Although high vacuum environment is required during MBE deposition to allow the direct injection of atom or molecular beam onto the single crystal substrate, the synthesized films usually have good crystal integrity, precise composition and excellent large-scale uniformity. It has been demonstrated that the topological features of MBE-grown thin films start to appear from the thickness of 2 quintuple layers due to the weaker inter-surface coupling between the top and bottom surface as compared with 1 quintuple layer film [12]. In 2009, Zhang et al. reported the direct observation of quantum interference by the surface states on the MBE-grown Bi_2Te_3 thin film [13]. The high-quality single crystal Bi_2Te_3 film limited the nonuniform morphology, and the atomically flat film has been demonstrated to be a good platform to study the scattering of the topologically non-trivial surface states which are quantum mechanically protected by the time-reversal symmetry confirmed by the imaging of standing waves of the surface states using STM [13]. In 2012, Liu et al. synthesized Bi2Te3, Bi2Se3 films and their alloys on GaAs (001) substrates by using MBE [14]. Reflection high-energy electron diffraction (RHEED), atomic force microscopy (AFM), X-ray diffraction (XRD), high-resolution transmission electron microscopy (HRTEM), Raman spectroscopy and mapping characterizations have been used to examine the topological insulator thin film qualities. As shown in Figures 1a and 2b, the Raman mapping patterns suggest that the position difference of the E2g Raman peaks are less than 1 cm^{-1} within a scan area of 15 μm × 15 μm, indicating good film uniformity in such a relatively small area. Figure 1c,d show the HRTEM images of the cross section of the grown topological insulator films on GaAs (001) substrate. Clear layered crystal features have been observed indicating good crystallinity [14]. However, quite few local and extended defects can also be observed which might be due to the instable growth process, and the imperfect interface between the topological insulator and the GaAs substrate is owing to the symmetry mismatch between the hexagonal lattice of Bi_2Te_3/Bi_2Se_3 and the cubic symmetry of GaAs (001) surface. Nevertheless, the topological insulator films synthesized by MBE is still of high crystalline quality, and further improvement of the interface can be expected through effective substrate engineering.

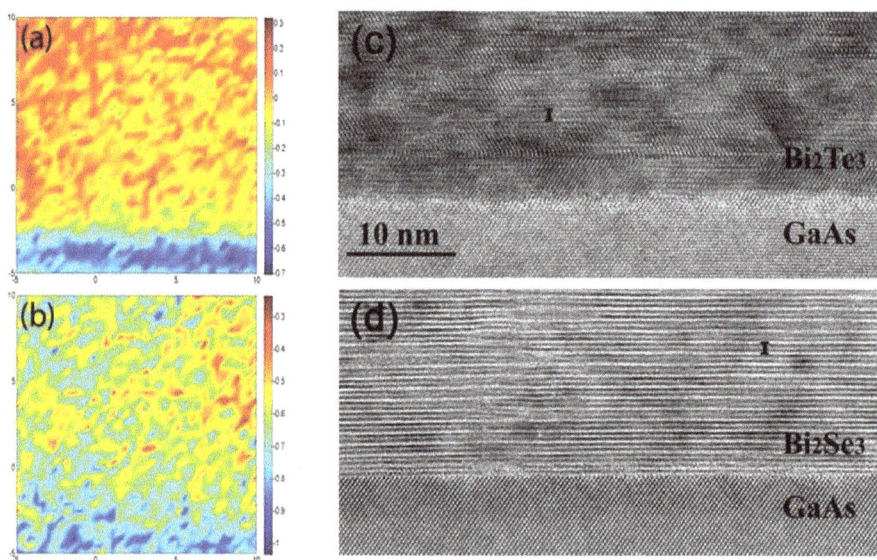

Figure 1. Raman mapping patterns (the position differences of the E^2_g peak) of (**a**) Bi_2Te_3 and (**b**) Bi_2Se_3 with the thickness of 136 nm and 150 nm, respectively. The area of measurement is 15 μm × 15 μm. High-resolution transmission electron microscopy (HRTEM) images of (**c**) Bi_2Te_3 and (**d**) Bi_2Se_3 topological insulators grown on a GaAs substrate [14]. Reproduced with permission from [14], Copyright AIP Publishing, 2012.

Referring to the MBE method, 3D topological insulators in planar geometry have been explored extensively using different synthetic methods. Chemical vapor deposition (CVD) is one of the most commonly used approach to grow topological insulator thin films in recent years. In a CVD process, chemical reactant evaporates and is gas-transferred to the targeted surface where chemical reaction occurs generating desired film materials. CVD is advantageous in fast-speed and large-area synthesis, and more importantly, the deposition can usually be carried out at low temperature which makes it very attractive in electronic device fabrication and integration compatible with current semiconductor technology. In 2016, Wang et al. reported the synthesis of ultra-large $Pb_{1-x}Sn_xTe$ topological insulator nanoplates by using low-cost and efficient CVD process [15]. The prominent characteristic of topological surface transport indicated that these nanoplates can be great candidates for low-dissipation transistors. The angle and temperature dependence of the magneto-conductance revealed the 2D nature of the weak antilocalization effect in them [15]. Besides, as shown in Figure 2, the peculiar 2D geometry of obtained nanoplates which can be applied to functional devices lay a foundation for the investigation of mirror symmetry–induced topological surface transport properties [15]. Metal-organic chemical vapor deposition (MOCVD) is a derivative technique of conventional CVD and has been implemented in the synthesis of topological insulator thin films recently. MOCVD uses metal organic compounds and hydrides as source material and topological insulator single crystal compounds grow on the substrate in the way of gas phase epitaxy through thermal decomposition reaction. MOCVD can enable high-purity, high-uniformity, large-scale and repeatable films, as well as precise control over the film thickness. In 2014, Bendt et al. demonstrated the layer-by-layer growth of smooth Sb_2Te_3 films on c-oriented Al_2O_3 substrates using Et_2Te_2 and i-Pr_3Sb as precursors [16]. The obtained high-quality films allowed for the measurement of the topological surface state for MOCVD grown Sb_2Te_3 by ARPES for the first time. The results illustrated the high-quality topological surface state which is even comparable to the optimized bulk single crystals Sb_2Te_3 films and detailed dispersions of the bulk valence band. In terms of electrical sheet resistivity, the characteristic of increasing monotonically with rising temperature was also found [16].

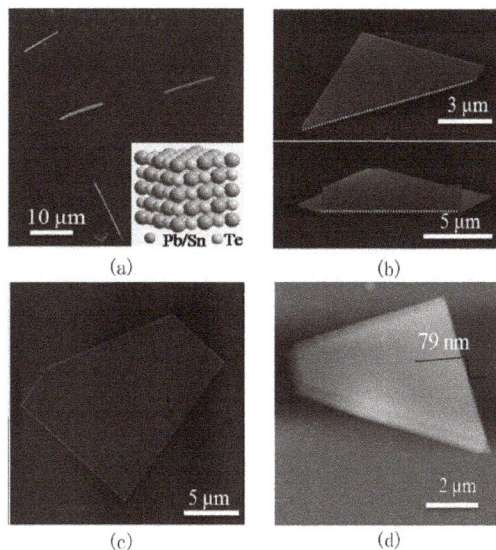

Figure 2. (a) SEM image of the vertically oriented $Pb_{1-x}Sn_xTe$ nanoplates. Inset: crystal structure model of $Pb_{1-x}Sn_xTe$. (b) 15°-tilted SEM images of the vertically oriented $Pb_{1-x}Sn_xTe$ nanoplates. (c) SEM and (d) AFM images of the planar $Pb_{1-x}Sn_xTe$ nanoplate [15]. Reproduced with permission from [15], Copyright John Wiley and Sons, 2015.

In addition to the topological insulator thin films, non-planar topological insulator nanostructures with reduced sample size have also become more and more intriguing such as nanowires, nanoribbons, nanosheets, and so forth. These nanostructures are expected to significantly enhance the surface conduction by suppressing the contribution of bulk carriers with their large surface-to-volume ratio. Typically, the quasi-one-dimensional nanostructures are synthesized by following the catalytic vapor–liquid–solid (VLS) method which has been widely used on the synthesis of silicon nanowires [17,18]. Au nanoparticles are commonly used as catalyst to stimulate the growth of semiconductor nanowire or nanoribbon through the precipitation from the supersaturated Au/semiconductor alloy. In 2010,

Peng et al. reported the quantum interference effect in Bi_2Se_3 nanoribbon grown by the Au-catalyzed VLS process [19]. The cross-sectional area of the nanoribbon is 6.6×10^{-15} m^2, which is largely determined by the size of catalyst nanoparticle (Figure 3a–d). Due to the large surface-to-volume ratio, the bulk conduction has been greatly limited, better manifesting the quantum interference effects of the surface states. Furthermore, such catalytical synthesis of topological insulator nanostructures can be applied in the fabrication of various heterogeneous structures. In 2016, Liu et al. reported single-crystalline topological insulator Bi_2Se_3 nanowires synthesized via Au-catalyzed VLS method, as shown in Figure 3e–h, which were further transferred and used to fabricate high-performance Bi_2Se_3/Si heterostructure photodetectors. The photodetectors exhibited excellent optoelectronic properties which were attributed to the high crystal quality of the Bi_2Se_3 nanowires and the high build-in electric field at the Bi_2Se_3/Si heterostructure interface [20]. This further evidenced the great prospect of high-quality topological insulator nanowires synthesized by catalytic VLS method.

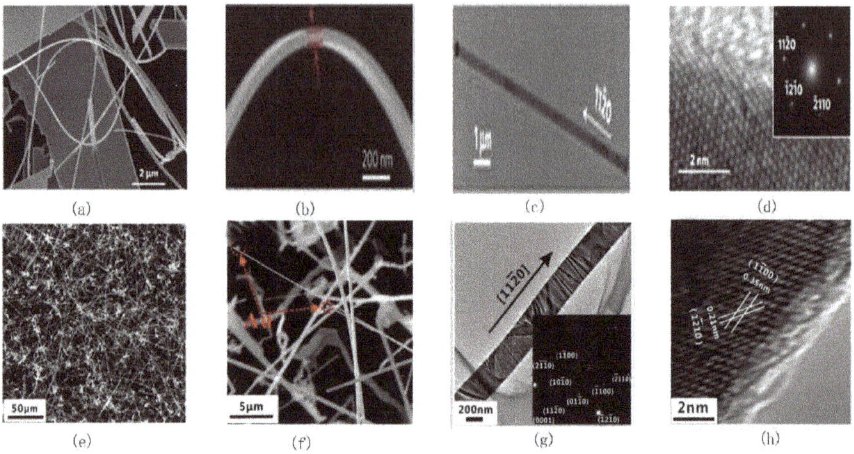

Figure 3. (**a,b**) SEM images of the as-grown Bi_2Se_3 nanoribbons by using VLS method [19]. (**c**) TEM image of a Bi_2Se_3 nanoribbon with Au nanoparticle at the tip. (**d**) HRTEM image of the edge of the Bi_2Se_3 nanoribbon with smooth surface. Inset: the selected-area electron diffraction (SAED) pattern shows the single-crystal characteristic of Bi_2Se_3 nanoribbon [19]. SEM images of the Bi_2Se_3 nanowires with (**e**) low magnification and (**f**) high magnification [20]. (**g**) Low-resolution TEM image of a single Bi_2Se_3 nanowire. (**h**) HRTEM image of the Bi_2Se_3 nanowire [20]. Reproduced with permission from [19], Copyright Springer Nature, 2009. Reproduced with permission from [20], Copyright Royal Society of Chemistry, 2016.

Despite the methods introduced above, other synthesis methods such as solvothermal synthesis have also been used to prepare topological insulator nanostructures. For example, Xiu et al. synthesized Bi_2Te_3 nanoribbons by heating the polyvinylpyrrolidone (PVP) solvent added with Bi_2O_3, Te and ethylenediamine tetraacetic acid (EDTA) powders [21]. However, such chemical approaches involving organic solvent usually include harvesting and multiple cleaning processes, which will inevitably introduce contaminates and deteriorate the surface condition of the synthesized nanostructure. Actually, the major bottleneck of the synthetical topological insulator materials still lies in the crystal properties, specifically the intrinsic defects such as S or Se vacancies and interstitial defects. These defects are responsible for the bulk conduction, and gas molecules can occupy these vacancies forming different types of doping if the topological insulator surface is not passivated or effectively protected [22]. As a result, a lot of attention has been paid to the optimization of the synthesis process and the device structure engineering for in-depth investigation on the material properties and device applications. The following section briefly introduces the device applications of synthetic topological insulator nanostructures, including field-effect transistors, optoelectronic devices, magnetoelectric devices and so forth.

3. Device Applications

3.1. Field-Effect Transistor

The field-effect transistor (FET), based on semiconductor nanostructures, has been regarded as one of the fundamental building blocks for future nanoelectronic device and circuit technologies. Topological insulator FET is also the critical device architecture for further optoelectronics and spintronics applications. Although a variety of FET devices based on exfoliated or MBE-grown topological insulator thin films have been fabricated [10,23–27], yet, up to now, high-performance topological insulator FET devices such as the analog of metal-oxide-semiconductor field-effect-transistor (MOSFET) have been rarely reported. For conventional Si-based MOSFET devices, the surface conduction of Si channel is protected by the thermal SiO_2 with optimized inversion characteristics allowing for better transistor performance. Similarly, the gapless surface states of 3D topological insulators have protected and robust conduction properties which are derived from the intrinsic material properties. Considering the current wide use and preference of silicon over other semiconductors, the integration of topological insulator nanostructures as the conduction channel replacing silicon in MOSFETs will be very attractive in future nanoelectronic device and circuit technologies.

Synthetic topological insulator nanostructures, especially those in non-planar geometry usually have larger surface-to-volume ratio, which is very advantageous in highlighting the unique topological features of the metallic surface states. Furthermore, the integration of such topological insulator nanostructures in FET devices with tunable electrical gate control, the transport properties of the carriers in the bulk and on the surface can be further investigated with possible electrical manipulation. For example, in 2013, Zhu et al. synthesized high-quality single crystal Bi_2Se_3 nanowires by using the VLS mechanism, and integrated them in a high-performance FET device through a self-alignment process (Figure 4a) [28]. The Bi_2Se_3 nanowires were grown from pre-patterned Au catalyst, enabling the direct device fabrication without nanowire harvesting and manual positioning. This approach not only enables the batch fabrication of homogeneous nanowire FETs but also limits the steps in which contamination might be introduced degrading the topological insulator surface properties. Excellent electrical performance has been achieved, including very sharp turn-on, near-zero cutoff current, large On/Off ratio (over 10^8), and well-saturated output current (Figure 4b,c). As compared with the conventional Bi_2Se_3 thin film transistor, the Bi_2Se_3 nanowire FET exhibited much better electrostatic gate control and subthreshold behaviors [29]. More significantly, the surface states were protected by high-quality high-k dielectric, and the surrounding-gate device geometry has greatly enhanced the gate control over the channel allowing for the full depletion of electrons from the nanowire. Therefore, the temperature dependent off-state current is due to the thermal excitation across the bandgap of the bulk, which means that the off-state current obtained above 240 K is contributed from the bulk conduction [28]. On the other hand, the linear dependence of the saturation current on the over-threshold voltage suggests the drift current model instead of the diffusion current model such as in conventional MOSFET. Such metal-like behavior exactly indicates that the on-state current is dominated by the contribution from the metallic surface conduction. And due to the very sharp turn-on performance of the FET, the separation of the surface and bulk conduction can be effectively achieved within a range of a few volts (~2 V) of gate voltage [28]. Such controlling and manipulation over the surface conduction and insulating switch-off through electrical approaches open up a suite of potential application in novel nanoelectronic and spintronic devices.

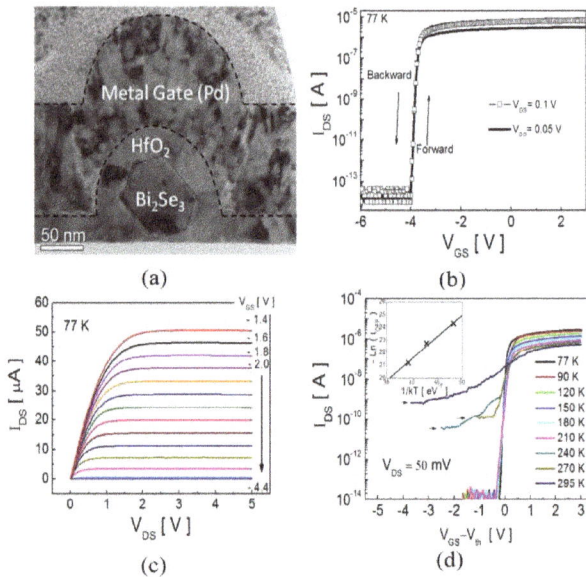

Figure 4. (**a**) TEM image of the cross-section of a Bi_2Se_3 nanowire field-effect transistor (FET); (**b**) I_{DS}-V_{GS} of Bi_2Se_3 nanowire FET at 77 K; (**c**) Linear-scale I_{DS}-V_{DS} curves for V_{GS} from −4.4 V to −1.4 V at 77 K. (**d**) The illustration of I_{DS}-V_{GS} at temperature from 77 K to 295 K with V_{DS} = 50 mV. Inset: the relationship between $\ln(I_{DS})$ at Off state and $1/kT$ [28]. Reproduced with permission from [28], Copyright Electrochemical Society, 2014.

Although the synthetic nanowire and nanoribbon geometry can enable advanced device structure to demonstrate the conduction from bulk and surface separately by electrical gating, these nanostructures typically have a width (cross-sectional) of tens of nanometers. On the contrary, some unique and interesting topological phase transition such as that between a trivial insulator and a nontrivial topological phase can only be achieved in ultrathin systems, which however has not aroused sufficient attention. In 2017, Liu et al. reported ultrathin $(Bi_{1-x}Sb)_2Se_3$ field effect transistor with On/Off ratio reaching ~25,000% as shown in Figure 5a,b [30]. The ultrathin film (4.2 nm) was synthesized by MBE and the varying Sb doping level has been proved to be also effective in tuning the transport properties in FETs in addition to the electrical gating. In such a combined approach, a large On/Off ratio can be achieved by tuning the Fermi level in the regime around the surface gap. It has been reported that the top and bottom surface states in Bi_2Se_3-based topological insulators will be hybridized and will form a surface gap when the film thickness is below a critical value of 6 nm [30,31]. The In doping in Bi_2Se_3 can also tune the transport properties as well as enabling the phase transition from a nontrivial topological metal to a trivial band insulator (the so-called metal-insulator transition). Both the film thinning and Sb-doping enhance the electric field across the Bi_2Se_3 channel, and the spin-orbit coupling strength is reduced due to substitution of Bi with lighter Sb element. Eventually, the bulk bandgap is reduced and the penetration depth of the surface states is increased becoming comparable to the film thickness, leading to the opening of a surface gap [30]. This is the first experimental observation of a large On/Off ratio in ultrathin topological insulator FET.

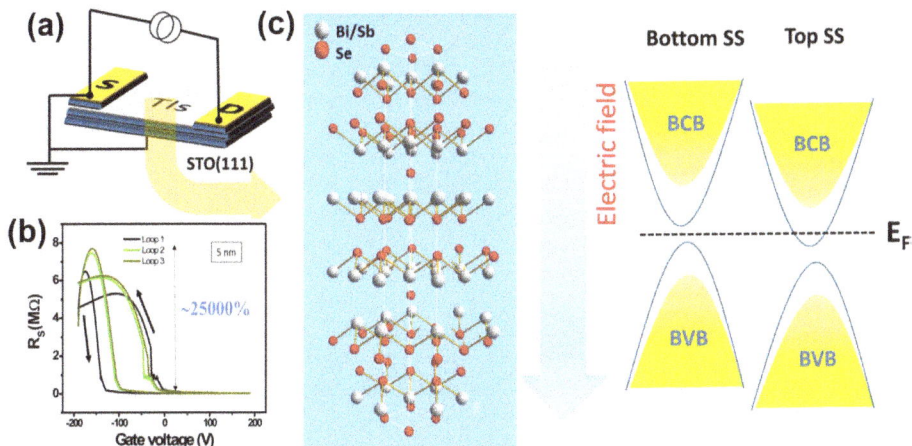

Figure 5. (**a**) The schematic of ultrathin $(Bi_{1-x}Sb_x)_2Se_3$ field effect transistor on $SrTiO_3$ substrate. (**b**) Rs vs. Vg of the FET based on 5 nm $(Bi_{1-x}Sb_x)_2Se_3$ film. (**c**) Mechanism of the large On/Off ratio: the Fermi level in the ultra-thin topological insulator film can be tuned within or close to the surface gap in both bottom and top surface states [31]. Reproduced with permission from [31], Copyright American Chemical Society, 2017.

As seen in most high-performance topological insulator FET research, the topological insulator surface channels are in direct contact with dielectric material in high-k/metal gate device structure. On the one hand, it is the prerequisite to implement the top gate FET geometry with topological insulator nanostructure as the channel. On the other hand, it has been found that topological insulator materials usually become heavily doped when exposed to air, so that the deposition of high-k dielectric can effectively prevent the N, O elements, and water molecules penetrating into the surface lattice. However, like other semiconductors without dangling bonds at the surface such as graphene and MoS_2, it might be hard to realize a uniform deposition of high-k dielectric on the surface of topological insulator by using atomic layer deposition (ALD) [23,32,33]. This is because of the absence of dangling bonds on the surface resulting in no chemical adsorption and nucleation sites in the initial deposition cycles. Liu et al. have studied the ALD Al_2O_3 deposition on Bi_2Te_3 for a dual-gate FET device fabrication by testing different ALD parameters [34]. Different ALD precursors (trimethylaluminum (TMA) and H_2O or TMA and O_3) have been tested and compared. In the case of graphene and MoS_2, in order to avoid the island or cluster morphology, some surface pretreatments are usually incorporated before the ALD deposition to generate nucleation sites for uniform dielectric deposition. Surprisingly, it is different for topological insulators as Liu et al. found that the surface is not as smooth as graphene which might be attributed to the oxidation of the Te-terminated surface by oxygen or water molecules. This oxidation facilitates the ALD process by generating nucleation sites for precursor adsorption [34]. However, as compared with the back-gate control, the top-gate control is less effective at the same electric field which is largely due to the degraded dielectric/Bi_2Te_3 interface during the ALD deposition. This is because of the introduced defects and impurities degrading the interface quality can act as trapping and de-trapping sites for the charge carriers leading to instable FET device performance. Nevertheless, like FETs based on other novel semiconductor materials such as transition metal dichalcogenides, more in-depth experimental and theoretical studies are still on-going to improve the high-k dielectric/topological insulator interface and optimize the ALD deposition process. In addition to the topological insulator/dielectric interface, the electron dynamics of topological insulator-based semiconductor-metal interface and interface engineering methods have also been investigated to provide reference for further device fabrications [35,36].

Similar to other novel semiconductors, the Fermi level in topological insulators can be effective tuned by constructing FET device and applying gate voltage [37–41]. In this way, the bulk carriers in topological insulators nanostructures can be easily depleted, achieving larger contribution from the surface conduction. However, sometimes the position of the Fermi level cannot be properly tuned only by using electrical gating to the same or the adjacent regime to the intrinsic Fermi level, because it might fall into the valence band under negative voltages. Elemental doping has been confirmed to be an effective approach to assist the manipulation of Fermi level in topological insulator nanostructures [42–44]. In 2012, Wang et al. reported the Na elemental doping in Bi_2Te_3 nanoplates synthesized by solvothermal method [45]. The ~0.8% Na doping was expected to tune the Fermi level of Bi_2Te_3 nanoplates toward the middle of the bandgap. By integrating such doped Bi_2Te_3 topological insulator nanoplates in a back-gate FET device (Figure 6a), the tuned Fermi level in the Bi_2Te_3 channel material enables the transition between different conduction types by using gate voltages. As shown in Figure 6b, the Hall measurement results indicated that the gate voltage can change the dominating conduction channels, and when the conduction contribution from the surface electrons and bulk electrons is larger than that from bulk holes, the Bi_2Te_3 FET exhibited n-type behavior. On the contrary, if the bulk holes dominated the conduction, p-type behavior was observed [45]. Moreover, through the magnetotransport measurement on the Na-doped Bi_2Te_3 nanoplate FET, distinct quantum interference effects under various gate voltage have been observed. As shown in Figure 6c,d, the frequency of the Shubnikov-de Hass (SdH) oscillation exhibited a clear increasing trend with increasing gate voltage, indicating higher Landau levels. The parameters estimated from the SdH oscillation such as the Fermi wavevector for different gate voltages also indicate that the oscillation is originated from the surface states [45]. Utilizing electrical gating to achieve the control and manipulation over the surface states of the intrinsic or doped topological insulator nanostructures has provided an efficient and effective approach to study the transport properties of the materials and dissipationless electronic and spintronic device applications.

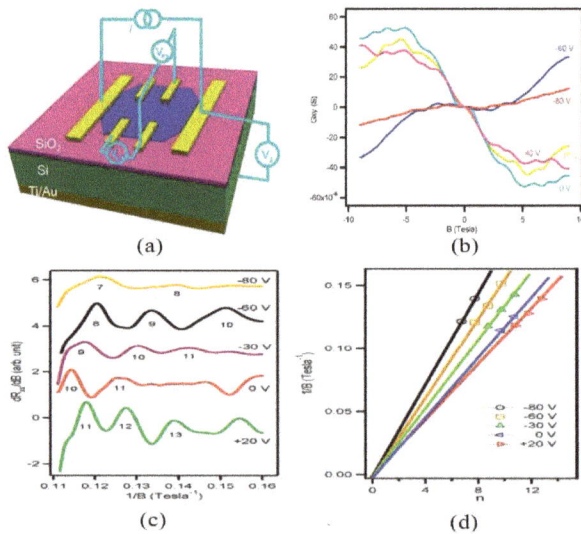

Figure 6. (**a**) A schematic of back-gate FET based on 40-nm-thick Bi_2Te_3 nanoplate with ~1.2 μm channel length. (**b**) Hall measurement results showing the type transition from n-type to p-type by changing gate voltage at 1.9 K. (**c**) SdH oscillations indexed by different Landau levels under increasing gate voltage at 1.9 K. (**d**) Landau level as a function of 1/B under different voltages [43]. Reproduced with permission from [43], Copyright Springer Nature, 2012.

It is worth mentioning here that synthetic topological insulator nanostructures have attracted more attention in recent years due to their high-quality single crystal nature and enhanced surface-to-volume ratio as compared with bulk materials. This has provided excellent geometries for probing the transport properties of the surface states, such as the research reported in References [19]. However, although these works took advantages of the geometry of the synthetic topological insulator nanostructures, there lacks effective gate control to further manifest the topological features of the surface states, which is very important to realize future electrically accessible devices. Up to now, full understanding over the transport properties of 3D topological insulator by interpreting the electrical behavior under various circumstances through careful device structure design and engineering has not been seen yet.

3.2. Optoelectronic Device

Optoelectronic devices can generate or sense light, or convert it into electric signal that can be processed by electronic devices and apparatus. Typical optoelectronic devices include lasers, light emitting diode (LED), photodetector, solar cell, and so forth. Up until now, various semiconductors have been studied extensively for different kinds of optoelectronic applications. For example, Si-based optoelectronic devices are advantageous in the CMOS compatibility which can be directed integrated on Si substrates. III–V compound-based optoelectronic devices are also attractive in the applications of optical fiber, infrared and visible LEDs and high efficiency solar cells. Novel 2D semiconductors such as MoS_2 and black phosphorus are also interesting candidates in realizing ultra-scaled optoelectronic devices due to their 2D nature and thickness-depended band structure. On the other hand, topological insulators also have intrinsic attractive optical properties such as high bulk refractive index and broadband surface plasmon resonance. In addition, topological insulators usually have a narrow bandgap which can improve the conductivity and also enables high transparency in near-infrared frequency regime. The Dirac-like surface states of topological insulator can bring strong optical absorption, enabling high-performance broadband photodetection capabilities. It has been theoretically proposed that topological insulators can electrically response to light signals, due to their novel and intriguing quantum phase of matter with spin-polarized surface states. For example, this effect has been preliminarily characterized by the photo-induced quantum phase transitions between conventional insulator and topological insulators in 2D electronic system [46], as well as the generation of helicity-dependent direct current by circularly polarized light [47]. In 2012, McIver et al. have experimentally observed a photocurrent which is originated from the topological helical Dirac fermions from an exfoliated Bi_2Se_3 thin flake [11]. More interestingly, the direction of the photocurrent can be reversed by reversing the helicity of the light. This is the first experimental demonstration of the circular and linear photogalvanic effects on topological insulators where the Rashba spin-split valence and conduction bands provide the asymmetric spin distribution. However, the observed polarization-dependent photocurrent always coexists with bulk photocurrent. This is because the depopulation of the Dirac cone using polarize light will generate bulk-like excited states. The contribution of photocurrent arising from the bulk can be expected to be eliminated by using samples with more insulating bulk and lower light energy. Nevertheless, the interference from the bulk state has been one of the major bottlenecks to further investigate the optoelectronic properties of the surface states in topological insulators.

Synthetic topological insulator nanostructures have also been tested in various optoelectronic device structures to study the photo-sensing and photo-detecting properties of the surface states. For example, in 2015, Zheng et al. reported a near infrared light photodetector based on Sb_2Te_3 ultra-thin film synthesized by MBE [48]. As shown in Figure 7, they found that the Sb_2Te_3 photodetector was sensitive to 980 nm light illumination, and the responsivity was 21.7 A/W, which was much better than most reported topological insulator-based photodetectors. However, both bulk and surface states contributed to the photocurrent and responsivity, with possible dominating contribution from the bulk due to the electron-hole pairs generated by photons. The interesting finding is that the bulk and surface contribution exhibited different temperature dependent responsive behaviors with the bulk state has a higher responsivity at room temperature while the surface state has higher responsivity at lower temperature [48]. This can lead to novel photo-detecting applications in complicated environment. Another shortcoming of the topological insulator back-gate FET based photodetector lies in the relatively slow response speed (Figure 7f) which is largely due to the degradation of the exposed surface states without protection. Dangling bonds or absorbents can be generated and introduced on the surface acting as the trapping centers thus reducing the response speed. Nevertheless, the contribution from the surface states still confirms the formation of a spin-polarized electrical signal under the polarized light illumination, and the relatively high responsivity is due to the location of the Dirac point near the Fermi level in the topological insulator film [49]. In order to improve the immunity of the optoelectronic properties of topological insulator devices, surface passivation by the ALD deposition of high-k dielectric is a useful and straightforward approach. However, as described in the previous section, uniform formation of oxide film on the surface of topological insulator also involves the deterioration of the surface conditions due to the adsorption of oxygen or water. The oxidation of topological insulators such as Bi_2Te_3 can shift the Fermi level towards up or down with respect to the Dirac point leading to degradation of photoconductivity in optoelectronic devices. More theoretical and experimental work focusing on the surface passivation or device structure engineering are still necessary to rule out the factors from external ambient for more accurate evaluation of the optoelectronic properties of topological insulators.

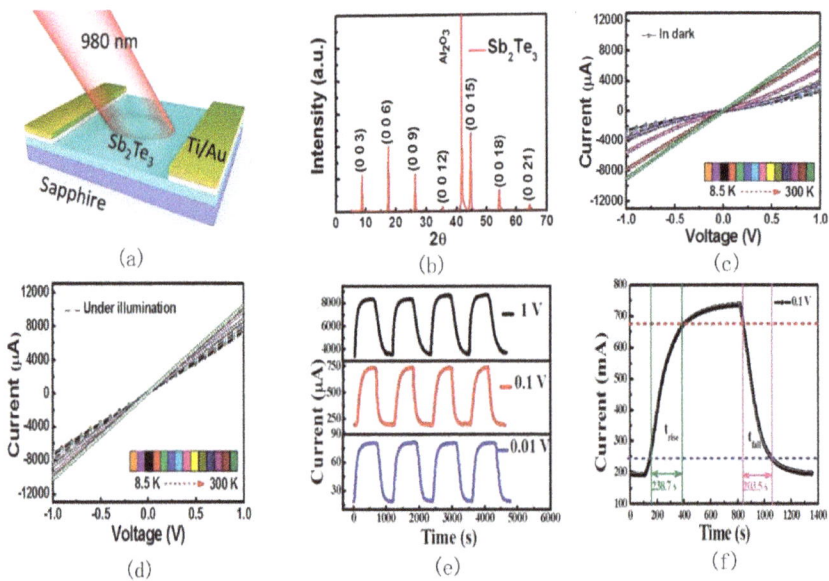

Figure 7. (**a**) The schematic illustration of Sb_2Te_3-based near infrared (NIR) light photodetector. (**b**) The X-ray diffraction image of Sb_2Te_3 film grown by MBE. (**c,d**) are the I–V characteristics of Sb_2Te_3 film in dark and under illumination, respectively, at the temperature range of 8.5–300 K. (**e**) Photoresponse of the Sb_2Te_3-based photodetector at 45 K under voltages of 0.01 V, 0.1 V and 1 V. (**f**) One cycle of the photoresponse at a temperature of 45 K and 0.1 V voltage [48]. Reproduced with permission from [48], Copyright Royal Society of Chemistry, 2015.

An alternative approach to enhance the surface contribution of the photocurrent is to use non-planar synthetic topological insulator nanostructures with larger surface-to-volume ratio such as nanowires and nanoribbons. For example, very recently in 2017, broad spectral photodetection from ultra-violet to near-infra-red was reported based on a Bi_2Te_3 nanowire photodetector [50]. Device performance was compared between the exfoliated Bi_2Te_3 nanosheets and the Bi_2Te_3 nanowires which were fabricated by milling the nanosheet using focused ion beam (FIB). The nanosheets of Bi_2Te_3 exhibited ultra-high photoresponsivity of 74 A/W at 1550 nm. More significantly, the nanowires transformed by FIB milling showed about one order enhancement in photoresponsivity. It is worth to mention that the FIB fabrication of nanowires involves harsh exposure environment with inevitable Ga ions implantation and even sample deformation during the milling process. The enhancement in photoresponsivity is exactly due to the nano-confinement effects of topological insulator surface states and more electron-hole pair generation for effective incident photons. Further aging tests showed that the devices still have robust detectivity and photoconductivity even after a period of four months' time. The above experimental results explicitly demonstrate the efficient carrier transport through the robust topological surface states which is resistant to material deformation and non-magnetic impurities. Promising improvement in topological insulator-based optoelectronic device can also be expected with future advanced synthetic nanomaterials and device engineering.

Another superiority of the synthetic nanostructures over the exfoliated flakes is the feasibility to fabricate high-quality topological insulator-based heterostructures [51–55]. The interface in the synthetic topological insulator-based heterostructure is much better than that of the exfoliated flake system. This is very helpful to enhance the effective generation and transfer of the photocarriers at the interface, enlarging the photocurrent without sacrificing the detecting spectral width. In 2015, Yao et al. reported a photodetector built with Bi_2Te_3/Si heterostructure which exhibited a prominent characteristic of ultra-broadband photodetection (370–118,000 nm) (Figure 8a–c) [56]. Besides, the responsivity of 1 A/W and detectivity of 2.5×10^{11} cm·Hz$^{1/2}$·W^{-1} have been achieved as well as excellent device stability even after long-time high-energy exposure and acidic treatment. However, the Bi_2Te_3 thin film in this heterostructure device was deposited by using pulsed laser deposition (PLD) technique, which is not favorable in forming high-quality single-crystal films. The intrinsic topological surface states and the Bi_2Te_3/Si interface will be deteriorated in the as-synthesized polycrystalline Bi_2Te_3 film, degrading the photodetection performance. In 2016, Zhang et al. reported a high-performance Bi_2Se_3/Si heterostructure photodetector in which the single crystal Bi_2Se_3 was deposited by using CVD method, which is superior in forming high-quality single-crystal topological insulator samples as compared with PLD method [57]. As shown in Figure 8d–f, high light responsivity (24.28 A/W), high detectivity (4.39×10^{12} cm·Hz$^{1/2}$·W^{-1}) and fast response speed (microseconds approximately) have been obtained. Broadband detection was also realized (300–1100 nm). Comparing the two reported work above, the similar band structure of Bi_2Te_3 and Bi_2Se_3 topological insulators actually demonstrated the significance of the crystalline quality and resulted interface of topological insulator thin film in optimizing the optoelectronic characteristics by using different crystal synthesis approaches. Later in 2016, Liu et al. investigated the photodetector built with Bi_2Se_3 nanowire/Si heterostructure [20]. The Au-catalyzed Bi_2Se_3 nanowires were transferred on pre-patterned Si substrate area to form the heterostructure. As shown in Figure 8g–i, the device showed an extremely high responsivity reaching 10^3 A/W which is much higher than the above Bi_2Se_3/Si heterostructures [20]. In addition, a broad spectral ranging from 380 to 1310 nm and fast response speed of tens of milliseconds were also achieved. Such high performance was attributed to the superior nanowire quality, and the strong built-in electric field at the heterostructure interface which prohibited the recombining of electron-hole pairs. It should be mentioned here that the Bi_2Se_3/Si heterostructure was formed simply by stacking the nanowires on Si substrate, this is not beneficial to form a high-quality interface with least defects. Unfortunately, no ideal heterostructure such as Bi_2Se_3 nanowire/Si nanowire by using VLS synthesis has yet been reported.

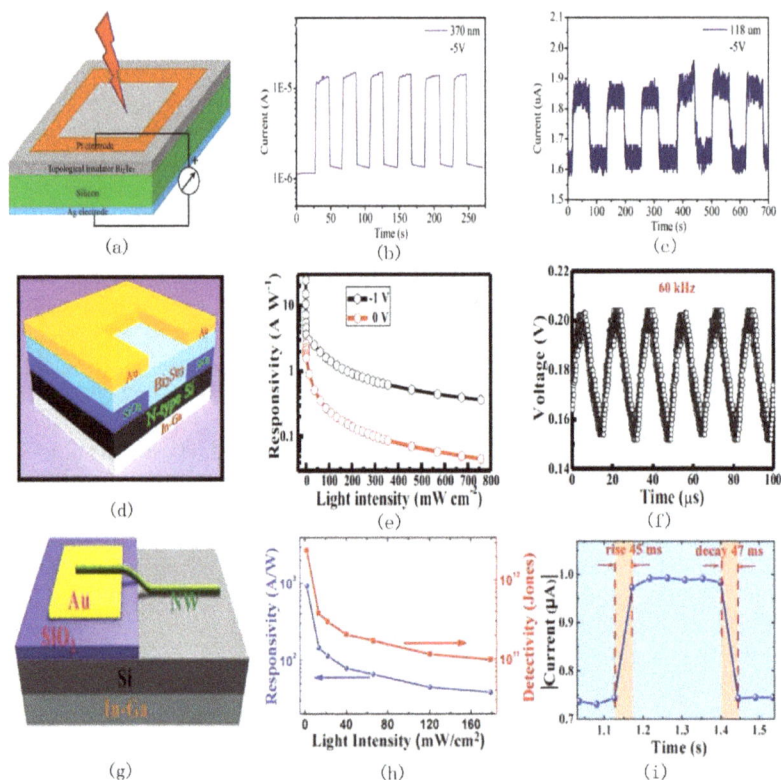

Figure 8. (**a**) Schematic structure of the Bi$_2$Te$_3$/Si-based photodetector. The time-dependent switching behavior under illumination of (**b**) 370 nm and (**c**) 118.8 μm at room temperature [56]. (**d**) The schematic illustration of the Bi$_2$Se$_3$/Si heterostructure device with Au and In−Ga electrodes [57]. (**e**) Photoresponsivity of the Bi$_2$Se$_3$/Si heterostructure photodetector as a function of light intensity at −1 V and 0 V at room temperature. (**f**) Transient response of the photodetector to the high-frequency (60 KHz) light signal at room temperature [57]. (**g**) The schematic illustration of Bi$_2$Se$_3$ nanowire/Si heterostructure photodetector [21]. (**h**) The responsivity and detectivity of Bi$_2$Se$_3$ nanowire/Si heterostructure photodetector versus light intensity at room temperature. (**i**) The response time (rise and decay times) investigation of the photodetector with magnified current response at room temperature [21]. Reproduced with permission from [21], Copyright Royal Society of Chemistry, 2016. Reproduced with permission from [56], Copyright RSC, 2015. Reproduced with permission from [57], Copyright American Chemical Society, 2016.

In 2017, Zhang et al. fabricated thin film-based SnTe/Si vertical heterostructure photodetectors with direct p-SnTe vapor deposition on n-Si surface. High-performance near-infrared detection has been achieved. With the assistance from the built-in electric field in SnTe/Si interface, the absorption efficiency and the separation of photogenerated carriers were greatly improved and the SnTe/Si heterostructure photodetector can have a responsivity of 2.36 AW^{-1}, a high detectivity of 1.54×10^{14} Jones, and a large bandwidth of 10^4 Hz in the near-infrared wavelength, which were shown in Figure 9a–c [58]. Later in 2017, Gu et al. also adopted SnTe/Si heterostructures but by CVD method to fabricate photovoltaic detectors. This self-driven detector realized broadband detection (254 nm–1550 nm), ultrafast response speed (~8 μs) and high detectivity (8.4×10^{12} Jones) as shown in Figure 9d–f [59]. Both work utilized SnTe as the p-type semiconductor to form a heterogeneous structure with Si. Sn Te is well known as the topological crystalline insulator (TCI), which is a new subset of topological insulator materials. TCI possesses exotic surface state properties which are protected by mirror symmetry. This is different from the regular topological insulators which are protected from the time-reversal symmetry. Because of the Dirac metal properties of the surface states, the TCI SnTe can naturally form a Schottky junction with Si at the interface, significantly improving the separation efficiency of photo-generated carriers [59].

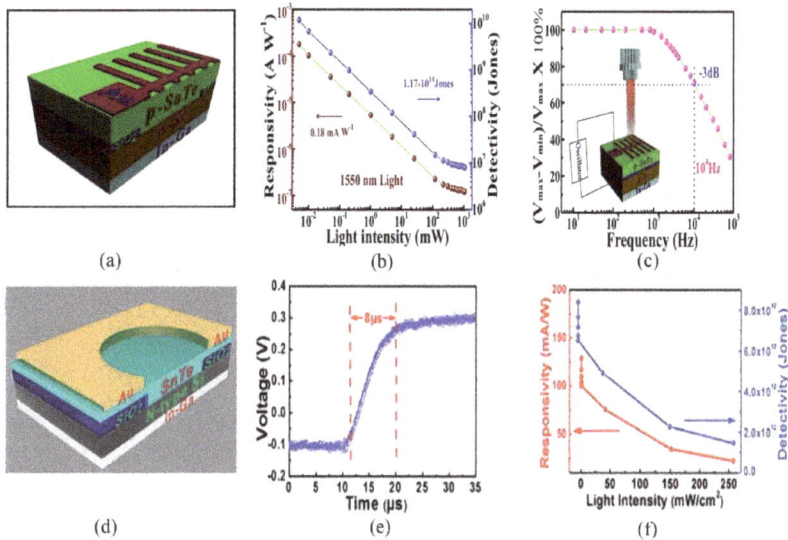

Figure 9. (**a**) The cross-sectional schematic diagram of SnTe/Si vertical heterostructure photodetector with Au top electrode and In-Ga bottom electrode. (**b**) The photoresponsivity and detectivity vs. light intensity curves under 1064 nm light illumination at zero bias voltage. (**c**) The relative balance $(V_{max} - V_{min})/V_{max} \times 100\%$ of the photovoltage of the device as a function of frequency. Inset: schematic diagram of measurement setup [58]. (**d**) Schematic illustration of the SnTe/Si heterostructure photovoltaic detectors. (**e**) The magnified photoresponse demonstrating the rise time of 8 µs. (**f**) The responsivity and detectivity as a function of light intensity at zero bias voltage and room temperature [59]. Reproduced with permission from [58], Copyright American Chemical Society, 2017. Reproduced with permission from [59], Copyright Royal Society of Chemistry, 2017.

More recently, the combinations of layered materials to fabricate heterostructures have received tremendous attention due to their intrinsic quasi-2D nature and proper lattice mismatch, which can be very attractive in optoelectronic applications [60–64]. Typical layered semiconductors including graphene, transition metal dichalcogenides, and topological insulators are promising candidates and have been widely studied. In 2015, Qiao et al. reported a broadband photodetector based on graphene-Bi_2Te_3 heterostructure [65]. The Bi_2Te_3 nanocrystals were epitaxially grown on a graphene template using CVD technique. Since there is no dangling bond on either Bi_2Te_3 or graphene surface, such growth forms a novel van der Waals heterostructure with an atomic gapless interface. As shown in Figure 10a–c, due to the small lattice mismatch between Bi_2Te_3 and graphene, the photo-excited carriers can be effectively transferred and separated at the interface. High photoresponsivity of 35 A/W at 532 nm wavelength and an expanded detection spectral range have been successfully achieved. In 2016, Yao et al. published their work on the synthesis of Bi_2Te_3/WS_2 heterostructure by using PLD deposition [66]. Although PLD is not favorable in preparing single-crystal films as mentioned above, the idea of integrating topological insulator and transition metal dichalcogenide to form 2D heterostructure is a significant advance in the next-generation photodetection. As demonstrated in Figure 10d–f, responsivity of 30.7 A/W with a pronounced detectivity of 2.3×10^{11} cm·Hz$^{1/2}$·W^{-1}, as well as a fast response within 20 ms have been obtained. Recently, a similar Bi_2Te_3/SnS 2D heterostructure photodetector has been reported by the same research team [67]. As shown in Figure 10g–i, it reached a high responsivity of 115 A/W, a large external quantum efficiency of $3.9 \times 10^4\%$ and a superior detectivity of 4.1×10^{11} $cm \cdot Hz^{1/2} \cdot W^{-1}$, which are among the best figures-of-merit of 2D photodetectors so far. Such decent performance is originated from the effective carrier separation at the Bi_2Te_3/chalcogenide interface and carrier transport with high mobility at the topological insulator surface. With the wide spreading of the novel 2D transition metal dichalcogenide in various research fields, its combination with the vast infrastructure of topological insulator will definitely open the new pathways for advanced optoelectronic applications.

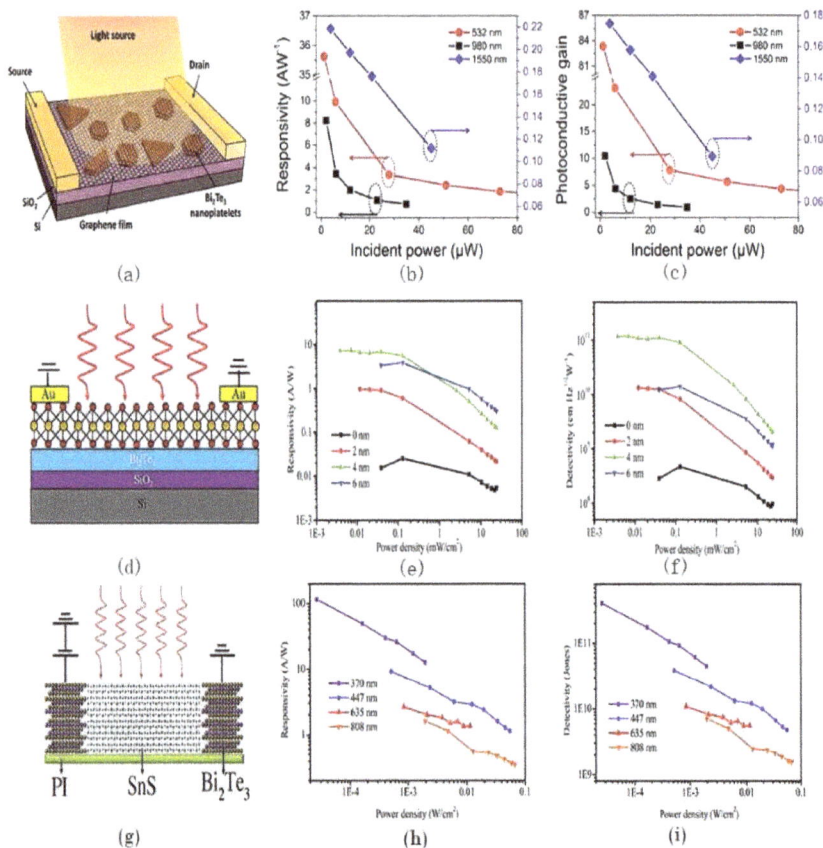

Figure 10. (**a**) The schematic structure of the graphene-Bi_2Te_3 heterostructure phototransistor [65]. (**b**) Photoresponsivity and (**c**) photoconductive gain of graphene-Bi_2Te_3 photodetector as a function of incident power at 532, 980, and 1550 nm, respectively [65]. (**d**) The cross-sectional view of WS_2/Bi_2Te_3 photodetector [66]. (**e**) Power-dependent responsivity and (**f**) detectivity of WS_2/Bi_2Te_3 photodetectors with different Bi_2Te_3 thicknesses [66]. (**g**) The schematic illustration of all-2D Bi_2Te_3-SnS-Bi_2Te_3 photodetector [67]. (**h**) Responsivity and (**i**) detectivity of the Bi_2Te_3-SnS-Bi_2Te_3 photodetector under the illumination of 370, 447, 635 and 808 nm, respectively [67]. Reproduced with permission from [65], Copyright American Chemical Society, 2015. Reproduced with permission from [66], Copyright Royal Society of Chemistry, 2016. Reproduced with permission from [67], Copyright John Wiley and Sons, 2018.

3.3. Magnetoelectric Device

Topological insulators have conductive surface states which are protected by time-reversal symmetry. Meanwhile, these states are related to the spin and the characteristic of QSH effect. Similar to the conventional QH effect, the topological structure in 3D topological insulators will lead to quantized electromagnetic response coefficients [68]. This means that when the time-reversal symmetry on the surface is broken, the quantized electromagnetic response will occur with such characteristics as induced magnetization by electric field or vice versa. This topologically new phenomenon is known as topological magnetoelectric effect (TME) [68,69].

Theoretically, one of the easiest ways to generate a surface symmetry breaking field is by coating the topological insulator surface with magnetic impurities such as a layer of ferromagnetic material, and tuning the chemical potential near the Dirac point [69]. In this way, the heterojunction of topological insulator and ferromagnetic material can enable the demonstration or even controlling of the magnetization through external approaches. This could be very interesting in a lot of physical investigations such as the realization of magnetic monopole effect. Although the theoretical and simulation work of TME have been initiated since about 10 years ago [70–72], experimental work verifying or even using the theory for practical device application is still rarely seen. For example, Fujita et al. proposed a novel 3D topological insulator cell with ferromagnetic doping on the surface breaking the time-reversal symmetry [73]. Long-range magnetic order was thus induced and opened up an energy gap at the Dirac point. They claimed that this new type of device can be utilized in novel memory and magnetic sensor, since the readout process was protected by the quantized hall effect, so that the magnetic storage should be robust against perturbations such as edge roughness, impurities and defects. However, the experimental realization of this proposed memory cell is difficult in maintaining the proper Fermi level position and electrical programming/erasing or readout operations at the same time. This problem was partly solved by Fan et al. in 2014 [74]. In their work, they use MBE to epitaxially synthesize $(Bi_{0.5}Sb_{0.5})_2Te_3/(Cr_{0.08}Bi_{0.54}Sb_{0.38})_2Te_3$ bilayer films. Conventionally, spin current generated in the heavy metal adjacent to a ferromagnet material is based on the spin-Hall effect, and can apply spin torques to the ferromagnetic layer resulting in current-induced magnetization manipulation [74]. Magnetically doped topological insulators (such as the Cr doping in this reference work) have been demonstrated to be good platform to study the spin-orbit torque due to the large spin-orbit coupling to invert the band structure. In the Cr-doped BiSbTe bilayer system (Figure 11a), spin accumulation will occur in the Cr-doped layer when passing a charge current due to the spin-Hall effect and the spin polarization. The accumulated angular momentum of spin can be transferred to the magnetization and affect the dynamics [74]. Figure 11b showing the second-harmonic anomalous hall effect resistance was used to calibrate the effective spin-orbit torque induced field. The magnetization switching can be realized through the charge current-induced spin-orbit torque, as shown in Figure 11c. The relationship between the effect spin-orbit field and the current amplitude was demonstrated in Figure 11d. This type of switching can be regarded as a direct utilization of the topological magnetoelectric effect involving the magnetization manipulation. Although the underlying mechanism of spin-orbit torque is still debated, the in-plane current-driven spin-orbit torque which stimulates the magnetization switching is a new mechanism towards memory and logic device applications [74].

Later in 2017, Wang et al. fabricated spin-orbit torque-driven magnetization switching in $Bi_2Se_3/NiFe$ heterostructures at room temperature. The charge-to-spin conversion efficiency in Bi_2Se_3 films reached ~1–1.75 while the needed current density for the magnetization switching was just ~6×10^5 A cm^{-2} which was orders of magnitude lower than the cases of heavy metals. This pave a new way for the topological insulator-based spintronic devices [75]. More recently, $Bi_{0.9}Sb_{0.1}$-based ultra-low power spin-orbit-torque switching was reported in 2018 [76]. The $Bi_{0.9}Sb_{0.1}$ thin films on $Mn_{0.6}Ga_{0.4}$ surface have high electrical conductivity (σ ~2.5×10^5 Ω^{-1} m^{-1}) and large spin Hall angle (θ_{SH} ~52) which can generate a spin-orbit field of 2770 Oe/(MA/cm^2) while the critical switching current density was as low as 1.5 MA/cm^2 in the bi-layers [76]. Although it was claimed that such BiSb system will be promising candidate for the industrial applications with topological insulators-based spin-orbit torque memory, the process integration of these novel materials still remains the challenging problem considering the realization of large spin Hall effect at room temperature and the controlling of surface orientation [76].

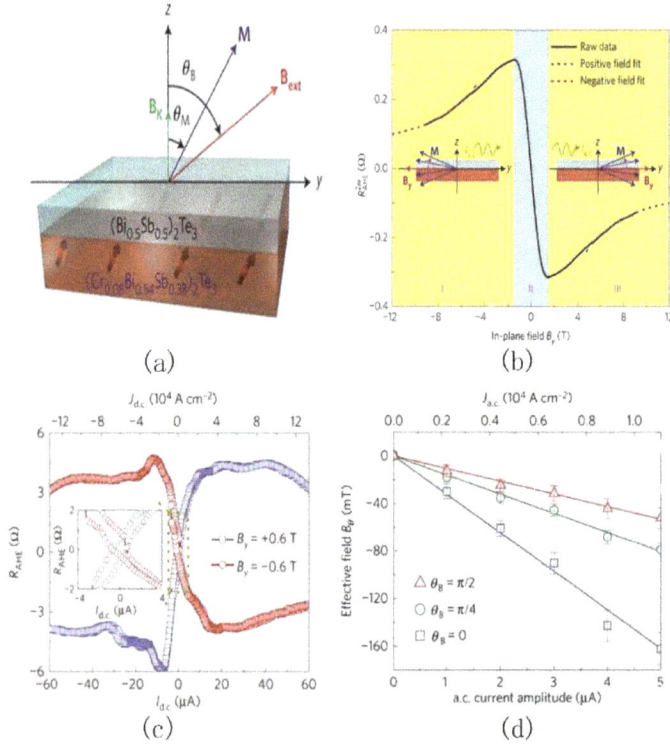

Figure 11. (**a**) The 3D schematic of the bilayer heterostructure consisting of 3-quintuple-layer $(Bi_{0.5}Sb_{0.5})_2Te_3$ at the top and 6-quintuple-layer $(Cr_{0.08}Bi_{0.54}Sb_{0.38})_2Te_3$ at the bottom. B_{ext} is the external magnetic field and M represents the magnetization of the bottom layer. B_K is the out-of-plane anisotropy field. (**b**) Second-harmonic anomalous Hall effect resistance versus the in-plane external magnetic field at room temperature. (**c**) Current-induced magnetization switching in the Hall bar device under in-plane external magnetic field at 1.9 K. (**d**) The effective spin–orbit field as a function of the a.c. current amplitude for $\theta_B = 0$, $\pi/4$ and $\pi/2$, respectively and the temperature is 1.9 K [74]. Reproduced with permission from [74], Copyright Springer Nature, 2014.

Compared to the above methods for magnetoelectric device applications, some other device design and mechanism have also been investigated. For example, Zhang et al. reported a novel magnetoresistance (MR) switching effect in 2012 by synthesizing Sn-doped Bi_2Te_3 thin films and integrated in a simple metal-insulator-metal device structure [77]. As shown in Figure 12, an external magnetic field was also applied to break the time-reversal symmetry. Different from the undoped Bi_2Te_3, in Sn-doped Bi_2Te_3, the bulk electrons will experience back scattering by Sn dopants, and the trajectory of the bulk electrons defined by the weak localization effect will be influenced by the time-reversal invariance, leading to a destroyed constructive interference of the electron's wave function. This will further decrease the resistance exhibiting an MR switching effect [77]. No matter how the magnetoelectric effect or the magnetoresistance switching behavior was realized, there is still an obvious gap between the theoretical prediction and practical device verification and application towards novel spintronic and magnetoelectric device applications.

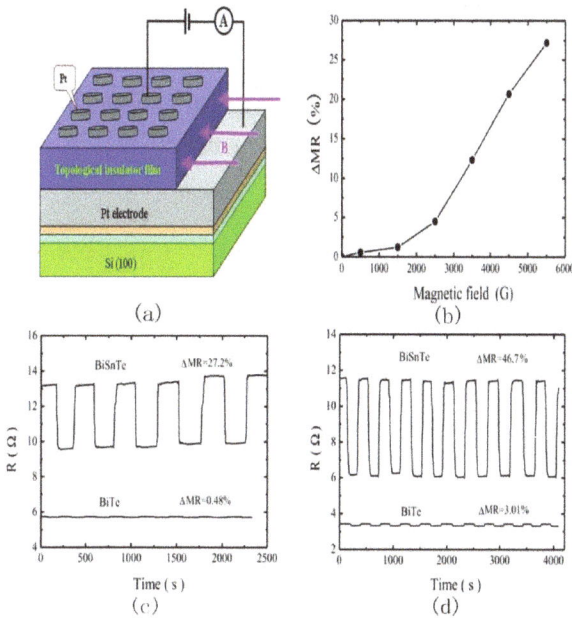

Figure 12. (**a**) Schematic illustration of the device with Pt electrodes. (**b**) The ΔMR of the Sn-doped Bi_2Te_3 film as a function of the magnetic field ranging from 0 to 5500 G at room temperature. (**c**) Transport properties of the Sn-doped Bi_2Te_3 and undoped Bi_2Te_3 films at room temperature. (**d**) The change in magnetoresistance of Sn-doped and undoped Bi_2Te_3 samples at the temperature of 80 K [77]. Reproduced with permission from [77], Copyright Springer Nature, 2018.

Considering back to the integration of topological insulator with ferromagnetic material, the spin injection and extraction through certain barriers are also interesting. This involves the spin-momentum locking of the 2D surface states in 3D topological insulators, which have great potential to realize dissipation-less transport and to achieve spin generators in spintronic devices. Existing theories and experiments have confirmed that the spin of electrons in the surface states of topological insulators is highly polarized, the energy band is reversed due to strong coupling of spin orbit and that the spin of electrons and the crystalline momentum are locked together called spin-momentum locking, forming a unique helical spin texture. Very recently, a 3D topological insulator spin valve has been reported with prominent switching behavior [78]. Bi_2Te_2Se nanoplatelets have been synthesized by using a catalyst-free vapor-solid method. A thin layer of insulating hexagonal boron nitride (hBN) was inserted separating the topological insulator and ferromagnetic electrode. By characterizing the contact resistance variation which is correlated with the spin valve switching, the functionality of the spin-polarized current carried by the helical surface states can be studied. The hysteresis in the switching curve reversed along with reversing the applied current when contact resistances were low while the polarity of the hysteresis was current-direction-independent under high contact resistances. This make it possible to modulate the spin exchange between the ferromagnetic material and topological insulator through the engineering of hBN layers [78].

The magnetoelectric device based on 3D topological insulator is an expanding subject with many theoretical and experimental issues remain uncertain. Further in-depth studies are needed to elucidate the mechanisms which can affect the topological magnetoelectric effect involving the detection, characterization and manipulation of magnetization, spin polarization, and spin filtering properties. Nevertheless, innovative spintronic devices utilizing various quantum effects can be expected in the future and it can be a big step in the advanced information technology.

3.4. Other Applications

Apart from the applications in field-effect transistor, optoelectronic device and magnetoelectric device described above, there are many other fields that topological insulator can be implemented in. For example, in 2008, Qi et al. employed the topological surface states to realize the detection of magnetic monopole [79]. In this theory, when an electric charge was approaching the surface of a topological insulator, a corresponding magnetic monopole charge can be induced acting as a mirror image. Hence, the detection of magnetic monopole can be interpreted by the measurement of the generated magnetic field. Besides, the conversion from a chiral Dirac fermion to a pair of neutral chiral Majorana fermions has also been realized, enabling the electrical detection of the interferometric signals [80]. Additionally, topological insulators can be used in batteries [81], gas sensors [82], solar cell [83] and memory devices [84] which have the superiority in quality factor and power consumption [73,84–86]. For example, Tian et al. observed the long-lived persistent electron spin polarization which was even unaffected with low temperature and removing current. This electrically controllable spin polarization makes it feasible to fabricate rechargeable spin battery and rewritable spin memory [81]. The sensing capabilities of topological insulators have been tested by Liu et al. by exposing the surface with conducting channels to various external environment [82]. A Bi_2Se_3-based organic polymer solar cell has been reported by Yuan et al. in 2015 with maximum photoelectric conversion efficiency of 4.37% [83], and a fast-speed non-volatile memory with write and read time as low as 20 nm has also been achieved based on 3D topological insulator materials [73]. Such applications with potential for significant innovations in future electronic devices are due to the attractive intrinsic properties of the topological insulators which are prominently distinguished from the conventional semiconductors. Nevertheless, great efforts are still needed for the synthetic topological insulator nanostructures to explore the strategies to realize new-concept and new-mechanism electronic devices.

4. Conclusions and Outlook

The scaling of Si-based CMOS technology has encountered greater challenges with more advanced technology node. On the other hand, topological insulators with the superiority of combing the spin and charge degrees of freedom has provided an alternative approach to overcome the physical limit. Low-power, high-performance spintronic devices can be achieved through the engineering over the spin-orbit coupling towards various applications such as nanoelectronics and optoelectronics [87]. Using different synthesis methods, novel topological insulator nanostructures can be effectively integrated in electronic devices leveraging the advantages afforded by the materials with current semiconductor technology. In this review, we briefly introduced synthetic approaches to grow topological insulator nanostructures and their device applications including field-effect transistor, optoelectronic device, magnetoelectric device and so forth. Through the development of 2D HgTe/CdTe quantum well to 3D topological insulators, the application of this series of materials have been extended to more theoretical and experimental research fields. The excellent device performance is originated from the unique properties of the surface states, and the enhanced topological feature in synthetic high-quality nanostructures with greater surface-to-volume ratio. So far, topological insulator nanowires, nanoribbons and nanoplates have attracted more interests because they will not only enable high-quality single crystal nanostructures but also facilitate advanced device architecture allowing for various characterization and manipulation methods with external means.

Topological insulators can have an enlarged scenario for future practical application, specifically in the interdisciplinary fields including condensed matter physics, microelectronics, information technology and so on. The intriguing metallic surface states make them promising candidate in new-concept spintronic devices, and the natural property that the surface has one Dirac point without spin degeneracy can give rise to the possibility in generating particles that can be utilized in quantum computing. Moreover, Majorana fermions which can be created in the interface between a topological insulator and a superconductor are just one step towards an error-prohibit topological quantum computer [88]. Actually, the current major bottleneck for topological insulators to be integrated in spintronic devices and quantum computing systems lies in the controllability over the combination with magnetic material. Recently, hybrid synthetic Bi_2Se_3 and EuS system has been reported in which EuS could maintain its stable magnetic properties up to room temperature, and the proximity-induced magnetism at the Bi_2Se_3/EuS interface can pave the way for energy-efficient topological control mechanism for future spin-based technologies [89]. So far, although the realization of surface state manipulation and heterogeneous integration can only be achieved preliminarily, further engineering on the synthetic process and device optimization can still be expected to open up a suite of potential application in novel nanoelectronics, optoelectronics, and spintronics.

Funding: This research was funded by the NSFC (61704030, 61376092 and 61427901), the Shanghai Pujiang Program (17PJ1400500), 02 State Key Project (2017ZX02315005), and Program of Shanghai Subject Chief Scientist (14XD1400900).

Conflicts of Interest: The authors declare no conflict of interest.

References

1. Klitzing, K.V.; Dorda, G.; Pepper, M. New Method for High-Accuracy Determination of the Fine-Structure Constant Based on Quantized Hall Resistance. *Phys. Rev. Lett.* **1980**, *45*, 494–497. [CrossRef]
2. Qi, X.L.; Zhang, S.C. The quantum spin Hall effect and topological insulators. *Phys. Today* **2010**, *63*, 33–38. [CrossRef]
3. Wang, G. The Symmetry Broken State of the 2D and 3D Topological Insulator. M.E. Thesis, Southeast University, Nanjing, China, 2013.
4. Hasan, M.Z.; Kane, C.L. Colloquium: Topological insulators. *Rev. Mod. Phys.* **2010**, *82*, 3045–3067. [CrossRef]
5. Bernevig, B.A.; Hughes, T.L.; Zhang, S.C. Quantum Spin Hall Effect and Topological Phase Transition in HgTe Quantum Wells. *Science* **2006**, *314*, 1757–1761. [CrossRef] [PubMed]
6. König, M.; Wiedmann, S.; Brüne, C.; Roth, A.; Buhmann, H.; Molenkamp, L.W.; Qi, X.L.; Zhang, S.C. Quantum spin hall insulator state in HgTe quantum wells. *Science* **2007**, *318*, 766–770.
7. Liang, F.; Kane, C.L. Topological Insulators with Inversion Symmetry. *Phys. Rev.* **2007**, *76*, 045302.
8. Hong, S.S.; Kundhikanjana, W.; Cha, J.J.; Lai, K.; Kong, D.; Meister, S.; Kelly, M.A.; Shen, Z.X.; Cui, Y. Ultrathin topological insulator Bi_2Se_3 nanoribbons exfoliated by atomic force microscopy. *Nano Lett.* **2010**, *10*, 3118–3122. [CrossRef] [PubMed]
9. Qu, D.X.; Hor, Y.S.; Xiong, J.; Cava, R.J.; Ong, N.P. Quantum oscillations and hall anomaly of surface states in the topological insulator Bi_2Te_3. *Science* **2010**, *329*, 821–824. [CrossRef] [PubMed]
10. Cho, S.; Butch, N.P.; Paglione, J.; Fuhrer, M.S. Insulating behavior in ultrathin bismuth selenide field effect transistors. *Nano Lett.* **2012**, *11*, 1925–1927. [CrossRef] [PubMed]
11. Mciver, J.W.; Hsieh, D.; Steinberg, H.; Jarillo-Herrero, P.; Gedik, N. Control over topological insulator photocurrents with light polarization. *Nat. Nanotechnol.* **2012**, *7*, 96–100. [CrossRef] [PubMed]
12. Li, Y.Y.; Wang, G.; Zhu, X.G.; Liu, M.H.; Ye, C.; Chen, X.; Wang, Y.Y.; He, K.; Wang, L.L.; Ma, X.C.; et al. Intrinsic Topological Insulator Bi2Te3 Thin Films on Si and Their Thickness Limit. *Adv. Mater.* **2010**, *22*, 4002–4007. [CrossRef] [PubMed]
13. Zhang, T.; Cheng, P.; Chen, X.; Jia, J.F.; Ma, X.; He, K.; Wang, L.; Zhang, H.; Dai, X.; Fang, Z.; et al. Experimental Demonstration of Topological Surface States Protected by Time-Reversal Symmetry. *Phys. Rev. Lett.* **2009**, *103*, 266803. [CrossRef] [PubMed]
14. Liu, X.; Smith, D.J.; Cao, H.; Chen, Y.; Fan, J.; Zhang, Y.; Pimpinella, R.E.; Dobrowolska, M.; Furdyna, J.K. Characterizations of Bi2Te3 and Bi2Se3 topological insulators grown by MBE on (100) GaAs substrates. *J. Vac. Sci. Technol. B* **2012**, *30*, 02B103. [CrossRef]
15. Wang, Q.; Cai, K.; Li, J.; Huang, Y.; Wang, Z.; Xu, K.; Wang, F.; Zhan, X.; Wang, F.; Wang, K.; et al. Rational Design of Ultralarge Pb1-x Snx Te Nanoplates for Exploring Crystalline Symmetry-Protected Topological Transport. *Adv. Mater.* **2016**, *28*, 617–623. [CrossRef] [PubMed]

16. Bendt, G.; Zastrow, S.; Nielsch, K.; Mandal, P.S.; Barriga, J.S.C.; Raderc, O.; Schulz, S. Stephan Deposition of topological insulator Sb2Te3 films by an MOCVD process. *J. Mater. Chem. A* **2014**, *2*, 8215–8222. [CrossRef]

17. Waner, R.S.; Ellis, W.C. Vaporliquid-solid mechanism of single crystal growth. *Appl. Phys. Lett.* **1964**, *4*, 89–90. [CrossRef]

18. Zhu, H.; Li, Q.; Yuan, H.; Baumgart, H.; Ioannou, D.E.; Richter, C.A. Self-aligned multi-channel silicon nanowire field-effect transistors. *Solid State Electron.* **2012**, *78*, 92–96. [CrossRef]

19. Peng, H.; Lai, K.; Kong, D.; Meister, S.; Chen, Y.; Qi, X.; Zhang, S.; Shen, Z.; Cui, Y. Aharonov-Bohm interference in topological insulator nanoribbons. *Nat. Mater.* **2009**, *9*, 225. [CrossRef] [PubMed]

20. Liu, C.; Zhang, H.; Sun, Z.; Ding, K.; Mao, J.; Shao, Z.; Jie, J. Topological Insulator Bi_2Se_3 Nanowire/Si Heterostructure Photodetector with Ultrahigh Responsivity and Broadband Response. *J. Mater. Chem. C* **2016**, *4*, 5648–5655. [CrossRef]

21. Xiu, F.; He, L.; Wang, Y.; Cheng, L.; Chang, L.; Lang, M.; Huang, G.; Kou, X.; Zhou, Y.; Jiang, X.; et al. Manipulating surface states in topological insulator nanoribbons. *Nat. Nanotechnol.* **2011**, *6*, 216–221. [CrossRef] [PubMed]

22. Schlenk, T.; Bianchi, M.; Koleini, M.; Eich, A.; Pietzsch, O.; Wehling, T.O.; Frauenheim, T.; Balatsky, A.; Mi, J.L.; Iversen, B.B.; et al. Controllable magnetic doping of the surface state of a topological insulator. *Phys. Rev. Lett.* **2013**, *110*, 1–10. [CrossRef] [PubMed]

23. Steinberg, H.; Gardner, D.R.; Lee, Y.S.; Jarillo-Herrero, P. Surface state transport and ambipolar electric field effect in Bi_2Se_3 nanodevices. *Nano. Lett.* **2010**, *10*, 5032. [CrossRef] [PubMed]

24. Chen, Y.P. Graphene and topological insulator based transistors: Beyond computing applications. In Proceedings of the Device Research Conference, University Park, TX, USA, 18–20 June 2012; pp. 37–38.

25. Yang, F.; Taskin, A.A.; Sasaki, S.; Segawa, K.; Ohno, Y.; Matsumoto, K.; Ando, Y. Dual-Gated Topological Insulator Thin-Film Device for Efficient Fermi-Level Tuning. *ACS Nano* **2015**, *9*, 4050–4055. [CrossRef] [PubMed]

26. Son, J.; Banerjee, K.; Brahlek, M.; Koirala, N.; Lee, S.; Ahn, J.; Oh, S.; Yang, H. Conductance modulation in topological insulator Bi2Se3 thin films with ionic liquid gating. *Appl. Phys. Lett.* **2013**, *103*, 016801. [CrossRef]

27. Zhang, Z.; Feng, X.; Guo, M.; Li, K.; Zhang, J.; Ou, Y.; Feng, Y.; Wang, L.; Chen, X.; He, K.; et al. Electrically tuned magnetic order and magnetoresistance in a topological insulator. *Nat. Commun.* **2014**, *5*, 4915. [CrossRef] [PubMed]

28. Zhu, H.; Zhao, E.; Richter, C.A.; Li, Q. Topological Insulator Bi_2Se_3 Nanowire Field Effect Transistors. *Ecs Trans.* **2014**, *64*, 51–59. [CrossRef]

29. Chang, J.; Register, L.F.; Banerjee, S.K. Topological insulator Bi_2Se_3 thin films as an alternative channel material in metal-oxide-semiconductor field-effect transistors. *J. Appl. Phys.* **2012**, *112*, 3045–3067. [CrossRef]

30. Liu, Y.H.; Chong, C.W.; Fanchiang, C.M.; Huang, J.C.A.; Han, H.C.; Li, Z.J.; Qiu, H.L.; Li, Y.C.; Liu, C.P. Ultrathin (Bi1-xSbx)2Se3 field effect transistor with large ON/OFF ratio. *Acs Appl. Mater. Interfaces* **2017**, *9*, 12859–12864. [CrossRef] [PubMed]

31. Xu, S.Y.; Neupane, M.; Liu, C.; Zhang, D.; Richardella, A.; AndrewWray, L.; Alidoust, N.; Leandersson, M.; Balasubramanian, T.; Sánchez-Barriga, J.; et al. Hedgehog Spin Texture and Berry's Phase Tuning in a Magnetic Topological Insulator. *Nat. Phys.* **2012**, *8*, 616–622. [CrossRef]

32. Kou, M.X.; Lang, Y.; Fan, Y.; Jiang, T.; Nie, J.; Zhang, W.; Jiang, Y.; Wang, Y.; Yao, L.H. Interplay between Different Magnetisms in Cr-Doped Topological Insulators. *ACS Nano* **2013**, *10*, 9205–9212. [CrossRef] [PubMed]

33. Zhu, H.; Richter, C.A.; Zhao, E.; Bonevich, J.E.; Kimes, W.A.; Jang, H.J.; Yuan, H.; Li, H.; Arab, A.; Kirillov, O.; et al. Topological insulator Bi_2Se_3 nanowire high performance field-effect transistors. *Sci. Rep.* **2013**, *3*, 1757. [CrossRef]

34. Liu, H.; Ye, P.D. Atomic-layer-deposited Al2O3 on Bi2Te3 for topological insulator field-effect transistors. *Appl. Phys. Lett.* **2011**, *99*, 3045. [CrossRef]

35. Wray, L.A.; Xu, S.; Neupane, M.; Xia, Y.; Hsieh, D.; Qian, D.; Fedorov, A.V.; Lin, H.; Basak, S.; Hor, Y.S.; et al. Electron dynamics in topological insulator based semiconductor-metal interfaces (topological p-n interface based on Bi2Se3 class). *arXiv* **2011**, arXiv:1105.4794.

36. Koirala, N.; Brahlek, M.; Salehi, M.; Wu, L.; Dai, J.; Waugh, J.; Nummy, T.; Han, M.G.; Moon, J.; Zhu, Y.; et al. Record surface state mobility and quantum Hall effect in topological insulator thin films via interface engineering. *Nano Lett.* **2015**, *8*, 3045. [CrossRef] [PubMed]

37. Hao, G.; Qi, X.; Xue, L.; Cai, C.; Li, J.; Wei, X.; Zhong, J. Fermi level tuning of topological insulator Bi$_2$(SexTe$_{1-x}$)$_3$ nanoplates. *J. Appl. Phys.* **2013**, *113*, 024306. [CrossRef]

38. Sacépé, B.; Oostinga, J.B.; Li, J.; Ubaldini, A.; Couto, N.J.; Giannini, E.; Morpurgo, A.F. Gate-tuned normal and superconducting transport at the surface of a topological insulator. *Nat. Commun.* **2011**, *2*, 575. [CrossRef] [PubMed]

39. Zhang, T.; Levy, N.; Ha, J.; Kuk, Y.; Stroscio, J.A. Scanning Tunneling Microscopy of Gate Tunable Topological Insulator Bi2Se3 Thin Films. *Phys. Rev. B* **2013**, *87*, 1504–1509.

40. Brinkman, A.; Snelder, M.; Brocks, G.H.L.A. Towards Controlling the Fermi Energy in Topological Materials. Master's Thesis, University of Twente, Enschede, The Netherlands, 2014.

41. Taskin, A.A.; Legg, H.F.; Yang, F.; Sasaki, S.; Kanai, Y.; Matsumoto, K.; Rosch, A.; Ando, Y. Planar Hall effect from the surface of topological insulators. *Nat. Commun.* **2017**, *8*, 1340.

42. Walsh, L.A.; Green, A.J.; Addou, R.; Nolting, W.; Cormier, C.R.; Barton, A.T.; Mowll, T.R.; Yue, R.; Lu, N.; Kim, J.; et al. Fermi Level Manipulation through Native Doping in the Topological Insulator Bi2Se3. *ACS Nano* **2018**, *12*, 6310–6318. [CrossRef] [PubMed]

43. Hong, S.S.; Cha, J.J.; Kong, D.; Cui, Y. Ultra-low carrier concentration and surface-dominant transport in antimony-doped Bi2Se3 topological insulator nanoribbons. *Nat. Commun.* **2012**, *3*, 757. [CrossRef] [PubMed]

44. Hor, Y.S.; Richardella, A.; Roushan, P.; Xia, Y.; Checkelsky, J.G.; Yazdani, A.; Hasan, M.Z.; Ong, N.P.; Cava, R.J. p-type Bi2Se3 for topological insulator and low temperature thermoelectric applications. *Phys. Rev. B* **2009**, *79*, 195208. [CrossRef]

45. Wang, Y.; Xiu, F.; Cheng, L.; He, L.; Lang, M.; Tang, J.; Kou, X.; Yu, X.; Jiang, X.; Chen, Z.; et al. Gate-Controlled Surface Conduction in Na-Doped Bi2Te3 Topological Insulator Nanoplates. *Nano Lett.* **2012**, *12*, 1170–1175. [CrossRef] [PubMed]

46. Inoue, J.; Tanaka, A. Photoinduced transition between conventional and topological insulators in two-dimensional electronic systems. *Phys. Rev. Lett.* **2010**, *105*, 017401. [CrossRef] [PubMed]

47. Hosur, P. Circular photogalvanic effect on topological insulator surfaces: Berry-curvature-dependent response. *Phys. Rev. B* **2011**, *83*, 426–432. [CrossRef]

48. Zheng, K.; Luo, L.B.; Zhang, T.F.; Liu, Y.H.; Yu, Y.Q.; Lu, R.; Qiu, H.L.; Li, Z.J.; Huang, J.C.A. Optoelectronic characteristics of a near infrared light photodetector based on a topological insulator Sb2Te3 film. *J. Mater. Chem. C* **2015**, *3*, 9154–9160. [CrossRef]

49. Zhang, H.; Liu, C.-X.; Qi, X.-L.; Dai, X.; Fang, Z.; Zhang, S.-C. Topological insulators in Bi2Se3, Bi2Te3 and Sb2Te3 with a single Dirac cone on the surface. *Nat. Phys.* **2009**, *5*, 438–442. [CrossRef]

50. Sharma, A.; Srivastava, A.K.; Senguttuvan, T.D.; Husale, S. Robust broad spectral photodetection (UV-NIR) and ultra high responsivity investigated in nanosheets and nanowires of Bi2Te3under harsh nano-milling conditions. *Sci. Rep.* **2017**, *7*, 17911. [CrossRef] [PubMed]

51. Fei, F.; Wei, Z.; Wang, Q.J.; Lu, P.; Wang, S.; Qin, Y.; Pan, D.; Zhao, B.; Wang, X.; Sun, J.; et al. Solvothermal synthesis of lateral heterojunction Sb2Te3/Bi2Te3 nanoplates. *Nano Lett.* **2015**, *15*, 5905–5911. [CrossRef] [PubMed]

52. Chong, S.K.; Han, K.B.; Nagaoka, A.; Tsuchikawa, R.; Liu, R.; Liu, H.; Vardeny, Z.V.; Pesin, D.A.; Lee, C.; Sparks, T.D.; et al. Topological Insulator-Based van der Waals Heterostructures for Effective Control of Massless and Massive Dirac Fermions. *arXiv* **2018**, arXiv:1805.09478.

53. Wang, X.B.; Cheng, L.; Wu, Y.; Zhu, D.P.; Wang, L.; Zhu, J.; Yang, H.; Chia, E.E.M. Topological-insulator-based terahertz modulator. *Sci. Rep.* **2017**, *7*, 13486. [CrossRef] [PubMed]

54. Yoshimi, R.; Tsukazaki, A.; Kikutake, K.; Checkelsky, J.G.; Takahashi, K.S.; Kawasaki, M.; Tokura, Y. Dirac electron states formed at the, heterointerface between a topological, insulator and a conventional semiconductor. *Nat. Mater.* **2014**, *13*, 253. [CrossRef] [PubMed]

55. Pang, M.Y.; Li, W.S.; Wong, K.H.; Surya, C. Electrical and optical properties of bismuth telluride/gallium nitride heterojunction diodes. *J. Non-Cryst. Solids* **2008**, *354*, 4238–4241. [CrossRef]

56. Yao, J.; Shao, J.; Wang, Y.; Zhao, Z.; Yang, G. Ultra-broadband and high response of the Bi2Te3-Si heterojunction and its application as a photodetector at room temperature in harsh working environments. *Nanoscale* **2015**, *7*, 12535–12541. [CrossRef] [PubMed]

57. Zhang, H.; Zhang, X.; Liu, C.; Lee, S.T.; Jie, J. High-Responsivity, High-Detectivity, Ultrafast Topological Insulator Bi2Se3/Silicon Heterostructure Broadband Photodetectors. *ACS Nano* **2016**, *10*, 5113. [CrossRef] [PubMed]

58. Zhang, H.; Man, B.; Zhang, Q. Topological Crystalline Insulator SnTe/Si Vertical Heterostructure Photodetectors for High-performance Near-infrared Detection. *Acs Appl. Mater. Interfaces* **2017**, *9*, 14067–14077. [CrossRef] [PubMed]

59. Gu, S.; Ding, K.; Pan, J.; Shao, Z.; Mao, J.; Zhang, X.; Jie, J. Self-driven, broadband and ultrafast photovoltaic detectors based on topological crystalline insulator SnTe/Si heterostructures. *J. Mater. Chem. A* **2017**, *5*, 11171–11178. [CrossRef]

60. An, X.; Liu, F.; Jung, Y.J.; Kar, S. Tunable grapheme-silicon heterojunctions for ultrasensitive photodetection. *Nano Lett.* **2017**, *13*, 909–916. [CrossRef] [PubMed]

61. Zhang, K.; Fang, X.; Wang, Y.; Wan, Y.; Song, Q.; Zhai, W.; Li, Y.; Ran, G.; Ye, Y.; Dai, L. Ultrasensitive near-infrared photodetectors based on graphene-MoTe2-graphene vertical van der Waals heterostructure. *ACS Appl. Mater. Interfaces* **2017**, *9*, 5392–5398. [CrossRef] [PubMed]

62. Liang, F.X.; Gao, Y.; Xie, C.; Tong, X.W.; Li, Z.J.; Luo, L.B. Recent advances in the fabrication of graphene–ZnO heterojunctions for optoelectronic device applications. *J. Mater. Chem. C* **2018**, *6*, 3815–3833. [CrossRef]

63. Huo, N.; Yang, Y.; Li, J. Optoelectronics based on 2D TMDs and heterostructures. *J. Semicond.* **2017**, *38*, 2–10. [CrossRef]

64. Sun, M.; Xie, D.; Sun, Y.; Li, W.; Teng, C.; Xu, J. Lateral multilayer/monolayer MoS2 heterojunction for high performance photodetector applications. *Sci. Rep.* **2017**, *7*, 4505. [CrossRef] [PubMed]

65. Qiao, H.; Yuan, J.; Xu, Z.; Chen, C.; Lin, S.; Wang, Y.; Song, J.; Liu, Y.; Khan, Q.; Hoh, H.Y.; et al. Broadband photodetectors based on graphene-Bi2Te3 heterostructure. *ACS Nano* **2015**, *9*, 1886–1894. [CrossRef] [PubMed]

66. Yao, J.; Zheng, Z.; Yang, G. Layered-material WS2/topological insulator Bi2Te3 heterostructure photodetector with ultrahigh responsivity in the range from 370 to 1550 nm. *J. Mater. Chem. C* **2016**, *4*, 7831–7840. [CrossRef]

67. Yao, J.; Yang, G. Flexible and High-Performance All-2D Photodetector for Wearable Devices. *Small* **2018**, *14*, 1704524. [CrossRef] [PubMed]

68. Qi, X.L.; Zhang, S.C. Topological insulators and superconductors. *Rev. Mod. Phys.* **2011**, *83*, 1057–1110. [CrossRef]

69. Qi, X.L. Topological field theory of time-reversal invariant insulators. *Phys. Rev. B* **2008**, *78*, 2599–2604. [CrossRef]

70. Crosse, J.A. Theory of topological insulator waveguides: Polarization control and the enhancement of the magneto-electric effect. *Sci. Rep.* **2017**, *7*, 43115. [CrossRef] [PubMed]

71. Nomura, K.; Nagaosa, N. Surface-quantized anomalous Hall current and the magnetoelectric effect in magnetically disordered topological insulators. *Phys. Rev. Lett.* **2011**, *106*, 166802. [CrossRef] [PubMed]

72. Grushin, A.G.; Neupert, T.; Chamon, C.; Mudry, C. Enhancing the stability of a fractional Chern insulator against competing phases. *Phys. Rev. B* **2012**, *86*, 205125. [CrossRef]

73. Fujita, T.; Jalil, M.B.A.; Tan, S.G. Topological Insulator Cell for Memory and Magnetic Sensor Applications. *Appl. Phys. Exp.* **2011**, *4*, 544–548. [CrossRef]

74. Fan, Y.; Upadhyaya, P.; Kou, X.; Lang, M.; Takei, S.; Wang, Z.; Tang, J.; He, L.; Chang, L.T.; Montazeri, M.; et al. Magnetization switching through giant spin-orbit torque in a magnetically doped topological insulator heterostructure. *Nat. Mater.* **2014**, *13*, 699–704. [CrossRef] [PubMed]

75. Wang, Y.; Zhu, D.; Wu, Y.; Yang, Y.; Yu, J.; Ramaswamy, R.; Mishra, R.; Shi, S.; Elyasi, M.; Teo, K.L.; Wu, Y.; Yang, H.; et al. Room temperature magnetization switching in topological insulator-ferromagnet heterostructures by spin-orbit torques. *Nat. Commun.* **2017**, *8*, 1364. [CrossRef] [PubMed]

76. Khang, N.H.D.; Ueda, Y.; Hai, P.N. A conductive topological insulator with large spin Hall effect for ultralow power spin–orbit torque switching. *Nat. Mater.* **2018**, *17*, 808–813. [CrossRef] [PubMed]

77. Zhang, H.B.; Yu, H.L.; Bao, D.H.; Li, S.W.; Wang, C.X.; Yang, G.W. Magnetoresistance Switch Effect of a Sn-Doped Bi2Te3 Topological Insulator. *Adv. Mater.* **2012**, *24*, 132. [CrossRef] [PubMed]

78. Vaklinova, K.; Polyudov, K.; Burghard, M.; Kern, K. Spin filter effect of hBN/Co detector electrodes in a 3D topological insulator spin valve. *J. Phys.* **2018**, *30*, 105302. [CrossRef] [PubMed]

79. Qi, X.L.; Li, R.D.; Zang, J.D.; Zhang, S.C. Inducing a magnetic monopole with topological surface states. *Science* **2009**, *323*, 1184–1187. [CrossRef] [PubMed]

80. Akhmerov, A.R.; Nilsson, J.; Beenakker, C.W.J. Electrically detected interferometry of Majorana fermions in a topological insulator. *Phys. Rev. Lett.* **2009**, *102*, 216404. [CrossRef] [PubMed]

81. Tian, J.; Hong, S.; Miotkowski, I.; Datta, S.; Chen, Y.P. Observation of current-induced, long-lived persistent spin polarization in a topological insulator: A rechargeable spin battery. *Sci. Adv.* **2017**, *3*, e1602531. [CrossRef] [PubMed]

82. Liu, B.; Xie, W.; Li, H.; Wang, Y.; Cai, D.; Wang, D.; Wang, L.; Liu, Y.; Li, Q.; Wang, T. Surrounding sensitive electronic properties of Bi2Te3 nanoplates-potential sensing applications of topological insulators. *Sci. Rep.* **2014**, *4*, 4639. [CrossRef] [PubMed]

83. Yuan, Z. The Application of Novel Interfacial Materials and Structure in Organic-Inorganic Solar Cells. Master's Thesis, Soochow University, Soochow, China, 2015.

84. Wang, Y.; Yu, H. Design exploration of ultra-low power non-volatile memory based on topological insulator. In Proceedings of the IEEE/ACM International Symposium on Nanoscale Architectures (NANOARCH), Amsterdam, The Netherlands, 4–6 July 2012; Volume 8474, pp. 30–35.

85. Paudel, H.P.; Leuenberger, M.N. A 3D topological insulator quantum dot for optically controlled quantum memory and quantum computing. In Proceedings of the APS March Meeting, American Physical Society, Baltimore, MD, USA, 18–22 March 2013; pp. 4049–4056.

86. Reza, A.K.; Fong, X.; Azim, Z.A.; Roy, K. Modeling and Evaluation of Topological Insulator/Ferromagnet Heterostructure-Based Memory. *IEEE Trans. Electron. Dev.* **2016**, *63*, 1359–1367. [CrossRef]

87. Wang, K.L.; Kou, X.; Upadhyaya, P.; Fan, Y.; Shao, Q.; Yu, G.; Amiri, P.K. Electric-Field Control of Spin-Orbit Interaction for Low-Power Spintronics. *Proc. IEEE* **2016**, *104*, 1974–2008. [CrossRef]

88. Moore, J.E. The birth of topological insulators. *Nature* **2010**, *464*, 194–198. [CrossRef] [PubMed]

89. Katmis, F.; Lauter, V.; Nogueira, F.S.; Assaf, B.A.; Jamer, M.E.; Wei, P.; Satpati, B.; Freeland, J.W.; Eremin, I.; Heiman, D. A high-temperature ferromagnetic topological insulating phase by proximity coupling. *Nature* **2016**, *533*, 513–516. [CrossRef] [PubMed]

© 2018 by the authors. Licensee MDPI, Basel, Switzerland. This article is an open access article distributed under the terms and conditions of the Creative Commons Attribution (CC BY) license (http://creativecommons.org/licenses/by/4.0/).

electronics

MDPI

Article

Gallium Nitride Normally-Off Vertical Field-Effect Transistor Featuring an Additional Back Current Blocking Layer Structure

Huolin Huang [1,2,*], Feiyu Li [1], Zhonghao Sun [1], Nan Sun [1], Feng Zhang [3], Yaqing Cao [1], Hui Zhang [1] and Pengcheng Tao [1]

[1] School of Optoelectronic Engineering and Instrumentation Science, Dalian University of Technology, Dalian 116024, China; lifeiyu@mail.dlut.edu.cn (F.L.); sunzhonghao@mail.dlut.edu.cn (Z.S.); dgsunnan@mail.dlut.edu.cn (N.S.); cyqtmxk@mail.dlut.edu.cn (Y.C.); zhanghui21841009@mail.dlut.edu.cn (H.Z.); pctao@dlut.edu.cn (P.T.)
[2] Key Laboratory for Micro/Nano Technology and System of Liaoning Province, Dalian University of Technology, Dalian 116024, China
[3] Key Laboratory of Semiconductor Materials Science & Beijing Key Laboratory of Low Dimensional Semiconductor Materials and Devices, Institute of Semiconductors, Chinese Academy of Sciences, Beijing 100083, China; fzhang@semi.ac.cn
* Correspondence: hlhuang@dlut.edu.cn; Tel.: +86-0411-8470786

Received: 31 December 2018; Accepted: 18 February 2019; Published: 20 February 2019

Abstract: A gallium nitride (GaN) semiconductor vertical field-effect transistor (VFET) has several attractive advantages such as high power density capability and small device size. Currently, some of the main issues hindering its development include the realization of normally off operation and the improvement of high breakdown voltage (BV) characteristics. In this work, a trenched-gate scheme is employed to realize the normally off VFET. Meanwhile, an additional back current blocking layer (BCBL) is proposed and inserted into the GaN normally off VFET to improve the device performance. The electrical characteristics of the proposed device (called BCBL-VFET) are investigated systematically and the structural parameters are optimized through theoretical calculations and TCAD simulations. We demonstrate that the BCBL-VFET exhibits a normally off operation with a large positive threshold voltage of 3.5 V and an obviously increased BV of 1800 V owing to the uniform electric field distribution achieved around the gate region. However, the device only shows a small degradation of on-resistance (R_{ON}). The proposed scheme provides a useful reference for engineers in device fabrication work and will be promising for the applications of power electronics.

Keywords: vertical field-effect transistor (VFET); back current blocking layer (BCBL); gallium nitride (GaN); normally off power devices

---

1. Introduction

With the rapid development of the power electronics industry, Si- or GaAs-based devices are approaching their material limit. The wide-bandgap semiconductors (known as third-generation semiconductors) have been widely employed and developed owing to their excellent physical properties such as large bandgap and high critical breakdown electric field. Among them, gallium nitride (GaN) is regarded as one of the most promising candidates for application in next-generation power devices [1,2], which is mainly due to the existence of high-density two-dimensional electron gas (2-DEG) in the AlGaN/GaN heterojunction interface induced by the strong polarization effect in wurtzite GaN [3,4]. Hence, the on-resistance (R_{ON}) and switching frequency of the power devices are improved dramatically.

GaN-based lateral high electron mobility transistor (HEMT) devices have been extensively demonstrated and have made great progress in the past few decades [5–8]. However, these lateral devices still encounter a few issues such as power density limit and output current collapse. For example, to improve the breakdown voltage (BV), the gate-to-drain distance of the HEMT has to be increased; this results in an obvious increase of the R_{ON}. Meanwhile, the peak electric field is located at the drain-side gate corner of the HEMT surface, which also leads to a limiting of the device output characteristics and a serious reliability issue when operated under high voltage [9,10]. Another challenge in lateral GaN-based HEMT is how to make the normally off device. Actually, normally off operation is strongly desired for safety and for efficient power switching in the integrated circuits (ICs) or systems for better compatibility. However, a standard AlGaN/GaN HEMT is naturally normally on because of the presence of the 2-DEG channel. Recently, various approaches such as gate-recess etching [5,7], fluorine-ion implantation [10], and p-cap layer deposition [6] have been explored to realize the normally off operation. However, it is still a big challenge to simultaneously obtain a high positive threshold voltage (V_{th}) and a large drain output current for an HEMT.

To overcome the mentioned issues, GaN vertical field-effect transistor (VFET) devices have been proposed and developed recently [11–15]. The drain electrode of the device is moved to the back of the wafer and, hence, the main electric field and conducting current flow are turned to the vertical direction. The peak point of the electric field is transferred into the bulk where the defect density is less and the crystal quality is better. Furthermore, the field distribution is more uniform. Therefore, the device current collapse is alleviated and the BV characteristics can be improved. Recently, some work has been conducted on normally on VFET devices and significant progress has been achieved. However, there is still a lack of sufficient research on the normally off VFET devices and, in particular, on their design and demonstration to improve the high voltage characteristics. Therefore, GaN power devices featuring both vertical structure and normally off operation will be very promising in the future applications of power electronics, and novel structure design and effective process technology are thus needed.

In this work, an additional back current blocking layer (BCBL) is introduced and inserted into the GaN normally off VFET to improve the BV characteristics. The electrical characteristics of the proposed device (called BCBL-VFET in this work) are investigated and the structural parameters are optimized through theoretical calculations and TCAD simulation work. Firstly, the pn-junction device was fabricated, and the measured I–V characteristics were employed to calibrate the physical parameters of the proposed device in the theoretical calculation and simulation work. Then the distribution characteristics of the electric field and impact ionization concentration were analyzed. Improved BV and R_{ON} were achieved after optimizing the thickness, depth, and spatial location of the current blocking layer in the GaN BCBL-VFET.

2. The Device Structure

Figure 1 shows the cross-sectional schematic of the proposed GaN BCBL-VFET. The key feature which is different from the conventional one is the introduction of the BCBL below the normal current blocking layer under the source electrode. The specifics of the device structure are listed in Table 1. In the conventional VFET, the breakdown points usually occur at the top pn-junction or at the etched gate corner under high-voltage operation because of the presence of high peak electric field. The additional insertion of the insulated BCBL will distribute the vertical voltage drop and also alleviate the crowding of the electric field around the gate corner. Hence, the BV of the proposed device will be improved significantly. Considering that the device still remains a good vertical current aperture, the output characteristics will not be degraded seriously. The structural parameters of the proposed BCBL such as the thickness, depth, and spatial location will be optimized by the theoretical calculation and simulation work, in detail.

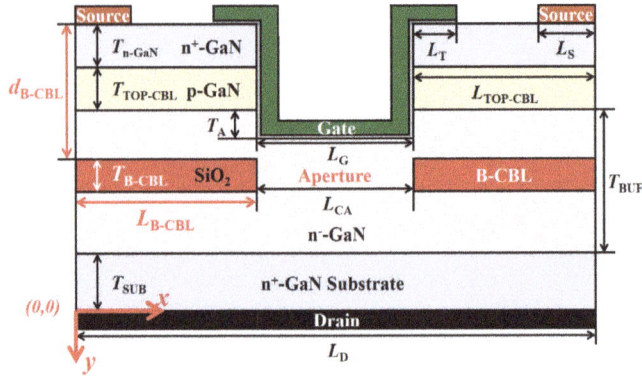

Figure 1. Cross-sectional schematic of the proposed GaN BCBL-VFET (back current blocking layer—vertical field-effect transistor).

Table 1. Specifics of the proposed GaN BCBL-VFET.

Parameters	Units	Values	Parameter Captions
L_D	μm	16.0	Drain length
L_{CA}	μm	5.0	Length of vertical current aperture
L_G	μm	5.0	Length of the recessed gate
$L_{TOP-CBL}$	μm	5.5	Length of the top p-GaN
$T_{TOP-CBL}$	μm	0.75	Thickness of the top p-GaN
T_{BUF}	μm	10.0	Thickness of the buffer layer

3. Fabrication Work and Parameter Calibration

3.1. Fabrication Process of the PN-Junction Devices

The GaN pn-junction was grown on a 350 μm GaN free-standing substrate using a metal organic chemical vapor deposition (MOCVD) system. The doping concentration of n+-GaN substrate was 5×10^{18} cm^{-3}. The epitaxial structure consisted of a 6 μm n−-GaN layer with a background n-type carrier concentration of 5×10^{16} cm^{-3} and a 500 nm p-GaN layer with an Mg doping concentration of 3×10^{19} cm^{-3}. The schematic cross-section, process flow, and top view of the fabricated GaN pn-junction device are shown in Figure 2. The Ti/Al/Ni/Au n-type ohmic contact was first formed on the back of the substrate by electron beam evaporation and annealing treatment under 800 °C for 60 s in N$_2$ atmosphere. Edge termination was realized by mesa etching using an inductively coupled plasma (ICP) system under an ambience consisting of Cl$_2$ and BCl$_3$ mixed gas. The etching process was continued for 100 s. The Ni p-type ohmic contact was then formed on the p-GaN epitaxial layer also by electron beam evaporation. The annealing temperature was 500 °C. The diameter of the anode pad was around 400 μm and the photo is shown in Figure 2c.

(a)

(b)

(c)

(d)

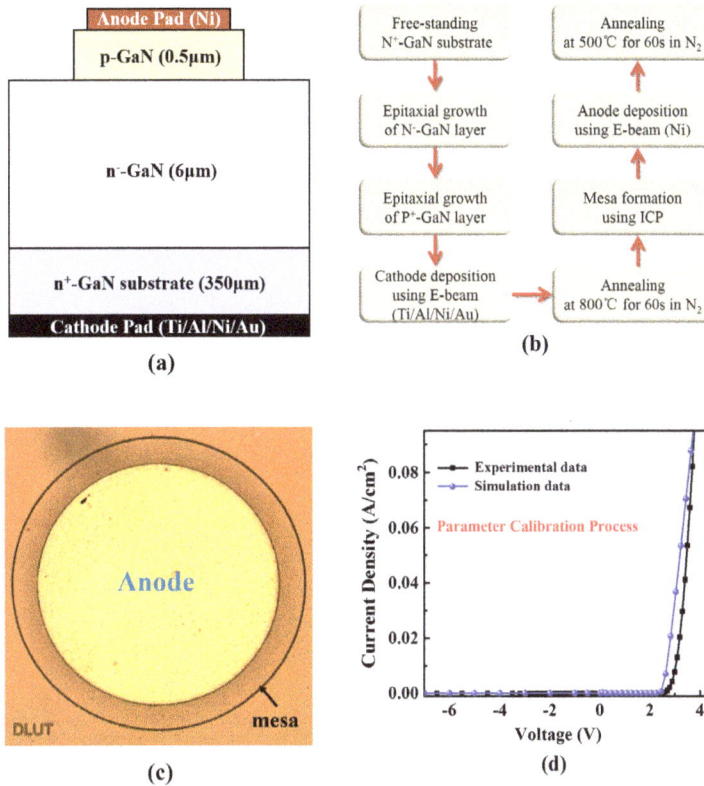

Figure 2. (**a**) Schematic cross-section, (**b**) fabrication process flow, (**c**) top view of the fabricated GaN pn-junction device, and (**d**) parameter calibration process by benchmarking the simulated I–V characteristics of GaN pn-junction device with the experimental data.

3.2. Parameter Calibration Process

The I–V characteristics of the pn-junction device were measured by a semiconductor characterization system (Keithley 4200) at room temperature for the purpose of the subsequent parameter calibration. In the TCAD simulation section, similar device configuration and structural parameters were employed. The physical parameters used in the simulations were calibrated by benchmarking the simulated I–V characteristics of the GaN pn-junction device with the experimental data. Figure 2d shows the comparisons of I–V curves which clearly indicates a good data match. Table 2 lists the employed physical parameters after calibration.

Table 2. Physical parameters adopted in the simulations after calibration.

Symbols	Units	Values	Parameter Descriptions
E_g	eV	3.4	Band gap of GaN at 300 K
ε_r	–	9.4	Relative permittivity of GaN
λ_S	eV	3.4	Affinity of GaN
μ_n	cm^2/Vs	600	Electron mobility of GaN
μ_p	cm^2/Vs	20	Hole mobility of GaN
v_{sat}	cm/s	1.2×10^7	Electron saturation velocity of GaN
E_m	V/cm	3.5×10^6	Critical electric field of GaN
N_C	cm^{-3}	2.2×10^{18}	Conduction band state density

To accurately simulate the device behaviors, appropriate physical models must be selected. In this work, the Shockley–Read–Hall (SRH) model was employed to govern the charge-trapping behavior. The Arora model was employed to determine the doping-dependent mobility for the low-field case. Furthermore, the Canali model was used for the high-field case considering that the carrier drift velocity is no longer proportional to the electric field, instead, the velocity saturates to a finite speed under high voltage. To simulate the off-state breakdown characteristics of power devices, the van Overstraeten–de Man model was used to generate the avalanche induced electron–hole pairs.

4. Results and Discussion

4.1. Parameter Optimization of the BCBL

The pn-junction in the GaN BCBL-VFET acts as the top current blocking layer (TCBL) to sustain the main bias voltage between the drain and source electrodes, which is similar to the case in the conventional VFET. The doping concentration of 8×10^{17} cm^{-3} and thickness of 0.75 µm were employed for the p-GaN layer based on the previous optimal data in the VFET. The concentrations of 5×10^{18} and 1.75×10^{16} cm^{-3} were employed for the upper n$^+$-GaN and the lower n$^-$-GaN layers, respectively, and the thicknesses of 0.25 and 10 µm were kept for them in the BCBL-VFET for the following simulations.

There are three important structural parameters in the BCBL, i.e., d_{BCBL} (the distance between the source and the BCBL), L_{BCBL} (the length of the BCBL), and T_{BCBL} (the thickness of the BCBL). They will all be optimized systematically by analyzing the BV and R_{ON} characteristics of the BCBL-VFET. Considering the trade-off between the BV and R_{ON}, a figure of merit (FOM = BV^2/R_{ON}) value is employed to evaluate the comprehensive device performance. The optimization work for the parameters d_{BCBL}, L_{BCBL}, and T_{BCBL} is discussed as below.

4.1.1. Optimization of d_{BCBL}

The simulated BV and R_{ON} data and the calculated FOM values are shown in Figure 3a,b, respectively. It is clear that the BV is increased obviously when the BCBL approaches the gate. All the devices with additional BCBL exhibit obviously improved off-state characteristics. The BV value is more than 1380 V, while the conventional VFET with only TCBL shows a BV value around 1300 V. Moreover, the R_{ON} variations in the proposed devices are small even though the BCBL is very close to the gate (the reason will be explained and discussed in detail later). This proves the feasibility of the proposed scheme. Figure 3b shows the variation in FOM values with the change in d_{BCBL}. A high FOM value around 9.1×10^8 W/cm^2 is found at d_{BCBL} = 2.0 µm.

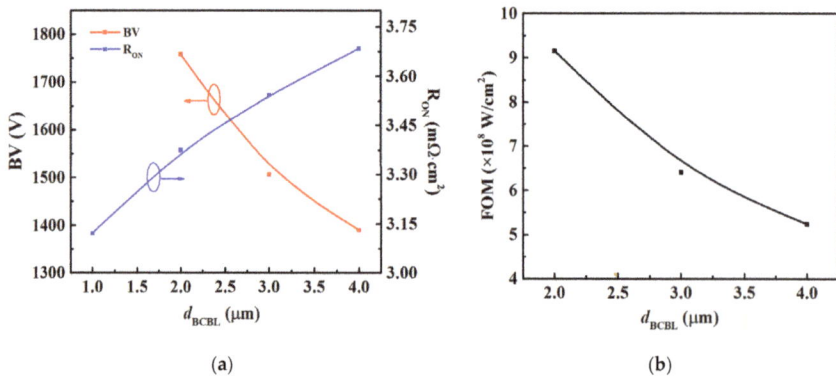

(a)   (b)

Figure 3. (a) Breakdown voltage (BV) and on-resistance (R_{ON}) versus d_{BCBL} (the distance between the source and the BCBL) and (b) figure of merit (FOM) values versus d_{BCBL}.

Figure 4 shows the distributions of the electric field and impact ionization concentration near the gate in the BCBL-VFET with the changed d_{BCBL}. The additional BCBL sustains part of the vertical voltage drop and hence improves the BV characteristic. The shared voltage drop is increased with the reduced d_{BCBL}, and the maximum BV is found at d_{BCBL} = 2.0 μm. However, it can be observed that the BV will be decreased when the d_{BCBL} is less than 2.0 μm since an additional high-field peak induced at the corner of the BCBL surpasses the critical electric field of the GaN material. The distributions of impact ionization concentration in Figure 4b confirm the inference. Note that the electron–hole pair production due to avalanche generation requires a certain threshold field strength and the possibility of acceleration, that is, a wide space charge region. More importantly, the ionization peak usually does not occur at the peak point of the electric field. The field peak positions of devices with various d_{BCBL} are different and the distributions of space charge regions are also different. That is why there is a possible inconsistency between the electric field and impact ionization concentration distributions in Figure 4a,b when the data are derived from the same position inside the device. The impact ionization level induced is much lower in the device with d_{BCBL} = 2.0 μm, compared with the other cases, although a small sharp field peak arises at the end of the BCBL.

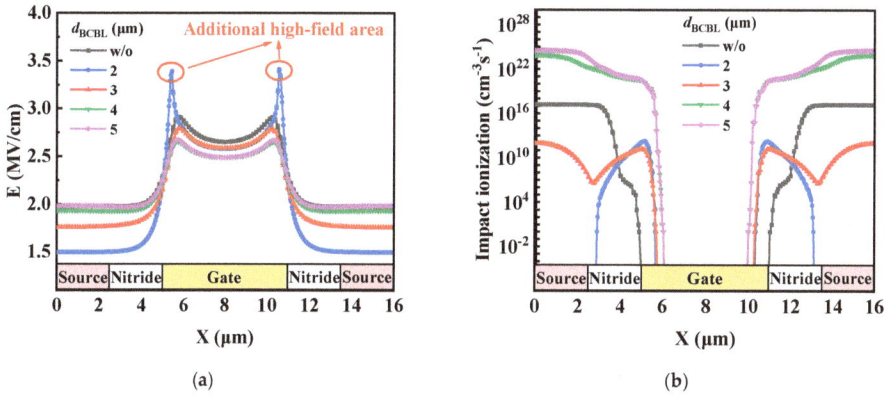

Figure 4. Distributions of (**a**) electric field and (**b**) impact ionization concentration near the gate in the BCBL-VFET with the changed d_{BCBL}.

Figure 5 shows the schematic view of the equivalent circuit in the conventional VFET and the proposed BCBL-VFET. The total resistance consists of source contact resistance (R_{CS}), source resistance (R_{n+}), channel resistance (R_{CH}), accumulation layer resistance (R_A), vertical current aperture resistance (R_{CA}), substrate resistance (R_{SUB}), and drain contact resistance (R_{CD}) [16]. Thus, the R_{ON} of the VFET can be written as:

$$R_{on} = R_{CS} + R_{n+} + R_{CH} + R_A + R_{CA} + R_{SUB} + R_{CD} \tag{1}$$

The R_{CA} which is the only difference between the conventional VFET (R_{CA}) and the BCBL-VFET (R'_{CA}) can be expressed respectively as below.

$$R_{CA} = R_{CA1} + R_{CA2} \tag{2}$$

$$R'_{CA} = R'_{CA1} + R'_{CA2} + R'_{CA3} \tag{3}$$

Here $R_{CA1} = \rho W_{CA1}/L_D Z$ and $R'_{CA1} = \rho W'_{CA1}/L_D Z$ are the uniform resistances in the current aperture, $R_{CA2} = 2\rho W_{CA2}/(L_D + L_{CA})Z$ and $R'_{CA2} = 2\rho W'_{CA2}/(L_D + L_{CA})Z$ are the ladder-shaped resistances, and $R'_{CA3} = \rho W'_{CA3}/L_{CA}Z$ is the resistance between two BCBLs. ρ is the material resistivity and Z is the device dimension perpendicular to the x–y plane.

Considering that W_{CA2} is approximately equal to W'_{CA2}, the R_{ON} variation induced by the insertion of the BCBL can be finally expressed as:

$$\Delta R_{on} = \frac{\rho W'_{CA3}}{Z}\left(\frac{1}{L_{CA}} - \frac{1}{L_D}\right) \tag{4}$$

It is clear that ΔR_{ON} has a functional dependence on the structural parameters L_{CA} and W'_{CA3}. However, this should not be notable because even the total R_{CA} makes up only a small percentage of the total R_{ON} in the BCBL-VFET.

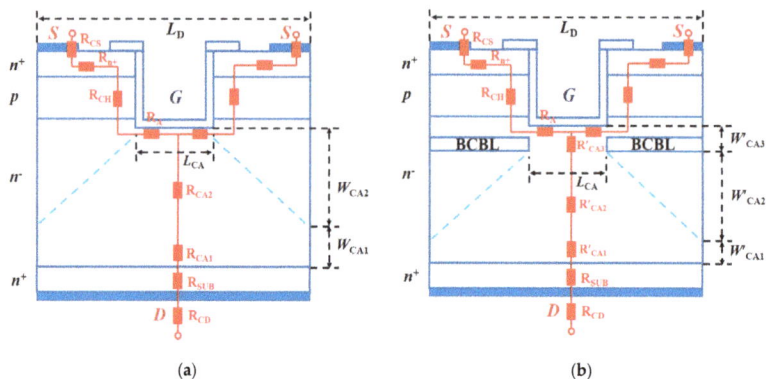

(a)   (b)

Figure 5. Schematic view of the equivalent circuit for the overall resistances between source and drain terminals in (**a**) the conventional vertical field-effect transistor (VFET) and (**b**) the proposed BCBL-VFET.

4.1.2. Optimization of L_{BCBL}

The simulated BV and R_{ON} data and the calculated FOM values with the changed L_{BCBL} are shown in Figure 6a,b, respectively. Both the BV and R_{ON} are increased obviously with the increasing L_{BCBL} mainly due to the narrowing of the vertical current aperture. Figure 6b shows a trade-off between the BV and R_{ON}, and then an optimal L_{BCBL} value at around 5.2 μm is determined. Figure 7 shows the distributions of the electric field and impact ionization concentration near the gate in the BCBL-VFET with the changed L_{BCBL}. The additional BCBL shares the voltage drop in the vertical direction which hence improves the BV characteristics of the devices. The average impact ionization concentration in the device with an L_{BCBL} around 5.2 μm is the lowest among all these devices which supports the above conclusions.

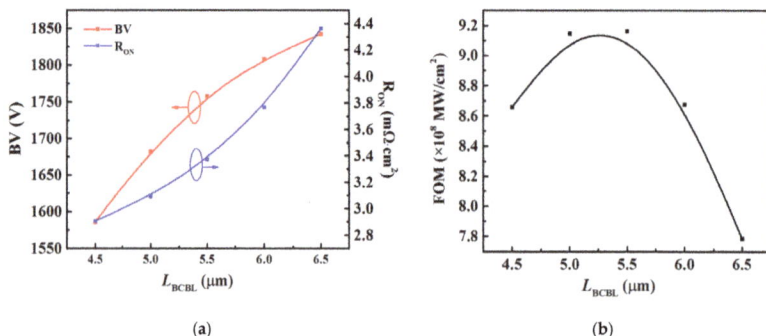

(a)   (b)

Figure 6. (**a**) BV and R_{ON} versus L_{BCBL} (the length of the BCBL) and (**b**) FOM values versus L_{BCBL}.

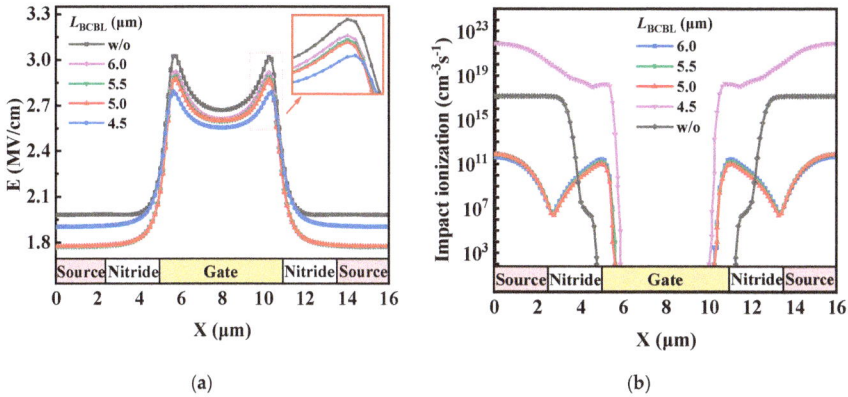

Figure 7. Distributions of (**a**) electric field and (**b**) impact ionization concentration near the gate in the BCBL-VFET with the changed L_{BCBL}.

4.1.3. Optimization of T_{BCBL}

The simulated BV and R_{ON} data and the calculated FOM values with the changed T_{BCBL} are shown in Figure 8a,b, respectively. The effects of T_{BCBL} variation on both the BV and R_{ON} are relatively small. Note that the BV characteristic is improved slightly when the T_{BCBL} increases from 0.1 to 0.4 µm and then the BV decreases; however, the reduction is small and kept within 100 V. Moreover, the R_{ON} variations in the proposed devices are also found to be small even though the thickness of the BCBL reaches 1.0 µm. This tells us that a relatively free value of T_{BCBL} can be employed in the BCBL-VFET. This is favorable in the device fabrication. Figure 9a shows the distributions of the electric field around the gate with the changed T_{BCBL}, which are found to be nearly the same among all these devices with a BCBL. However, a different distribution of impact ionization concentration is found in Figure 9b when the thickness of the BCBL reaches 1.0 µm. An additional peak of impact ionization starts to form around the BCBL corner and hence the BV is reduced. Therefore, the T_{BCBL} value should be kept at less than 0.6 µm.

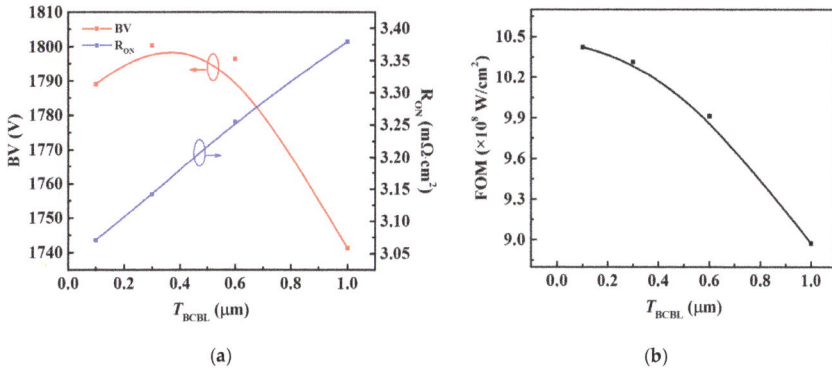

Figure 8. (**a**) BV and R_{ON} versus T_{BCBL} (the thickness of the BCBL) and (**b**) FOM values versus T_{BCBL}.

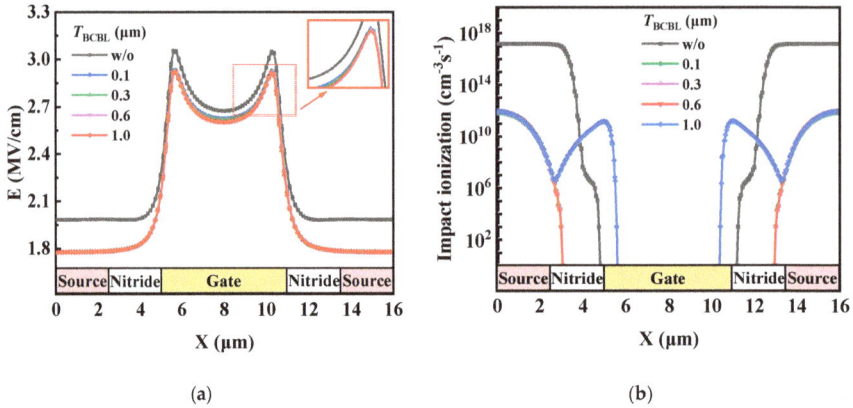

Figure 9. Distributions of (**a**) electric field and (**b**) impact ionization concentration around the gate in the BCBL-VFET with the changed T_{BCBL}.

4.2. Performances of the Optimized BCBL-VFET

The optimal parameters of d_{BCBL}, L_{BCBL}, and T_{BCBL} were determined to be 2.0, 5.2, and 0.6 µm, respectively. Figure 10 shows the comparisons of electric field distribution profiles between the optimized BCBL-VFET and the control device. It can be observed that the BCBL can shield the electric field from the pn-junction below the source electrode and hence efficiently protect the TCBL. Furthermore, the electric field distribution is made more uniform and the high-field area around the gate corner is reduced. The output dc I_d–V_d, transfer I_d–V_g, gate transconductance g_m, and BV characteristics of the optimized BCBL-VFET are shown in Figure 11. It can be seen clearly that the normally off operation with a large positive V_{th} (~3.5 V) is realized in the BCBL-VFET. Only a slight degradation of output I_d is found, although the V_{th} alters by about 1.0 V which might be due to the influence of the BCBL affecting the conducting channel. The degradation of drain current is relatively small at $V_{gt} < 1.5$ V and $V_d > 8.0$ V. To get the optimum performance, the device is recommended to work at about $V_{gt} = 1.5$ V and $V_d = 10.0$ V. The g_m peak is around 800 S/cm^2 at $V_g = 4.5$ V and remains nearly the same in comparison to that in the conventional VFET. The large g_m value achieved in the BCBL-VFET ensures a high switching frequency of the device. Furthermore, note that the BV of the proposed device is improved significantly compared with the control one.

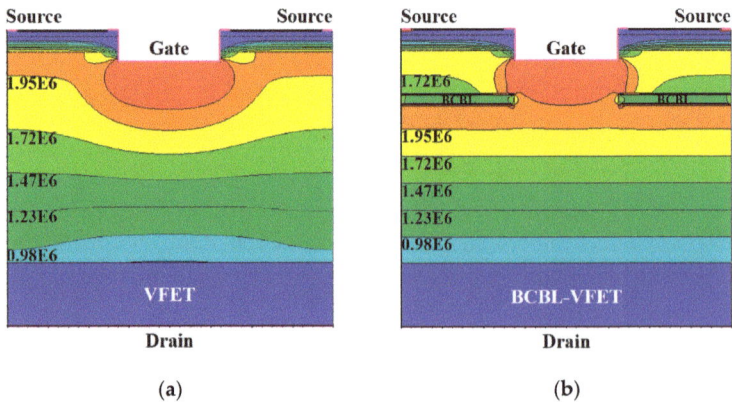

Figure 10. Comparisons of electric field distribution profiles between (**a**) the conventional VFET and (**b**) the proposed BCBL-VFET device.

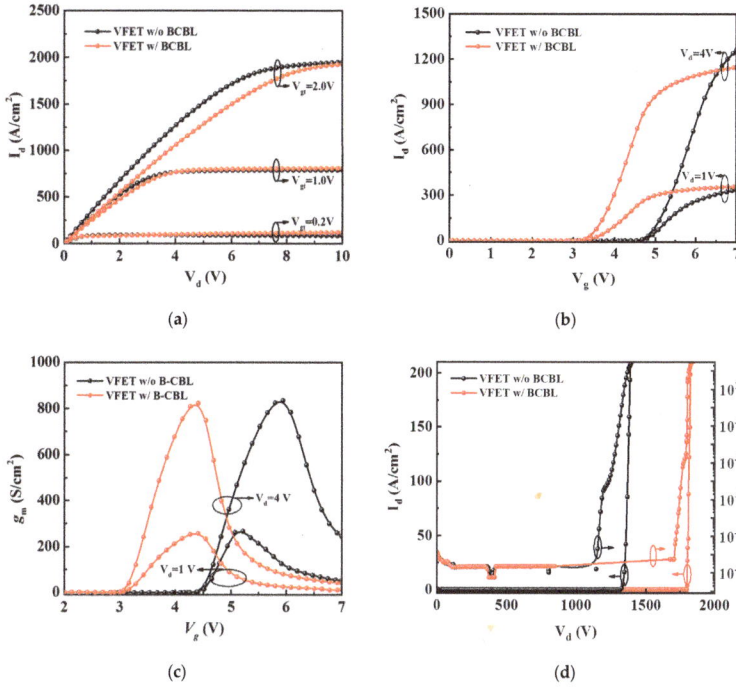

Figure 11. Comparisons of the typical (**a**) output dc I_d–V_d curves, (**b**) transfer I_d–V_g curves, (**c**) transconductance g_m, and (**d**) BV characteristics between the proposed BCBL-VFET and conventional VFET devices. For a fair comparison, here V_{gt} (= V_g – V_{th}) is employed.

Figure 12 summarizes the performance comparisons between the proposed BCBL-VFET and other reported GaN-based power devices including VFET, metal-oxide-semiconductor field-effect transistor (MOSFET), and HEMT devices [17–25]. It can be seen that the performance of the BCBL-VFET is closer to the GaN limit, compared with the control device and other work. This demonstrates that the proposed scheme would be very promising in the applications of power electronics.

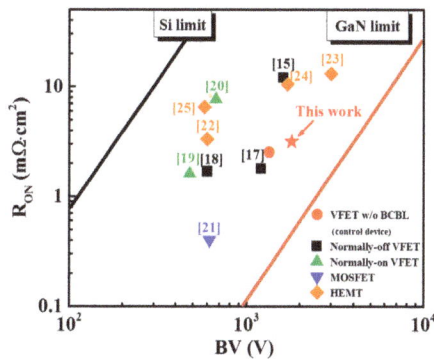

Figure 12. Performance comparisons among the state-of-the-art GaN-based power devices. The red star represents this work (BCBL-VFET).

The simulation work is significant and it helps people better understand the physical mechanism in the VFET and save experimental time. The structural parameters are optimized and provided to people for a better device fabrication. However, the actual operating conditions including the operation temperature, switching process, and matching circuit should be also considered seriously for practical applications [26–30]. It is noteworthy that the optimal device structures in this work are based on the dc characteristics. Due to the thermal and trapping effects under actual operating conditions, the dynamic R_{ON} and BV can be quite different from the static ones [10,29]. Actually, they are usually worse than the theoretical dc performances, and this needs to be considered in the fabrication work. Nevertheless, the proposed device structure and the optimized results are still constructive and meaningful, and the theoretical trend of device performance changes with the structural parameters should be similar between the dynamic and static ones.

The g_m value in the proposed BCBL-VFET is large enough to improve the device switching rate. Since a GaN semiconductor has high electron mobility and saturation rate, the on/off switching speed of the BCBL-VFET should be fast enough after having a stable fabrication technology and a matching gate drive circuit. In this work, the designed BV rating is more than 1800 V which can be applied in an electric vehicle with 600 V rating even taking into account the deviations from the realistic structural parameters and processing technology. Moreover, GaN-based power devices are very promising in the applications in high-temperature environments. The output current–voltage characteristics can remain stable after repeated high-temperature operations. Based on the previous simulation and measurement data [29], only a small V_{th} deviation of around 0.1 V was found in the metal-insulator-semiconductor (MIS)-gate GaN-based power devices after several rounds of high-temperature measurements from 25 to 150 °C. More nonlinear dynamic behaviors and actual operating conditions will be considered and taken into account in the following work.

5. Conclusions

A GaN normally off VFET device with an additional BCBL is introduced and demonstrated. The BCBL can lower the peak electric field around the gate and hence make the electric field distribution uniform, which leads to an obvious improvement of the BV characteristics. The structural parameters of the proposed BCBL-VFET have been optimized through theoretical calculations and TCAD simulations. An improved BV of 1800 V and a slightly increased R_{ON} of 3.14 m$\Omega \cdot$cm^2 are achieved in the BCBL-VFET, compared with 1300 V and 2.52 m$\Omega \cdot$cm^2 in the control VFET device. The FOM value of the proposed device can be up to 1.03 GW/cm^2, which is 52.5% higher than the control one.

Author Contributions: Writing—Original Draft, F.L.; Methodology, Z.S.; Investigation, N.S. and H.Z.; Software, F.Z.; Data Curation, Y.C.; Validation, P.T.; Writing—Review & Editing, H.H.

Funding: This research was funded by the National Science Foundation of China (grant number 51607022), the Fundamental Research Funds for the Central Universities (grant number DUT17LK13), the Open Project Program of the Key Laboratory of Semiconductor Materials Science (grant number KLSMS-1610), and the Open Project Program of the Key Laboratory of Nanodevices and Applications (grant number 18JG02) from the Chinese Academy of Sciences.

Acknowledgments: The authors are grateful for the support from the Computer Centre of the National University of Singapore.

Conflicts of Interest: The authors declare no conflict of interest.

References

1. Chen, K.J.; Haberlen, O.; Lidow, A.; Tsai, C.L.; Ueda, T.; Uemoto, Y.; Wu, Y.F. GaN-on-Si Power Technology: Devices and Applications. *IEEE Trans. Electron. Devices* **2017**, *64*, 779–795. [CrossRef]
2. Kachi, T. Recent progress of GaN power devices for automotive applications. *Jpn. J. Appl. Phys.* **2014**, *53*, 100210. [CrossRef]
3. Ambacher, O.; Foutz, B.; Smart, J.; Shealy, J.R.; Weimann, N.G.; Chu, K.; Murphy, M.; Sierakowski, A.J.; Schaff, W.J.; Eastman, L.F.; et al. Two dimensional electron gases induced by spontaneous and piezoelectric polarization in undoped and doped AlGaN/GaN heterostructures. *J. Appl. Phys.* **2000**, *87*, 334–344. [CrossRef]

4. Miao, M.S.; Weber, J.R.; Van de Walle, C.G. Oxidation and the origin of the two-dimensional electron gas in AlGaN/GaN heterostructures. *J. Appl. Phys.* **2010**, *107*, 123713. [CrossRef]

5. Oka, T.; Nozawa, T. AlGaN/GaN Recessed MIS-Gate HFET With High-Threshold-Voltage Normally-Off Operation for Power Electronics Applications. *IEEE Electron. Device Lett.* **2008**, *29*, 668–670. [CrossRef]

6. Lee, F.; Su, L.; Wang, C.; Wu, Y.; Huang, J. Impact of Gate Metal on the Performance of p-GaN/AlGaN/GaN High Electron Mobility Transistors. *IEEE Electron. Device Lett.* **2015**, *36*, 232–234. [CrossRef]

7. Jiang, H.; Tang, C.W.; Lau, K.M. Enhancement-Mode GaN MOS-HEMTs With Recess-Free Barrier Engineering and High-k ZrO_2 Gate Dielectric. *IEEE Electron. Device Lett.* **2018**, *39*, 405–408. [CrossRef]

8. Dai, S.; Zhou, Y.; Zhong, Y.; Zhang, K.; Zhu, G.; Gao, H.; Sun, Q.; Chen, T.; Yang, H. High f_T AlGa(In)N/GaN HEMTs Grown on Si With a Low Gate Leakage and a High ON/OFF Current Ratio. *IEEE Electron. Device Lett.* **2018**, *39*, 576–579. [CrossRef]

9. Anderson, T.J.; Tadjer, M.J.; Hite, J.K.; Greenlee, J.D.; Koehler, A.D.; Hobart, K.D.; Kub, F.J. Effect of Reduced Extended Defect Density in MOCVD Grown AlGaN/GaN HEMTs on Native GaN Substrates. *IEEE Electron. Device Lett.* **2016**, *37*, 28–30. [CrossRef]

10. Huang, H.L.; Liang, Y.C. Formation of combined partially recessed and multiple fluorinated-dielectric layers gate structures for high threshold voltage GaN-based HEMT power devices. *Solid-State Electron.* **2015**, *114*, 148–154. [CrossRef]

11. Ben-Yaacov, I.; Seck, Y.K.; Mishra, U.K.; DenBaars, S.P. AlGaN/GaN current aperture vertical electron transistors with regrown channels. *J. Appl. Phys.* **2004**, *95*, 2073–2078. [CrossRef]

12. Kanechika, M.; Sugimoto, M.; Soejima, N.; Ueda, H.; Ishiguro, O.; Kodama, M.; Hnyashi, E.; Itoh, K.; Uesugi, T.; Kachi, T. A vertical insulated gate AlGaN/GaN heterojunction field-effect transistor. *Jpn. J. Appl. Phys.* **2007**, *46*, L503–L505. [CrossRef]

13. Ozbek, A.M.; Baliga, B.J. Planar Nearly Ideal Edge-Termination Technique for GaN Devices. *IEEE Electron. Device Lett.* **2011**, *32*, 300–302. [CrossRef]

14. Chowdhury, S.; Wong, M.H.; Swenson, B.L.; Mishra, U.K. CAVET on Bulk GaN Substrates Achieved With MBE-Regrown AlGaN/GaN Layers to Suppress Dispersion. *IEEE Electron. Device Lett.* **2012**, *33*, 41–43. [CrossRef]

15. Oka, T.; Ueno, Y.; Ina, T.; Hasegawa, K. Vertical GaN-based trench metal oxide semiconductor field-effect transistors on a free-standing GaN substrate with blocking voltage of 1.6 kV. *Appl Phys Express* **2014**, *7*, 021002. [CrossRef]

16. Baliga, B.J. *Fundamentals of Power Semiconductor Devices*; Springer Science & Business Media: Berlin, Germany, 2008; pp. 359–365.

17. Oka, T.; Ina, T.; Ueno, Y.; Nishii, J. 1.8 mΩ·cm^2 vertical GaN-based trench metal–oxide–semiconductor field-effect transistors on a free-standing GaN substrate for 1.2-kV-class operation. *Appl. Phys. Express* **2015**, *8*, 054101. [CrossRef]

18. Li, R.; Cao, Y.; Chen, M.; Chu, R. 600 V/1.7 Ω Normally-Off GaN Vertical Trench Metal–Oxide–Semiconductor Field-Effect Transistor. *IEEE Electron Device Lett.* **2016**, *37*, 1466–1469. [CrossRef]

19. Shrestha, N.M.; Li, Y.M.; Chang, E.Y. Optimal design of the multiple-apertures-GaN-based vertical HEMTs with SiO_2 current blocking layer. *J. Comput. Electron.* **2016**, *15*, 154–162. [CrossRef]

20. Yaegassi, S.; Okada, M.; Saitou, Y.; Yokoyama, M.; Nakata, K.; Katayama, K.; Ueno, M.; Kiyama, M.; Katsuyama, T.; Nakamura, T. Vertical heterojunction field-effect transistors utilizing re-grown AlGaN/GaN two-dimensional electron gas channels on GaN substrates. *Phys. Status Solidi C* **2011**, *8*, 450–452. [CrossRef]

21. Kang, E.G.; Kim, Y.T. Design of Trench Gate GaN Power MOSFET using Al_2O_3 Gate Oxide. *J. Phys. Conf. Ser.* **2012**, *352*, 012025. [CrossRef]

22. Saito, W.; Takada, Y.; Kuraguchi, M.; Tsuda, K.; Omura, I.; Ogura, T. Design and demonstration of high breakdown voltage GaN high electron mobility transistor (HEMT) using field plate structure for power electronics applications. *Jpn. J. Appl. Phys.* **2004**, *43*, 2239–2242. [CrossRef]

23. Dogmus, E.; Zegaoui, M.; Medjdoub, F. GaN-on-silicon high-electron-mobility transistor technology with ultra-low leakage up to 3000 V using local substrate removal and AlN ultra-wide bandgap. *Appl. Phys. Express* **2018**, *11*, 034102. [CrossRef]

24. Zhao, Z.Q.; Zhao, Z.Y.; Luo, Q.; Du, J.F. High-voltage RESURF AlGaN/GaN high electron mobility transistor with back electrode. *Electron. Lett.* **2013**, *49*, 1638–1640. [CrossRef]

25. Huang, H.; Liang, Y.C.; Samudra, G.S.; Ngo, C.L.L. Normally-Off AlGaN/GaN-on-Si MIS-HEMTs Using Combined Partially Recessed and Fluorinated Trap-Charge Gate Structures. *IEEE Electron. Device Lett.* **2014**, *35*, 569–571. [CrossRef]

26. Campbell, C.F.; Balistreri, A.; Kao, M.Y.; Dumka, D.C.; Hitt, J. GaN Takes the Lead. *IEEE Microw. Mag.* **2012**, *13*, 44–53. [CrossRef]

27. Crupi, G.; Raffo, A.; Avolio, G.; Schreurs, D.M.M.P.; Vannini, G.; Caddemi, A. Temperature Influence on GaN HEMT Equivalent Circuit. *IEEE Microw. Wirel. Components* **2016**, *26*, 813–815. [CrossRef]

28. Nalli, A.; Raffo, A.; Crupi, G.; D'Angelo, S.; Resca, D.; Scappaviva, F.; Salvo, G.; Caddemi, A.; Vannini, G. GaN HEMT Noise Model Based on Electromagnetic Simulations. *IEEE Trans. Microw. Theory* **2015**, *63*, 2498–2508. [CrossRef]

29. Huang, H.; Li, F.; Sun, Z.; Cao, Y. Model Development for Threshold Voltage Stability Dependent on High Temperature Operations in Wide-Bandgap GaN-Based HEMT Power Devices. *Micromachines* **2018**, *9*, 658. [CrossRef] [PubMed]

30. Quaglia, R.; Camarchia, V.; Pirola, M.; Ghione, G. GaN Monolithic Power Amplifiers for Microwave Backhaul Applications. *Electronics* **2016**, *5*, 25. [CrossRef]

© 2019 by the authors. Licensee MDPI, Basel, Switzerland. This article is an open access article distributed under the terms and conditions of the Creative Commons Attribution (CC BY) license (http://creativecommons.org/licenses/by/4.0/).

electronics

MDPI

Article

AlGaN/GaN MIS-HEMT with PECVD SiN$_x$, SiON, SiO$_2$ as Gate Dielectric and Passivation Layer

Kuiwei Geng [1], Ditao Chen [1], Quanbin Zhou [1] and Hong Wang [1,2,3,*]

[1]   Engineering Research Center for Optoelectronics of Guangdong Province, School of Electronics and Information Engineering, South China University of Technology, Guangzhou 510640, China; gengkw@scut.edu.cn (K.G.); 201620108002@mail.scut.edu.cn (D.C.); zhouquanbin86@163.com (Q.Z.)

[2]   School of Physics and Optoelectronics, South China university of Technology, Guangzhou 510640, China

[3]   Zhongshan Institute of Modern Industrial Technology, South China University of Technology, Zhongshan 528437, China

*   Correspondence: phhwang@scut.edu.cn; Tel.: +86-136-0006-6193

Received: 5 October 2018; Accepted: 7 December 2018; Published: 10 December 2018

Abstract: Three different insulator layers SiN$_x$, SiON, and SiO$_2$ were used as a gate dielectric and passivation layer in AlGaN/GaN metal–insulator–semiconductor high-electron-mobility transistors (MIS-HEMT). The SiN$_x$, SiON, and SiO$_2$ were deposited by a plasma-enhanced chemical vapor deposition (PECVD) system. Great differences in the gate leakage current, breakdown voltage, interface traps, and current collapse were observed. The SiON MIS-HEMT exhibited the highest breakdown voltage and I$_{on}$/I$_{off}$ ratio. The SiN$_x$ MIS-HEMT performed well in current collapse but exhibited the highest gate leakage current density. The SiO$_2$ MIS-HEMT possessed the lowest gate leakage current density but suffered from the early breakdown of the metal–insulator–semiconductor (MIS) diode. As for interface traps, the SiN$_x$ MIS-HEMT has the largest shallow trap density and the lowest deep trap density. The SiO$_2$ MIS-HEMT has the largest deep trap density. The factors causing current collapse were confirmed by Photoluminescence (PL) spectra. Based on the direct current (DC) characteristics, SiN$_x$ and SiON both have advantages and disadvantages.

Keywords: gallium nitride; MISHEMT; dielectric layer; interface traps; current collapse; PECVD

1. Introduction

In the past decades, the wide bandgap semiconductor material, gallium nitride (GaN), attracted great attention due to its wide bandgap, high breakdown electric field, and excellent thermal properties [1]. Gate leakage current and current collapse are the main issues that limit the performance of AlGaN/GaN high-electron-mobility transistors (HEMTs). To overcome these problems, different dielectric materials have been proposed for the fabrication of metal-insulator-semiconductor (MIS) HEMTs, such as SiO$_2$ [2–6], SiN$_x$ [7–11], SiONe [12–14], ZrO$_2$ [15], Al$_2$O$_3$ [16–18], and HfO$_2$ [19], etc. Each material has advantages and disadvantages. Some groups studied stack dielectric layers like SiN$_x$/Al$_2$O$_3$ [20] and SiN$_x$/SiO$_2$ [21] to improve leakage current and stability. The Al$_2$O$_3$ and HfO$_2$ gate dielectric layer deposited by atomic layer deposition (ALD) has shown advantages in reducing gate leakage and eliminating current collapse [22,23]. However, Al$_2$O$_3$ and HfO$_2$ are not suitable as passivation layers due to the low deposition rate of ALD. The plasma-enhanced chemical vapor deposition (PECVD) is one of the key sectors in conventional GaN-based light emitting diode (LED) production lines and complementary metal–oxide–semiconductor (CMOS) production lines. Thus, lots of work has been done on PECVD-deposited silicon nitride, silicon oxide, and silicon oxynitride. Compared to SiN$_x$, SiO$_2$ has a larger conduction band offset with GaN, which is related to leakage current. However, SiN$_x$ has a relatively higher dielectric constant (~7), which contributes to better gate control of two dimensional electron gas (2DEG). As a trade-off, SiON can be modulated to retain some advantages from both SiN_x and SiO_2 and has been proved to be a good candidate for a gate dielectric [12]. Considering the passivation effect, these dielectric layers can reduce the surface states, modulate the strain, and improve the reliability [24–26]. Although some comparisons have been made on the above materials [13,27,28], some published data are often controversial and the overall result is still not sufficiently clear. This shows that many questions are still unanswered and a better understanding of the passivation effect on the device performance is required.

In this work, we have made comprehensive comparisons between MIS-HEMTs with PECVD-deposited SiN_x, SiON, and SiO_2 as a gate dielectric and passivation layer. The differences in direct current (DC) static characteristics and current collapse were investigated. The interface traps were studied by Capacitance versus Voltage (C-V) measurements and a pulse mode drain current versus gate-to-source voltage (Id-V_{gs}) test. The Photoluminescence (PL) spectra test was also applied to confirm the passivation effectiveness.

2. Materials and Methods

The AlGaN/GaN epilayer used in this work is grown on Si (111) substrate using metal–organic chemical vapor deposition (MOCVD). The epitaxial structure consists of a 3.5 µm GaN buffer layer, a 300 nm GaN channel layer, a 1 nm AlN interlayer, a 22 nm $Al_{0.23}Ga_{0.77}N$ barrier layer, and a 3 nm GaN cap layer. A 2DEG mobility of 1831 $cm^2/V \cdot s$ and a sheet carrier concentration of 8.3×10^{12} cm^{-2} are measured by Hall effect measurement.

The device fabrication of MIS-HEMT started with the cleaning of the epitaxial wafer with a standard solvent. Then, devices were isolated using BCl_3 and Cl_2 etching in an Inductively Coupled Plasma (ICP) system. Prior to the deposition of ohmic metal, surface treatment was performed by immersing in HCL for 60 s. The Ti/Al/Ni/Au metal stack was then deposited by E-beam evaporation. Rapid temperature annealing at 850 °C for 1 min in a N_2 environment was then performed to form ohmic contact. The contact resistance was 2.11 $\Omega \cdot mm$ and the specific contact resistance was 1.75×10^{-4} Ω cm^2, as extracted by a circular transmission line model. After that, SiN_x, SiON, and SiO_2 dielectric layers with a thickness of 20 nm were deposited separately on the surfaces of different AlGaN/GaN samples using PECVD. Additionally, a Si dummy wafer and an AlGaN/GaN dummy wafer were loaded, together with the sample, in each deposition process. The dielectric/AlGaN/GaN samples were used for PL spectra measurement. The thickness and refractive index of deposited thin films were measured for the dummy wafer using an ellipsometer. The deposition properties of SiN_x, SiON, and SiO_2 are listed in Table 1.

Table 1. Deposition properties of SiN_x, SiON, and SiO_2.

Sample	Pressure (millitorr)	RF Power(W)	SiH_4 [1] (sccm[2])	N_2O (sccm)	NH_3 (sccm)	TEM (°C)	Refractive Index [3]
SiN_x	650	50	150	0	25	300	1.82
SiON	500	75	25	20	40	300	1.56
SiO_2	650	50	100	1000	0	300	1.46

[1] SiH_4 (5%)/N_2; [2] Standard Cubic Centimeter per Minute; [3] Refractive Index at the wavelength of 632.8 nm.

The dielectric layers above drain and source electrodes were then removed by ICP. Finally, all samples were carried out using the same gate contact process. A Ni/Au (50/150 nm) gate metal was deposited by E-Beam evaporation. All samples have the same epitaxial structure and fabrication process, except for the type of dielectric layers. The MIS-HEMTs with different dielectric layers are labeled as SiN_x MIS-HEMT, SiON MIS-HEMT, and SiO_2 MIS-HEMT, respectively. Figure 1 shows the schematic cross-sectional view of MIS-HEMT. The gate length L_G, gate width W_G, gate to drain distance L_{GD}, and gate to source distance L_{GS} are 3, 150, 20, and 10 µm, respectively.

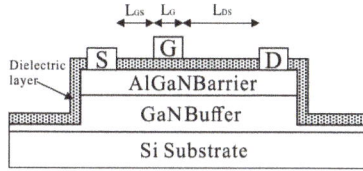

Figure 1. Metal–insulator–semiconductor high-electron-mobility transistors (MIS-HEMT) structure cross-section view.

3. Results and Discussion

Table 2 shows some selected properties of MIS-HEMTs. The typical DC output characteristics of SiN_x MIS-HEMT, SiON MIS-HEMT, and SiO_2 MIS-HEMT are shown in Figure 2a–c. The drain current densities at drain-to-source voltage Vds = 20 V and gate-to-source voltage V_{gs} = 8 V are 623 mA/mm, 590 mA/mm, and 620 mA/mm, respectively, for SiN_x MIS-HEMT, SiON MIS-HEMT, and SiO_2 MIS-HEMT. The specific on-resistance extracted at Vds = 3 V is 9.89 Ω·mm, 11.6 Ω·mm, and 11.4 Ω·mm, respectively. These output characteristics show that the SiN_x MIS-HEMT has the highest maximum drain current and lowest static on-resistance. Figure 2d shows the off-state breakdown characteristics measured at V_{gs} = −18 V. SiON MIS-HEMT exhibits a higher breakdown voltage compared with the other two samples. SiN_x performs slightly better in improving the saturated drain current of MIS-HEMT, and SiON can withstand a higher electric field strength.

Table 2. Selected properties of SiN_x, SiON, and SiO_2 MIS-HEMT.

Sample	I_{dmax} (mA/mm)	$g_m$max (mS/mm)	V_{th}	Gate Leakage [1] (mA/mm)	Off-State Breakdown Voltage (V)	%I [2]	Dynamic Ron/Static Ron [3]
SiN_x	623	62.7	−16.7	4.46 E-4	364	11.6%	1.18
SiON	590	55.3	−11.7	3.86 E-5	428	71.26%	5.64
SiO_2	620	81.3	−9.9	3.12 E-5	284	84.14%	24.5

[1] Gate leakage current density at two-terminal reverse voltage = −20 V; [2] Reduction of drain current and increase of Ron due to current collapse at off-state V_{gs} = −18 V, Vdstress = 50 V for 10 s; [3] Dynamic Ron at off-state Vds stress = 50 V and static Ron without stress.

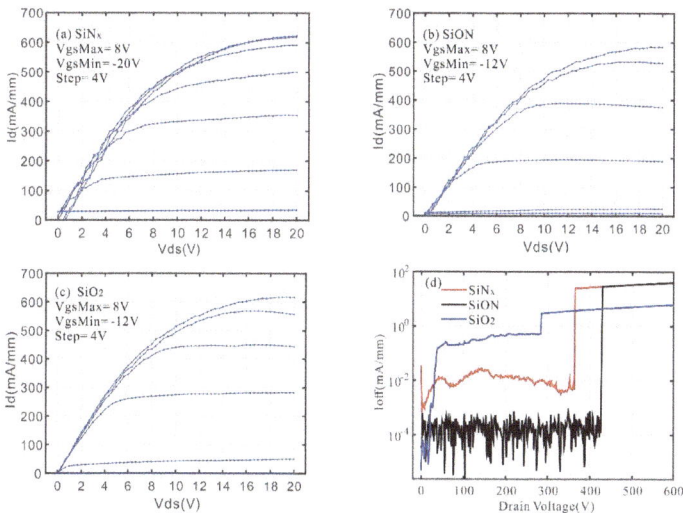

Figure 2. Output characteristic of (**a**) SiN_x MIS-HEMT, (**b**)SiON MIS-HEMT, and (**c**) SiO_2 MIS-HEMT. (**d**) off-state breakdown characteristic of the three samples, measured at V_{gs} = −18 V.

Figure 3d plots the gate-to-source two-terminal leakage current curve when V_{gs} changes from 5 to −40 V with drain electrode dangling. The gate-to-source leakage current (Igs) density of SiN_x MIS-HEMT is 4.46×10^{-4} mA/mm at $V_{gs} = 20$ V, which is 1 order larger than that of SiON or SiO_2 MIS-HEMT. This can be attributed to the lower conduction band offset of SiN_x from GaN [12]. As for SiO_2 MIS-HEMT, there is a rapid increase of leakage current when the gate voltage bias is lower than −35 V. This phenomenon means that the SiO_2 dielectric layers are more easily damaged than SiN_x and SiON.

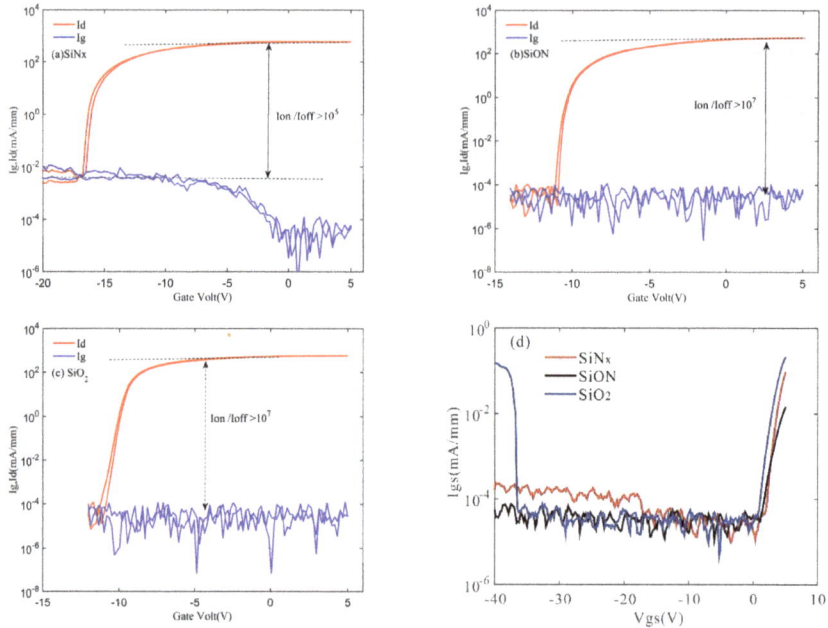

Figure 3. Transfer characteristic of (**a**) SiN_x MIS-HEMT, (**b**) SiON MIS-HEMT, and (**c**) SiO_2 MIS-HEMT, and (**d**) gate leakage current curve measured at two terminals for the three samples.

The transfer curves of the three samples are shown in Figure 3a–c. The drain voltage was fixed at 15 V, and the minimum gate voltage was −20 V, −14 V, and −12 V, respectively, for SiN_x MIS-HEMT, SiON MIS-HEMT, and SiO_2 MIS-HEMT. The drain current, Id, is almost equal to the gate current, Ig, in the off-state for all samples. This result reveals that the off-state drain leakage current is mainly from the gate electrode. As a result, the comparison of the off-state drain currents of these two samples agrees with the two-terminal gate leakage current. A high I_{on}/I_{off} ratio $> 10^7$ was observed on SiON MIS-HEMT and SiO_2 MIS-HEMT, which is two orders larger than SiN_x MIS-HEMT.

To investigate the interface condition of the three samples, a forward and backward C-V measurement with a frequency of 1 MHz was applied. As shown in Figure 4, each C-V curve has two rising slopes. The 1st slope represents the completed depletion of 2DEG and the 2nd slope indicates the electron transfer from AlGaN/GaN to dielectric/AlGaN interface [29,30]. A relatively low voltage corresponding to the 1st slope is observed in the SiN_x MIS structure. This may be attributed to the fixed charge in the dielectric/AlGaN interface. Some researches show that a large amount of fixed charges exist in the dielectric layer and dielectric/AlGaN interface [31–35], differing from interface traps analyzed in the work, these kinds of fixed charges are not modulated by the gate voltage and do not lead to voltage hysteresis. Therefore, they have a negligible effect on the CV hysteresis measurement. However, the positive fixed charge would cause a negative voltage shift of the flat band voltage (V_{FB}), and thus lead to a low threshold voltage, V_{th} [33]. More investigation concerning fixed charges is needed in the future. In the CV curve, the backward hysteresis and threshold voltage shifts are always attributed to interface traps. The inset of Figure 4 shows that there is little voltage shift on the 2nd slope while obvious hysteresis occurred on the 1st slope of the SiON and SiO$_2$ MIS sample. This phenomenon occurs because larger number of deep traps with long emission time constants appear in the SiON and SiO$_2$ MIS structure. We also used dielectric capacitance in series with the barrier capacitance model to extract the Cox of the three samples [36]. The Cox is 297.86, 277.82, and 364.82 nF/cm^{-2} for SiN$_x$, SiON, and SiO$_2$ MIS-HEMT, respectively. C-V measurement provides a rough comparison of the three structures but its precision is limited by the sweeping rates. The pulse-mode Id-V$_{gs}$ measurement was employed for a more accurate extraction of interface states.

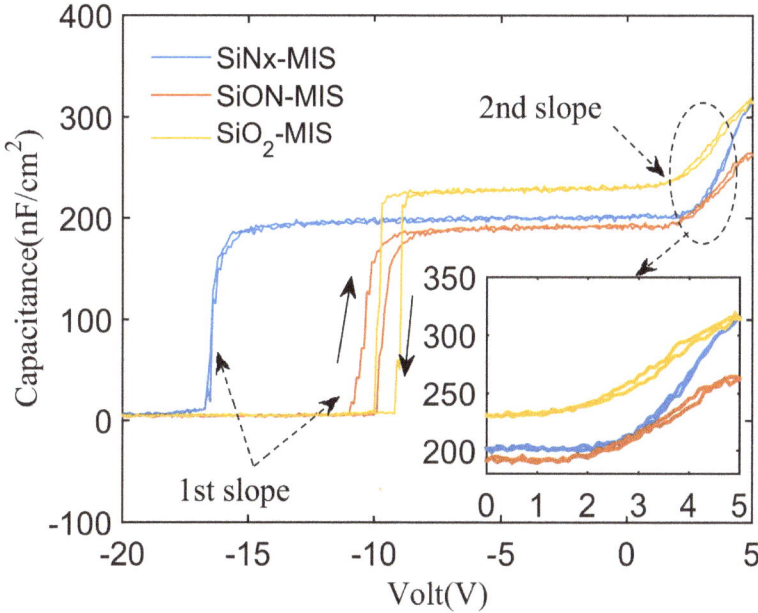

Figure 4. Forward and backward capacitance versus voltage (C-V) measurement with frequency of 1 MHz. Inset: magnified voltage range of 0~5 V.

In this work, the pulse period was fixed at 500 ms and the selected measured pulse widths were 100 μs, 1 ms, 10 ms, 100 ms, and 200 ms. Vds was kept at 1 V to reduce drain-to-gate field-assisted detrapping [29]. The inset of Figure 5b shows the forward sweep and backward sweep measurement conditions of V$_{gs}$. It is reported that the acceptor-like interface states were originally empty and would capture electrons during the forward sweep of V$_{gs}$ with a low V$_{gs}$ base [29]; therefore, the forward sweep curve was chosen to be the basic line. In the backward sweep of V$_{gs}$, interface traps with emission times longer than the measurement pulse width would remain occupied by electrons, which would lead to a positive shift of V$_{th}$. The detectable traps emission time τ is related to its energy using Shockley-Read-Hall statistics:

$$\tau = \frac{1}{v_{th}\sigma_n N_C}\exp\left(\frac{\Delta E}{kT}\right) \tag{1}$$

where v$_{th}$, σ$_n$, and N$_C$ are the electron thermal velocity, electron capture cross-section, and electron concentration at the effective density of states in the conduction band in GaN. $\Delta E = E_C - E_T$ is the energy gap between the conduction band and interface trap. k is the Boltzman constant and T is the temperature.

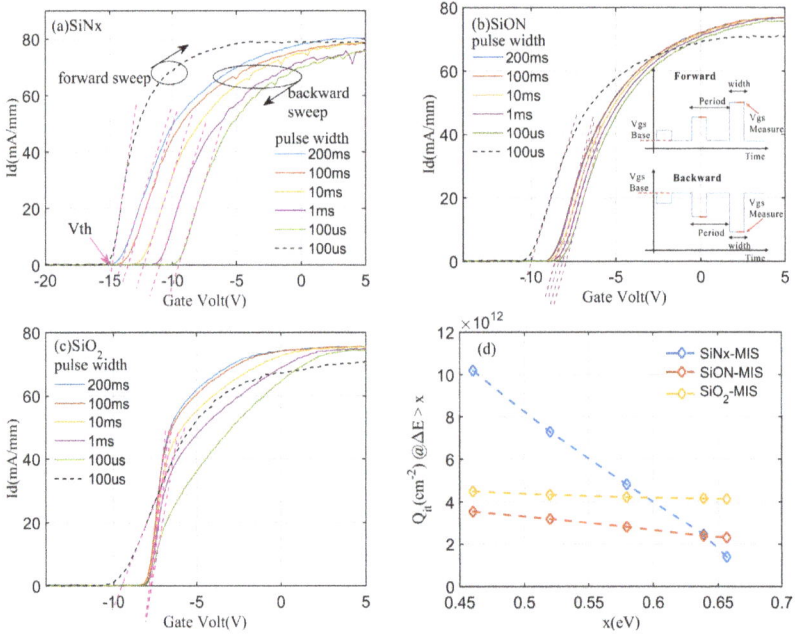

Figure 5. Pulse mode Id-V_{gs} curves with pulse width variations of 100 μs, 1 ms, 10 ms, 100 ms, and 200 ms. The pulse period is 500 ms. The Vds was kept at 1 VThe backward V_{gs} base is 5 V. (**a**) SiN$_x$ MIS-HEMT with forward V_{gs} base of −20 V (**b**) SiON MIS-HEMT with forward V_{gs} base of −14 V. Inset: pulse V_{gs} condition with forward sweep and backward sweep (**c**) SiO$_2$ MIS-HEMT with forward V_{gs} base of −14 V (**d**) Interface trap charge density Q_{it} at $\Delta E > x$(eV) of three samples.

For SiN$_x$ MIS-HEMT, the forward V_{gs} base was −20 V. For SiON and SiO$_2$ MIS-HEMT, the forward V_{gs} base was −14 V. The backward V_{gs} base was 5 V for all three samples. As shown in Figure 5a–c, SiN$_x$ MIS-HEMT shows a strong correlation between the measurement pulse width and the threshold voltage shift. SiO$_2$ MIS-HEMT shows little V_{th} change with different pulse widths. The corresponding interface-trapped charge density (Q_{it}) can be determined by:

$$Q_{it} = \frac{C_{ox} \cdot \Delta V_{th}}{q} \tag{2}$$

where C_{ox} values were extracted from C-V curves and the threshold voltage shift ΔV_{th} values were extracted from the results shown in Figure 5a–c.

Using Equations (1) and (2), Q_{it} with different ranges of ΔE were extracted and are shown in Figure 5d. At $\Delta E > 0.460$ eV, the Q_{it} is 1.02×10^{13} cm^{-2}, 3.54×10^{12} cm^{-2}, and 4.49×10^{12} cm^{-2} for SiN$_x$, SiON, and SiO$_2$ MIS-HEMT, respectively. At $\Delta E > 0.657$ eV, the Q_{it} turns out to be 1.38×10^{12} cm^{-2}, 2.31×10^{12} cm^{-2}, and 4.13×10^{12} cm^{-2}, respectively. SiN MIS-HEMT has the largest detected Q_{it} at $\Delta E > 0.460$ eV. These kinds of interface traps have an emission time longer than 100 μs. In the backward sweep of V_{gs}, these traps would remain occupied by an electron, and lead to the largest V_{th} shift observed in SiN$_x$ MIS-HEMT. In addition, the detectable Q_{it} density of the SiN$_x$ sample includes a number of 8.81×10^{12} cm^{-2} located at 0.460 eV $< \Delta E <$ 0.657 eV. These kind of traps (acceptor like) are regarded as shallow traps with short emission times. As for the SiO$_2$ MIS-HEMT, the difference between Q_{it} at $\Delta E > 0.460$ eV and Q_{it} at $\Delta E > 0.657$ eV is small, which indicates that most of its interface traps are deep traps. In conclusion, SiN$_x$ MIS-HEMT has the largest density of shallow interface traps and the lowest density of deep traps among the three samples. SiO$_2$ MIS-HEMT has the lowest density of shallow traps and largest density of deep traps. The performance of SiON MIS-HEMT is between the SiN$_x$ and SiO$_2$ sample. These results explain the difference in hysteresis in the 1st slope of the C-V curve. The larger deep trap densities of SiON and SiO$_2$ MIS-HEMT are responsible for the hysteresis in the 1st slope of the C-V curve. In addition, these calculated results show that shallow energy levels are more likely to be occupied by traps in the SiN$_x$/AlGaN interface than in the SiON/AlGaN interface and SiO$_2$/AlGaN interface. This offers chances for electron hopping, which could partially explain why SiN$_x$ MIS-HEMT has the highest gate leakage current and the lowest I$_{on}$/I$_{off}$ ratio among the three samples.

The Off-state Current collapse characteristic was measured by a slow switching test using an Agilent B1505A power device analyzer [37]. Various stress voltages from 5 V up to 50 V were applied on drain-to-source electrodes when V$_{gs}$ was fixed at −18 V to ensure that the channel was pinched off. After the stress situation for 10 s, V$_{gs}$ was changed to 0 V and the on-state Id-Vd curve was measured. In addition, the time interval was 10 ms between the two data points. This would lead to trap discharging and recover the current collapse to some extent. Therefore, the deep traps would be the major factor causing current collapse. Figure 6a–c shows that the degradation of the drain current was 11.6%, 71.26%, and 84.14%, respectively, for SiN$_x$ MIS-HEMT, SiON MIS-HEMT, and SiO$_2$ MIS-HEMT. As Figure 6d shows, the dynamic Ron increases more quickly along with off-state drain bias stress (Vdstress) for SiO$_2$ MIS-HEMT. After 50 V Vdstress was applied to the devices, the ratio of dynamic Ron and static Ron turns out to be 1.18, 5.64, and 24.5, respectively, for SiN$_x$ MIS-HEMT, SiON MIS-HEMT, and SiO$_2$ MIS-HEMT. Among these three samples, SiN$_x$ MIS-HEMT shows better performance with regard to suppressing the current collapse than the other two samples. Though SiN$_x$ MIS-HEMT has the largest detected interface trap density, most of them are relatively shallow traps located at 0.46 eV < ΔE < 0.657 eV. These kinds of shallow traps contribute less to current collapse. SiON and SiO$_2$ MIS-HEMT have higher trap density than SiN$_x$ MIS-HEMT at ΔE > 0.657 eV. This leads to more serious current collapse observed in SiON and SiO$_2$ MIS-HEMT. The current collapse performance coincides with the extracted results of C-V and pulse Id-V$_{gs}$ measurements for interface deep traps, which indicates that deep interface traps strongly influence the collapse characteristic.

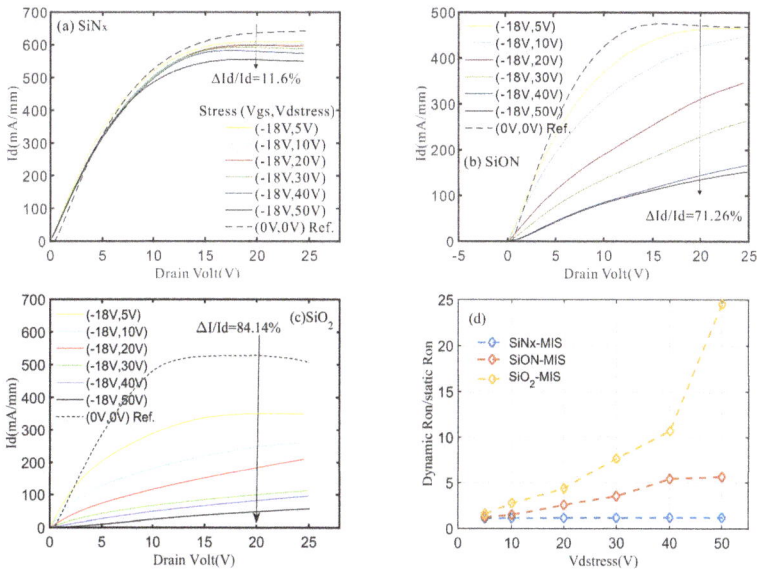

Figure 6. Id-Vds curves at V$_{gs}$ = 0 V of (**a**) SiN$_x$ MIS-HEMT (**b**) SiON MIS-HEMT (**c**) SiO$_2$ MIS-HEMT after off-state stress bias at V$_{gs}$ = −18 V, Vds stress at 5, 10, 20, 30, 40, and 50 V for 10 s. (**d**) The ratio of dynamic Ron and static Ron versus different off-state Vds stress.

To further investigate the difference in characteristics of SiN$_x$, SiON and SiO$_2$ MIS-HEMT, room temperature photoluminescence (PL) spectra were recorded. As shown in the inset of Figure 7, the relative intensities of the yellow band (wavelength at approximately 560 nm) of the three dielectric/AlGaN/GaN samples are quite different. It has been reported that Si and O impurity, which would act as shallow donors, can effectively impact the yellow luminescence (YL) [38,39]. The largest YL intensity observed on SiO$_2$/AlGaN/GaN sample implies that it has a maximum number of Si and O shallow donors among the three samples. During the deposition of SiON and SiO$_2$, the reactive gas N$_2$O would cause uncontrollable oxidation of the AlGaN interface and therefore generate several kinds of surface states [40]. The SiN$_x$ chemical deposition process involves NH$_3$ plasma treatment, which would be effective to suppress the N-vacancies-related surface defects at the AlGaN surface [41]. When the off-state stress is applied to the devices, the shallow donors, oxides, and defects mentioned above would capture electrons and cause the phenomenon of the virtual gate [42]. The formation of the virtual gate would increase the depletion region and thus cause the decrease of drain current. The above factors together lead to the different performance on current collapse of the three samples.

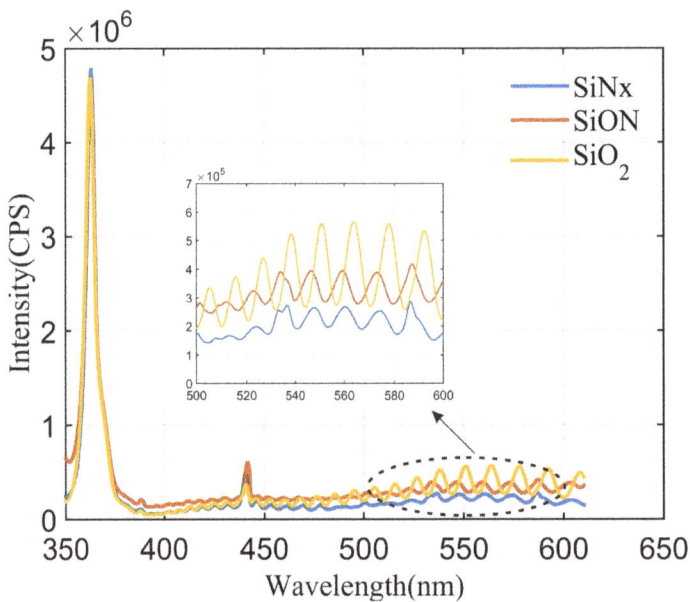

Figure 7. Photoluminescence (PL) spectra at room temperature of three samples.

4. Conclusions

In conclusion, we have fabricated AlGaN/GaN MIS-HEMTs with PECVD-deposited SiN$_x$, SiON, and SiO$_2$ as the gate dielectric and passivation layer. The DC static characteristics, interface traps, and current collapse of MIS-HEMTs with different dielectrics were comprehensively compared. The SiN$_x$ sample performs well with regard to suppressing the current collapse but suffers from high leakage current and high shallow trap density. The SiON MIS-HEMT exhibits a low gate leakage current of 3.86×10^{-5} mA/mm and a high breakdown voltage of 428 V, which indicates it is a great candidate as a gate dielectric and passivation layer. However, its deposition parameters need to be further optimized in order to enhance the reliability and stability.

Author Contributions: Conceptualization, K.G. and D.C.; Formal analysis, K.G. and D.C.; Investigation, Q.Z.; Methodology, K.G. and Q.Z.; Supervision, H.W.; Writing original draft, K.G. and D.C.; D.C. designed the experiment, prepared the samples and performed the measurements. K.G. contributed to the conception of the study and designed the experiment. Q.Z. contributed to the data analysis and wrote the manuscript. H.W. supervised the study and reviewed the manuscript.

Funding: This work was supported by Science and Technologies plan Projects of Guangdong Province (Nos. 2017B010112003, 2017A050506013), and Applied Technologies Research and Development Projects of Guangdong Province (Nos. 2015B010127013, 2016B010123004), and Science and Technologies plan Projects of Guangzhou City (Nos. 201504291502518, 201604046021, 201704030139), and Science and Technology Development Special Fund Projects of Zhongshan City (Nos. 2017F2FC0002, 2017A1009).

Conflicts of Interest: The authors declare no conflict of interest.

References

1. Chen, K.J.; Haberlen, O.; Lidow, A.; Tsai, C.L.; Ueda, T.; Uemoto, Y.; Wu, Y. GaN-on-Si Power Technology: Devices and Applications. *IEEE Trans. Electron. Dev.* **2017**, *64*, 779–795. [CrossRef]
2. Lee, J.; Kim, H.; Seo, K.; Cho, C.; Cha, H. High quality PECVD SiO_2 process for recessed MOS-gate of AlGaN/GaN-on-Si metal–oxide–semiconductor heterostructure field-effect transistors. *Solid State Electron.* **2016**, *122*, 32–36. [CrossRef]
3. Chakroun, A.; Jaouad, A.; Soltani, A.; Arenas, O.; Aimez, V.; Ares, R.; Maher, H. AlGaN/GaN MOS-HEMT Device Fabricated Using a High Quality PECVD Passivation Process. *IEEE Electr. Device Lett.* **2017**, *38*, 779–782. [CrossRef]
4. Kordoš, P.; Heidelberger, G.; Bernát, J.; Fox, A.; Marso, M.; Lüth, H. High-power SiO_2/AlGaN/GaN metal-oxide-semiconductor heterostructure field-effect transistors. *Appl. Phys. Lett.* **2005**, *87*, 143501. [CrossRef]
5. Arulkumaran, S.; Egawa, T.; Ishikawa, H. Studies of Electron Beam Evaporated SiO_2/AlGaN/GaN Metal—Oxide—Semiconductor High-Electron-Mobility Transistors. *Jpn. J. Appl. Phys.* **2005**, *44*. [CrossRef]
6. Chiou, Y. Photo-CVD SiO_2 Layers on AlGaN/GaN/AlGaN MOS-HFETs. *J. Electrochem. Soc.* **2004**, *151*, G863–G865. [CrossRef]
7. Jiang, H.; Liu, C.; Chen, Y.; Lu, X.; Tang, C.W.; Lau, K.M. Investigation of Insitu SiN as Gate Dielectric and Surface Passivation for GaN MISHEMTs. *IEEE Trans. Electron. Dev.* **2017**, *64*, 832–839. [CrossRef]
8. Tang, Z.; Jiang, Q.; Lu, Y.; Huang, S.; Yang, S.; Tang, X.; Chen, K.J. 600-V Normally Off SiN_x/AlGaN/GaN MIS-HEMT With Large Gate Swing and Low Current Collapse. *IEEE Electr Device Lett.* **2013**, *34*, 1373–1375. [CrossRef]
9. Zhang, Z.; Yu, G.; Zhang, X.; Deng, X.; Li, S.; Fan, Y.; Sun, S.; Song, L.; Tan, S.; Wu, D.; et al. Studies on High-Voltage GaN-on-Si MIS-HEMTs Using LPCVD Si_3N_4 as Gate Dielectric and Passivation Layer. *IEEE Trans. Electron. Dev.* **2016**, *63*, 731–738. [CrossRef]
10. Arulkumaran, S.; Hong, L.Z.; Ing, N.G.; Selvaraj, S.L.; Egawa, T. Influence of Ammonia in the Deposition Process of SiN on the Performance of SiN/AlGaN/GaN Metal–Insulator–Semiconductor High-Electron-Mobility Transistors on 4-in. Si(111). *Appl. Phys. Express* **2009**, *2*, 31001. [CrossRef]
11. Adivarahan, V.; Gaevski, M.; Sun, W.H.; Fatima, H.; Koudymov, A.; Saygi, S.; Simin, G.; Yang, J.; Khan, M.A.; Tarakji, A.; et al. Submicron gate Si_3N_4/AlGaN/GaN-metal-insulator-semiconductor heterostructure field-effect transistors. *IEEE Electr. Device Lett.* **2003**, *24*, 541–543. [CrossRef]
12. Kim, H.; Han, S.; Jang, W.; Cho, C.; Seo, K.; Oh, J.; Cha, H. Normally-Off GaN-on-Si MISFET Using PECVD SiON Gate Dielectric. *IEEE Electr. Device Lett.* **2017**, *38*, 1090–1093. [CrossRef]
13. Balachander, K.; Arulkumaran, S.; Egawa, T.; Sano, Y.; Baskar, K. A comparison on the Electrical Characteristics of SiO_2, SiON and SiN as the Gate Insulators for the Fabrication of AlGaN/GaN Metal–Oxide/Insulator–Semiconductor High-Electron Mobility-Transistors. *Jpn. J. Appl. Phys.* **2005**, *44*, 4911–4913. [CrossRef]
14. Balachander, K.; Arulkumaran, S.; Egawa, T.; Sano, Y.; Baskar, K. Demonstration of AlGaN/GaN metal-oxide-semiconductor high-electron-mobility transistors with silicon-oxy-nitride as the gate insulator. *Mater. Sci. Eng. B* **2005**, *119*, 36–40. [CrossRef]
15. Balachander, K.; Arulkumaran, S.; Ishikawa, H.; Baskar, K.; Egawa, T. Studies on electron beam evaporated ZrO_2/AlGaN/GaN metal-oxide-semiconductor high-electron-mobility transistors. *Phys. Status Solidi A* **2005**, *202*, R16–R18. [CrossRef]

16. Ye, P.D.; Yang, B.; Ng, K.K.; Bude, J.; Wilk, G.D.; Halder, S.; Hwang, J.C.M. GaN metal-oxide-semiconductor high-electron-mobility-transistor with atomic layer deposited Al_2O_3 as gate dielectric. *Appl. Phys. Lett.* **2005**, *86*, 63501. [CrossRef]

17. Liu, Z.H.; Ng, G.I.; Arulkumaran, S.; Maung, Y.K.T.; Teo, K.L.; Foo, S.C.; Sahmuganathan, V.; Xu, T.; Lee, C.H. High Microwave-Noise Performance of AlGaN/GaN MISHEMTs on Silicon with Al_2O_3 Gate Insulator Grown by ALD. *IEEE Electr. Device Lett.* **2010**, *31*, 96–98.

18. Asif, M.; Chen, C.; Peng, D.; Xi, W.; Zhi, J. Improved DC and RF performance of InAlAs/InGaAs InP based HEMTs using ultra-thin 15 nm $ALD-Al_2O_3$ surface passivation. *Solid State Electron.* **2018**, *142*, 36–40. [CrossRef]

19. Liu, C.; Chor, E.F.; Tan, L.S. Investigations of HfO_2/AlGaN/GaN metal-oxide-semiconductor high electron mobility transistors. *Appl. Phys. Lett.* **2006**, *88*, 173504. [CrossRef]

20. Anand, M.J.; Ng, G.I.; Vicknesh, S.; Arulkumaran, S.; Ranjan, K. Reduction of current collapse in AlGaN/GaN MISHEMT with bilayer SiN/Al_2O_3 dielectric gate stack. *Phys. Status Solidi C* **2013**, *10*, 1421–1425. [CrossRef]

21. Balachander, K.; Arulkumaran, S.; Sano, Y.; Egawa, T.; Baskar, K. Fabrication of AlGaN/GaN double-insulator metal-oxide-semiconductor high-electron-mobility transistors using SiO_2 and SiN as gate insulators. *Phys. Status Solidi A* **2005**, *202*, R32–R34. [CrossRef]

22. Miyazaki, E.; Goda, Y.; Kishimoto, S.; Mizutani, T. Comparative study of AlGaN/GaN metal-oxide-semiconductor heterostructure field-effect transistors with Al_2O_3 and HfO_2 gate oxide. *Solid State Electron.* **2011**, *62*, 152–155. [CrossRef]

23. Chang, Y.C.; Huang, M.L.; Chang, Y.H.; Lee, Y.J.; Chiu, H.C.; Kwo, J.; Hong, M. Atomic-layer-deposited Al_2O_3 and HfO_2 on GaN: A comparative study on interfaces and electrical characteristics. *Microelectron. Eng.* **2011**, *88*, 1207–1210. [CrossRef]

24. Green, B.M.; Chu, K.K.; Chumbes, E.M.; Smart, J.A.; Shealy, J.R.; Eastman, L.F. The effect of surface passivation on the microwave characteristics of undoped AlGaN/GaN HEMTs. *IEEE Electr. Device Lett.* **2000**, *21*, 268–270. [CrossRef]

25. Meneghesso, G.; Verzellesi, G.; Danesin, F.; Rampazzo, F.; Zanon, F.; Tazzoli, A.; Meneghini, M.; Zanoni, E. Reliability of GaN High-Electron-Mobility Transistors: State of the Art and Perspectives. *IEEE Trans. Device Mater. Reliab.* **2008**, *8*, 332–343. [CrossRef]

26. Arulkumaran, S.; Ng, G.I.; Liu, Z.H. Effect of gate-source and gate-drain Si_3N_4 passivation on current collapse in AlGaN/GaN high-electron-mobility transistors on silicon. *Appl. Phys. Lett.* **2007**, *90*, 173504. [CrossRef]

27. Arulkumaran, S.; Egawa, T.; Ishikawa, H.; Jimbo, T.; Sano, Y. Surface passivation effects on AlGaN/GaN high-electron-mobility transistors with SiO_2, Si_3N_4, and silicon oxynitride. *Appl. Phys. Lett.* **2004**, *84*, 613–615. [CrossRef]

28. Javorka, P.; Bernat, J.; Fox, A.; Marso, M.; Lüth, H.; Kordoš, P. Influence of SiO_2 and Si_3N_4 passivation on AlGaN/GaN/Si HEMT performance. *Electron. Lett.* **2003**, *39*, 1155–1157. [CrossRef]

29. Lu, X.; Yu, K.; Jiang, H.; Zhang, A.; Lau, K.M. Study of Interface Traps in AlGaN/GaN MISHEMTs Using LPCVD SiNx as Gate Dielectric. *IEEE Trans. Electron. Dev.* **2017**, *64*, 824–831. [CrossRef]

30. Mizue, C.; Hori, Y.; Miczek, M.; Hashizume, T. Capacitance-Voltage Characteristics of Al_2O_3/AlGaN/GaN Structures and State Density Distribution at Al_2O_3/AlGaN Interface. *Jpn. J. Appl. Phys.* **2011**, *50*. [CrossRef]

31. Capriotti, M.; Alexewicz, A.; Fleury, C.; Gavagnin, M.; Bethge, O.; Visalli, D.; Derluyn, J.; Wanzenböck, H.D.; Bertagnolli, E.; Pogany, D.; et al. Fixed interface charges between AlGaN barrier and gate stack composed of insitu grown SiN and Al_2O_3 in AlGaN/GaN high electron mobility transistors with normally off capability. *Appl. Phys. Lett.* **2014**, *104*, 113502. [CrossRef]

32. Hung, T.; Krishnamoorthy, S.; Esposto, M.; Neelim Nath, D.; Sung Park, P.; Rajan, S. Interface charge engineering at atomic layer deposited dielectric/III-nitride interfaces. *Appl. Phys. Lett.* **2013**, *102*, 72105. [CrossRef]

33. Zhu, J.; Ma, X.; Xie, Y.; Hou, B.; Chen, W.; Zhang, J.; Hao, Y. Improved Interface and Transport Properties of AlGaN/GaN MIS-HEMTs with PEALD-Grown AlN Gate Dielectric. *IEEE Trans. Electron. Dev.* **2015**, *62*, 512–518.

34. Son, J.; Chobpattana, V.; McSkimming, B.M.; Stemmer, S. Fixed charge in high-k/GaN metal-oxide-semiconductor capacitor structures. *Appl. Phys. Lett.* **2012**, *101*, 102905. [CrossRef]

35. Liu, S.; Huang, C.; Chang, C.; Lin, Y.; Chen, B.; Tsai, S.; Majlis, B.Y.; Dee, C.; Chang, E.Y. Effective Passivation with High-Density Positive Fixed Charges for GaN MIS-HEMTs. *IEEE J. Electron. Dev.* **2017**, *5*, 170–174. [CrossRef]

36. Yang, S.; Liu, S.; Lu, Y.; Liu, C.; Chen, K.J. AC-Capacitance Techniques for Interface Trap Analysis in GaN-Based Buried-Channel MIS-HEMTs. *IEEE Trans. Electron Dev.* **2015**, *62*, 1870–1878. [CrossRef]

37. Zhang, D.; Cheng, X.; Zheng, L.; Shen, L.; Wang, Q.; Gu, Z.; Qian, R.; Wu, D.; Zhou, W.; Cao, D.; et al. Effects of polycrystalline AlN film on the dynamic performance of AlGaN/GaN high electron mobility transistors. *Mater. Des.* **2018**, *148*, 1–7. [CrossRef]

38. Reshchikov, M.A.; Morkoç, H. Luminescence properties of defects in GaN. *J. Appl. Phys.* **2005**, *97*, 61301. [CrossRef]

39. Kaufmann, U.; Kunzer, M.; Obloh, H.; Maier, M.; Manz, C. Origin of defect-related photoluminescence bands in doped and nominally undoped GaN. *Phys. Rev. B Condens. Matter* **1999**, *59*, 5561. [CrossRef]

40. Chevtchenko, S.A.; Reshchikov, M.A.; Fan, Q.; Ni, X.; Moon, Y.T.; Baski, A.A.; Morkoç, H. Study of SiN_x and SiO_2 passivation of GaN surfaces. *J. Appl. Phys.* **2007**, *101*, 1139–1190. [CrossRef]

41. Hashizume, T.; Hasegawa, H. Effects of nitrogen deficiency on electronic properties of AlGaN surfaces subjected to thermal and plasma processes. *Appl. Surf. Sci.* **2004**, *234*, 387–394. [CrossRef]

42. Eller, B.S.; Yang, J.; Nemanich, R.J. Electronic surface and dielectric interface states on GaN and AlGaN. *J. Vacuum Sci. Technol. A Vacuum Surf. Films* **2013**, *31*, 50807. [CrossRef]

© 2018 by the authors. Licensee MDPI, Basel, Switzerland. This article is an open access article distributed under the terms and conditions of the Creative Commons Attribution (CC BY) license (http://creativecommons.org/licenses/by/4.0/).

electronics

MDPI

Article

Modeling of 2DEG characteristics of In$_x$Al$_{1-x}$N/AlN/GaN-Based HEMT Considering Polarization and Quantum Mechanical Effect

Jian Qin [1,2,3]**, Quanbin Zhou** [2,3]**, Biyan Liao** [2,3] **and Hong Wang** [2,3,*]

[1]   Department of Electronics and Communication Engineering, Guangzhou University, Guangzhou 510006, China; gzu_jian@gzhu.edu.cn

[2]   Engineering Research Center for Optoelectronics of Guangdong Province, South China University of Technology, Guangzhou 510641, China; zhouquanbin86@163.com (Q.Z.); LydiaLiao1018@163.com (B.L.)

[3]   Zhongshan Institute of Modern industrial Technology, South China University of Technology Zhongshan, Zhongshan 528437, China

*   Correspondence: phhwang@scut.edu.cn

Received: 23 October 2018; Accepted: 5 December 2018; Published: 8 December 2018

Abstract: A comprehensive model for 2DEG characteristics of In$_x$Al$_{1-x}$N/AlN/GaN heterostructure has been presented, taking both polarization and bulk ionized charge into account. Investigations on the 2DEG density and electron distribution across the heterostructure have been carried out using solutions of coupled 1-D Schrödinger-Poisson equations solved by an improved iterative scheme. The proposed model extends a previous approach allowing for estimating the quantum mechanical effect for a generic InAlN/GaN-based HEMT within the range of the Hartree approximation. A critical AlN thickness (~2.28 nm) is predicted when considering the 2DEG density in dependence on a lattice matched In$_{0.17}$Al$_{0.83}$N thickness. The obtained results present in this work provide a guideline for the experimental observation of the subband structure of InAlN/GaN heterostructure and may be used as a design tool for the optimization of that epilayer structure.

Keywords: 2DEG density; InAlN/GaN heterostructure; polarization effect; quantum mechanical

1. Introduction

Due to their unique material properties, GaN-based high electron mobility transistors (HEMTs) are the most promising candidates for high frequency and high power applications. Over past decades, Al$_x$Ga$_{1-x}$N/GaN has been the focus of research for HEMT [1–5]. It is well known that the epitaxial growth of AlGaN over GaN leads to the formation of 2DEG with sheet density order of 10^{13} cm^{-2}. However, when the fraction of Al exceeds 0.3 [6–8], the 2DEG transport properties would be seriously degraded and would finally limit further improvement of the device's performance. To avoid the problem of strain relaxation in AlGaN, a novel material system has been developed by replacing an AlGaN barrier with InAlN [9–14]. It is believed that the InAlN/GaN structure induces a higher sheet density and better carrier confinement as compared with the typical AlGaN/GaN structure. The ultrahigh polarization allows In$_x$Al$_{1-x}$N/GaN heterojunction featuring 2DEG density well above 2×10^{13} cm^{-2} but with a much thinner InAlN layer that is typically less than 15 nm. These superior features make InAlN/GaN heterostructure attractive, especially for the next generation of high power and millimeter wave applications.

Due to the large piezoelectric coefficients and lattice mismatch between InAlN and GaN, a strain induced polarization electric field is generated at the interface of InAlN/GaN with the order of 10^6 V/cm [15]. Such a high electric field gives rise to a sufficiently narrow triangular potential well and causes a profound quantization effect in the motion of the electron perpendicular to the interface. Although a significant process in InAlN/AlN/GaN has been reported [16–19], the detailed study of the electron quantization effect on these heterostructures is still less well known. Accurate models which are at the same time computationally efficient for optimization of epilayer structure engineering are also highly desirable. Medjdoub et al. [20] for the first time presented the capability of a InAlN/GaN heterostructure to deliver better electrical performance and to be less unstable under high temperature as compared with an AlGaN/GaN structure. Gonschorek et al. [21] presented the results in an $In_xAl_{1-x}N/GaN$ structure showing an enhanced 2DEG density with increasing barrier layer thickness. The model is based on reference [1] where the effect of bulk ionized charge is ignored and the relationship between gate voltage and sheet density of 2DEG was not given. Several other groups have made a successful attempt to describe the 2DEG transport characteristics of gallium nitride-based HEMT using a numerical method based on the effective mass approximation [22,23]. However, little direct information on the charge distribution could be obtained. The dependence of polarization charge and bulk ionized charge on the behavior of the electrostatic potential requires clarification for more detail. Contrary to this, we emphasize that the sheet density of polarization charge and quantum mechanical effect are the fundamental design issues for InAlN-GaN based HEMT, and that these effects can influence the device's electrical characteristics and must be carefully accounted for with a reliable description of those devices. Recent experimental works have reported capacitance enhancement through the density modulation of the light generated carrier [24], and the quantum mechanical description of the enhanced capacitance value was given in term of exchange and correlation energies of the interacting many-body system of 2DEG [25–27]. Strictly speaking, the most accurate quantum mechanical approach is to solve Schrödinger-Poisson system taking these many-body effects into account, however, from a device modeling perspective, the incorporation of electron-electron interaction highly complicates the numerical procedure. In this frame work, the self-consistent quantum mechanical calculation is discussed within the range of Hartree approximation [28]. As a trade-off between complexity and computation efficiency, this approximation is expected to work well for two-dimensional systems, especially when the electron concentration is sufficiently high.

In what follows, we develop a novel comprehensive model for 2DEG characteristics of $In_xAl_{1-x}N/AlN/GaN$ heterostructure, taking both bulk ionized dopant and polarization charge into account. An improved efficient scheme for the self-consistent solution of Schrödinger's equation and Poisson's equation has been presented. We report a detail study on the 2DEG density and corresponding charge distribution within the vicinity of the heterojunction. The calculations are carried out under the most general conditions and can be therefore easily extended to other III-IV based devices.

2. Modeling and Theory

2.1. Polarization Effect

The schematic of $In_xAl_{1-x}N/AlN/GaN$ heterostructure studied in this work is given in Figure 1. The heterostructure consists of a 500 nm thick unintentionally doped GaN layer followed by the growth of 10 nm thickness of an $In_{0.12}Al_{0.88}N$ epitaxial layer over it. A thin 1 nm AlN interlayer was inserted to provide sufficient separation of the electron wave from the InAlN. We assumed that the layer of heterostructure would grow on the thick GaN substrate pseudo-morphically along [001], denoted as the z direction in the following figure.

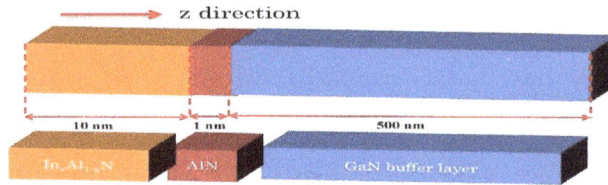

Figure 1. Schematic of the $In_xAl_{1-x}N/AlN/GaN$ heterostructure studied in this work.

Since the electronic transport properties change considerably due to piezoelectric polarization arising from material strain, the model starts with the calculation for diagonal strain components modeled by

$$\varepsilon_{xx} = (a_1 - a_2)/a_2 \tag{1}$$

$$\varepsilon_{zz} = -\varepsilon_{xx}(C_{13}/C_{33}) \tag{2}$$

where a_1, a_2 are the lattice parameters of the different layers, ε_{xx}, ε_{zz} are bi-axial strain and uni-axial strain, which refer to the direction of parallel and perpendicular to the grown interface respectively. As strain and piezoelectric physical constants are known, piezoelectric polarization at each heterointerface can be calculated with [29,30].

$$P^{PE} = 2\varepsilon_{zz}\left(e_{31} - e_{33}\frac{C_{13}}{C_{33}}\right) \tag{3}$$

where C_{13}, C_{33} are elastic constants. The ternary material of $In_xAl_{1-x}N$ is calculated with parameters given in Table 1 following Vegard's law.

$$P^{sp}_{InAlN}(x) = xP^{sp}_{InN} + (1-x)P^{sp}_{AlN} + b \cdot x(1-x) \tag{4}$$

where b is the bowing parameter ($b = 0.07$ C/m²) [31], x is the indium content.

For a generic $In_xAl_{1-x}N/GaN$ based HEMT structure, the indium content dependent sheet carrier density $n_s(x)$ is given by [32,33]

$$n_s(x) = \frac{\sigma(x)}{q} - \left(\frac{\varepsilon_0\varepsilon(x)}{d \cdot q^2}\right)\left[q\varphi_b + E_f(x) - \Delta E_c(x)\right] \tag{5}$$

Following the analysis of references [33,34], n_s may also be related to the gate and channel voltage through Gauss law as

$$q \cdot n_s = C_{eff}\left(V_g - V_{off} - \varphi_s(V_{ch})\right) \tag{6}$$

$$\varphi_s(V_{ch}) = V_{ch} + E_f(V_{ch}) \tag{7}$$

In Equations (5) to (7), σ is the combination of the spontaneous polarization and piezoelectric charge which is determined by Equations (3) and (4) respectively, ε is the dielectric constant, d is the InAlN barrier thickness, φ_b is the Schottky barrier height, E_f is the Fermi energy level with respect to the GaN conduction band, ΔE_c represents the conduction band discontinuity given by $\Delta E_{C,ABN}(x) = 0.7\left[E_{g,ABN}(x) - E_{g,ABN}(0)\right]$, the concept of effective band offset (Equation (3) in [21]) is used when the effect of AlN layer is taken into account, C_{eff} is effective capacitance (per unit area) between the gate and 2DEG, V_{off} is cut-off voltage below which it is under sub-threshold region. The surface potential φ_s is the Fermi level in the body where the channel voltage $V_{ch} = V_s$ at the source and $V_{ch} = V_{ds}$ at the drain respectively.

2.2. Self-Consistent Calculations

In the framework of the effective mass approximation, the following calculations of the Eigen-states for electron are taken into account:

$$\left[-\frac{\hbar^2}{2m^*}\frac{d^2\xi_i(z)}{dz^2}\right] + [E_c(z) - E_i]\xi_i(z) = 0 \tag{8}$$

$$\frac{d}{dz}\left(\varepsilon_0\varepsilon(z)\frac{dE_c(z)}{dz}\right) = -q[n_{2d}(z) + n_{3d}(z)] \tag{9}$$

where

$$E_c(z) = -q\varphi(z) + \Delta E_c \tag{10}$$

$$n_{2d}(z) = \sum_{i=0}^{+\infty}\frac{m^*kT}{\pi\hbar^2}\ln\left[1 + \exp\left(\frac{E_f - E_i}{kT}\right)\right] \cdot |\xi_i(z)|^2 \tag{11}$$

$$n_{3d}(z) = N_D^+(z) - N_A^-(z) \tag{12}$$

In Equations (8) to (12), k is the Boltzmann constant, $\hbar$ is the Planck constant, m^* is effective mass, $\varphi(z)$ is electrostatic potential. E_i is energy of the ith sub-band level and $\xi_i(z)$ is the corresponding wave function. $\varepsilon(z)$ is the position-dependent dielectric constant, N_D^+, N_A^- are the bulk densities of the ionized donors and acceptor, respectively. Note that the total sheet charge in this work consists of two parts: the first part is derived from the polarization induced charge calculated by the combination of Equations (3) and (4). Another part comes from the terms which are assumed to be insensitive to quantization, i.e., the ionized dopant concentrations N_D^+ and N_A^- as given in Equation (12). These two components are assumed to be independent of each other.

Our calculation is based on the solving the coupled Schrödinger and Poisson's equation set given in Equations (8) and (9). A finite iteration scheme is deployed similarly to the approach given in references [23,35]. For InAlN/AlN/GaN HEMT, the boundary conditions are given as follows: at the metal-semiconductor interface, the potential above the Fermi level is the Schottky barrier height. This barrier energy of the heterostructure is set as 1.2 eV relative to the ground energy is defined to the energy level close to the substrate.

At the beginning, one starts with an initial empirical potential $\varphi^0(z)$ then calculates the wave function $\xi_i(z)$ as well as the corresponding eigenenergies E_i using Schrödinger's equation. Thereafter, n_i is calculated by $nn_i = \frac{m^*kT}{\pi\hbar^2}\ln\left[1 + \exp\left(\frac{E_f - E_i}{kT}\right)\right]$ and the electron distribution is calculated using Equation (11). The new value of $n'(z)$ along with N_D^+ and N_A^- are used for calculating potential distribution through using of Equations (9) and (10). This updated value of $E_c'(z)$ will be used in Schrödinger's equation to find the new wave function and eigenenergies. The iteration process is repeated until convergence criterion is finally achieved. However, the abovementioned procedure does not converge all the time due to the high sensitivity of the eigenenergies on the potential and of the quantum charge n_{2d} on the corresponding energy levels. An improved iterative scheme is developed to get around this problem. In this frame work, for each i-th subband, at the n-th iterative step, we define a set of space-dependent energies $E_i^n(z)$ and effective state density $N_i^n(z)$ given as

$$E_i^n(z) = E_i^n - E_i^n(z) \tag{13}$$

$$N_i^n(z) = \frac{m^*kT}{\pi\hbar^2}|\xi_i^n(z)|^2 \tag{14}$$

The corresponding quantum charge is then defined as

$$n_{2d}^{n+1}(z) = \sum_{i=0}^{+\infty}N_i^n(z)\cdot\log\left[1 + \exp\left(\frac{E_f - E_c^{n+1}(z) - E_i^n(z)}{kT}\right)\right] \tag{15}$$

75

At step $n + 1$, Equation (15) will be included into the Poisson's equation (9), and this equation will be solved for determining the updated E_c^{n+1}. Note that we involve the term of space dependent $E_i^n(z)$ in Equation (13) so as to overcome the numerical oscillations during the feedback of iterative calculation. This dependence finally vanishes when the algorithm reaches convergence at $E_c^n = E_c^{n+1}$, and Equation (15) gives the same charge density as Equation (11).

On the other hand, in a classical system, one uses the concept of effective density of state N_c and the order $1/2$ Fermi-Dirac integral when accounting for the local dependence of electron concentration on the potential. In that case, the right term of Equation (9) can be replaced by

$$n = N_c \cdot F_{1/2}\Big((E_f - E_c(z))/kT\Big) \tag{16}$$

This approximate model is compared with our quantum effect calculation to reveal how these quantum mechanicals influence the InAlN/GaN-based HEMT performance.

3. Results and Discussion

The calculations were carried out at room temperature for the $In_x Al_{1-x} N/AlN/GaN$ structure given in Figure 1. In the following, the barrier indium content is assumed to be ranging from 0.03 to 0.23. The ionized dopant concentrations N_d^+ and residual N_A^- in the buffer layer are assumed to be 3×10^{18} cm^{-3} and 1×10^{13} cm^{-3} respectively unless otherwise stated. Other physical constants used in this calculation are summarized in Table 1.

Table 1. Physics parameters of InN, AlN, and GaN at room temperature [30,31].

Title 1	InN	AlN	GaN
a_x (Å)	3.545	3.112	3.189
P^{sp} (C/m^2)	−0.032	−0.081	−0.032
$e_{31} - (C_{31}/C_{33})e_{33}$ (C/m^2)	−0.86	−0.91	−0.69
m_e^* (m_0)	0.11	0.48	0.22
ε_x (ε_0)	13	8.5	10

Figure 2 shows the sketch of the band bending with (solid, red) and without (dash, black) the polarization effect, along with the calculated 2DEG density from the polarization induced sheet charge given in the inset. To incorporate the contribution of polarization effect, a bound sheet charge width approximately ~6 Å width [30] is added at the AlN/GaN interface in our self-consistent calculation. We use the indium fraction x as a variable parameter to alter the sheet carrier density so that the sheet density of electron satisfies both Equations (5) and (11). As seen from Figure 2, the polarization effect dominates the behavior of electrostatic potential within the vicinity of the heterojunction. The 2D channel becomes sharper and the confined electrons are pushed close to the interface. Note that the 2DEG density increases almost linearly with the decreasing of the In fraction (corresponding to increasing of Al fraction) as indicated in the inset of Figure 2. i.e., for a representative 10 nm InAlN barrier heterostructure, the sheet carrier density is calculated to be 2.55×10^{13} cm^{-2} for the In fraction of $x = 0.23$ and may be increased to 3.3×10^{13} cm^{-2} for x = 0.07. When the fraction of indium is around 17–18%, InAlN is assumed to be lattice matched with GaN material, the heterostructure is free of strain (effect of piezoelectric polarization vanishes, $P^{PE} = 0$) and the polarization charge for the interface is mainly determined by the spontaneous effect given by Equation (4).

Figure 3 presents the conduction band profile of the $In_{0.12}Al_{0.88}N/AlN/GaN$ heterostructure with barrier thickness of 10 nm and AlN interlayer thickness of 1 nm, along with the wave functions representing the probability distribution of electrons of the corresponding energy level. For the sake of simplicity, only the ground state and first exited state (i = 0, 1) are considered. The numbers given in Figure 3 indicate the probability occupation of the corresponding subband energy. The wave function within the ground state of a similar $Al_{0.2}Ga_{0.8}N/AlN/GaN$ heterostructure is also given for comparison. For a 10 nm $In_{0.12}Al_{0.88}N/AlN/GaN$ heterostructure, over 86.2% of the channel electrons are occupied by the ground state, suggesting the formation of 2DEG system at AlN/GaN interface. The variation of the ground and first exited subband (E_1–E_0) is around 179 meV at the room temperature as compared with 72 meV for that of $Al_{0.2}Ga_{0.88}N/GaN$. A larger energy separation implies a stronger electrical field and better 2DEG confinement, as shown in the result given in Figure 4.

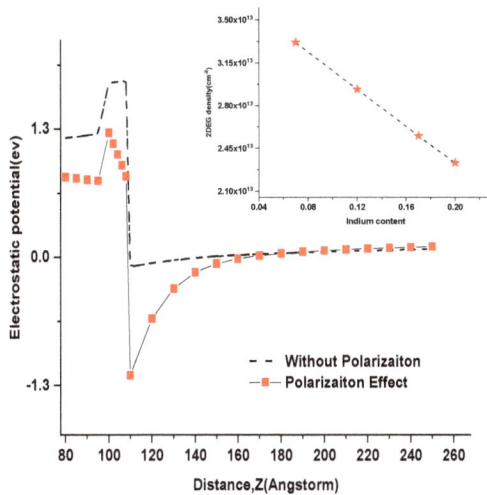

Figure 2. Sketch of the band bending with (solid, red) and without (dash, black) the polarization effect being taken into account. The calculated 2DEG density for a 10 nm InAlN/AlN/GaN heterostructure as a function of indium fraction is given in the inset.

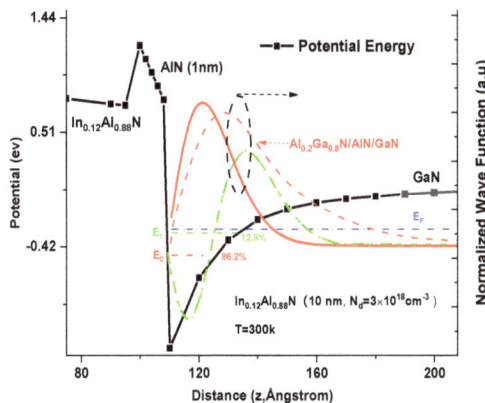

Figure 3. The conduction band profile of the $In_{0.12}Al_{0.88}N/AlN/GaN$ heterostructure and the wave function of the ground state ($i = 0$, red, line) and first excited ($i = 1$, green, dash) under room temperature. A normalized first subband wave function for an $Al_{0.2}Ga_{0.8}N/AlN/GaN$ (red, short dash) is given for comparison.

To quantify the 2DEG confinement characteristics, we approximately define the width of two-dimensional electron gas as the variation between the two depths at which the electron concentration shrunk to 10% of its maximum value. As seen from Figure 4, the 2DEG width is around 5.4 nm for a 10 nm $Al_{0.2}GaN_{0.8}N$ barrier, and this amount deceases to ~3.3 nm as observed in a comparable $In_{0.12}Ga_{0.88}N$ structure. An improved carrier confinement shows fundamental advantages in suppressing the short-channel effect, i.e., DIGBL, self-heating as suggested in references [36–38]. The reduction in short-channel effect helps to increase the frequency performance of GaN-based heterostructure devices.

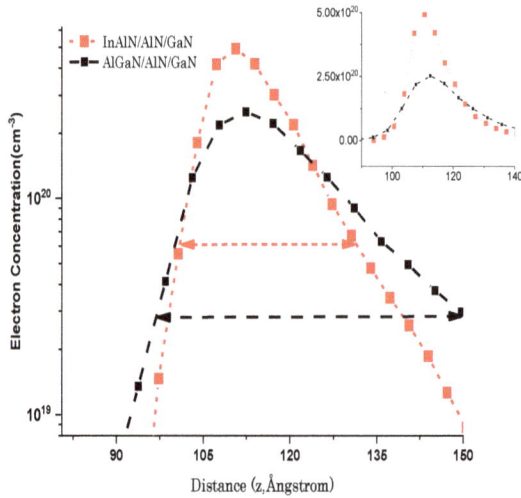

Figure 4. Comparison of the electron distribution of $In_{0.12}Al_{0.88}N/AlN/GaN$ and case of $Al_{0.2}Ga_{0.8}N/AlN/GaN$ for characteristics of 2DEG confinement.

The electron distribution of $In_{0.12}Al_{0.88}N/AlN/GaN$ depending on AlN thickness is shown in Figure 5. A simple $In_{0.12}Al_{0.88}N/GaN$ structure (AlN = 0 nm) is also given for comparison. For the InAlN/AlN/GaN with a 1 nm AlN interlayer insertion, the electron concentration at the interface increases slightly due to the enlarged band offset over the barrier. There is only an order of 5.3% of the electron density penetrated into the barrier with the penetration width of around 8.2 Å. For AlN interlayer thickness larger than 1 nm, little influence was observed on the wave function and charge distribution.

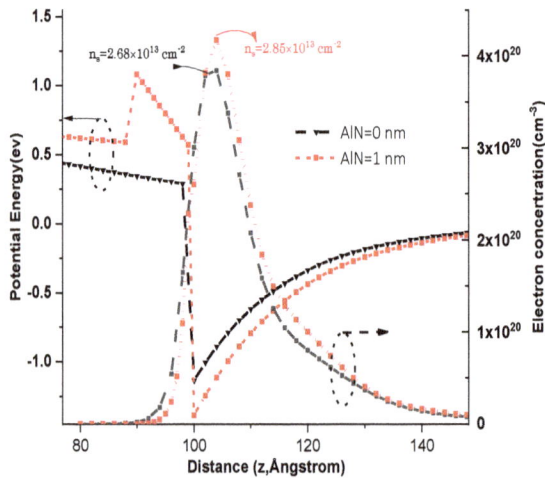

Figure 5. The electron concentration and conduction band distribution of $In_{0.12}Al_{0.88}N/AlN/GaN$ with different AlN thicknesses.

The 2DEG density in dependence on the barrier thickness for a lattice-matched In$_{0.17}$Al$_{0.83}$N/AlN/GaN heterostructure is given in Figure 6. The AlN thickness varies from 1 to 3 nm. The result of Figure 6 comes from the calculated maximum sheet density located at the interface of the AlN/GaN structure. Other parameters are kept the same as in Figure 5. The calculated results have been compared with the available experimental data given in references [9–11]. A reasonable agreement is achieved justifying the validation of this work. It was found that the 2DEG sheet density increases as AlN thickness increases. The insertion of the AlN layer tends to weaken the increasing dependence of 2DEG density on the barrier thickness. For 1 nm AlN layer, the sheet density of 2DEG increases with the barrier thickness and then tends to saturate. This trend is profoundly contrary to the case when increasing AlN thickness to 3 nm. According to the theoretical calculation, a critical AlN thickness (~2.28 nm for lattice-matched In$_{0.17}$Al$_{0.83}$N/AlN/GaN) may exist at which little variation occurs on the 2DEG density when increasing InAlN thickness, as shown in Figure 6. It is now clear that the InAlN barrier predominately induces the 2DEG for a thinner AlN layer, however, the AlN layer takes the dominant contribution to 2DEG formation when it is thick enough. A similar phenomenon was also observed in an AlN/GaN structure [22].

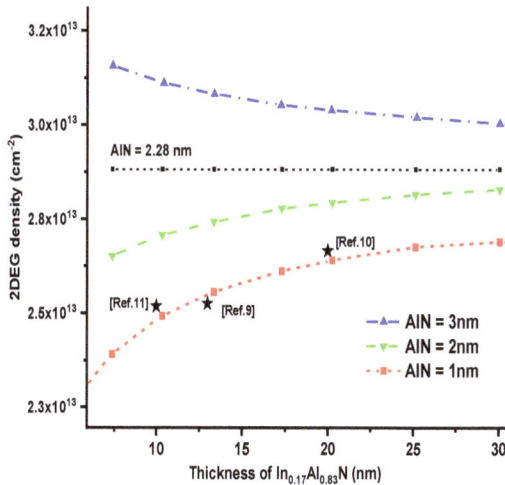

Figure 6. The 2DEG sheet density of In$_{0.17}$Al$_{0.83}$N/AlN/GaN versus InAlN barrier thickness with different AlN thicknesses, The obtained experimental data from [9–11] are given for comparison.

To clarify the impact of quantization effect on realistic devices, a generic In$_x$Al$_{1-x}$N/AlN/GaN-based HEMT is utilized to study the channel concentration dependence on the applied gate voltage by using different models. The result is shown in Figure 7 wherein the electron distribution at $V_G = 0\,V$ is given in the inset. The boundary conditions are given as following: the conduction band edge at gate on the InAlN is fixed to the Schottky-barrier height (1.2 eV) and the applied voltage. The channel is under grounded ($V_{ch} = 0$) and Fermi level is fixed to a sufficiently small value below $1 \times 10^{-4} \times (kT/q)$. Only the ground state and first exited state ($i = 0, 1$) are taken into account. The corresponding wave functions are assumed to be vanished at boundaries of the structure. Other calculation parameters are kept the same as with Figure 3.

As for the quantum mechanical results, the maximum of the electron concentration is found to be located at around several angstroms from the interface instead of the interface as predicted by the classical model. This result coincides with the charge width approximation given in Figure 2 previously. The cut off voltage V_{off} is found to be mainly determined by the polarization charge present at the heterointerface. It is worth noting that when channel carrier density does not exceed $\sim 6 \times 10^{12}\,cm^{-2}$, even the classical model simply through the density of state N_c using Equation (16) can evaluate the channel electron within a 5 percent error. The deviation starts to gradually enhance when sheet carrier density reach to 2.3×10^{13} cm^{-2} (corresponding to electric field order of ~3.8 MV/cm at interface). The result implies that the behavior of InAlN/GaN–based HEMT under the sub-threshold region may be described with good accuracy even by the classical approach using Fermi-Dirac statistics. As for increasing the applied gate voltage, the conduction band can no longer be treated as a continuous state due to the enlarged electric field. In this condition, a significant quantization effect may arise when dealing with transport properties of the 2DEG in these small-geometry devices.

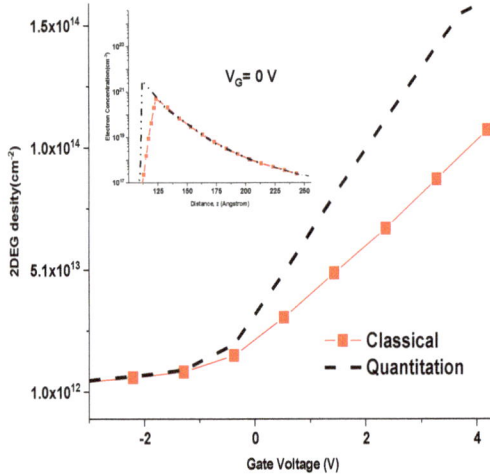

Figure 7. The gate voltage dependence of the 2DEG density for a generic InAlN/AlN/GaN using classical model (Equation (16)) and quantum mechanical (Equation (11)). The inset presents the electron distribution within the heterostructure at $V_g = 0$ V. The extracted parameters used in calculation are given as: $V_{off} = -3.3$ V, $C_{eff} = 8.36 \times 10^{-6}$ F/cm^2, $N_c = 2.3 \times 10^{18}$ cm^{-3}.

4. Conclusions

A comprehensive model of 2DEG characteristic of In$_x$Al$_{1-x}$N/AlN/GaN heterostructure has been given in this paper taking polarization and quantum mechanical effect into account. We report a detailed study on the 2DEG density and electron distribution across the heterostructure using a solution of Schrödinger and Poisson's equations solved through an improved iterative scheme.

It is found that that the polarization effect dominates the behavior of electrostatic potential within the vicinity of the heterojunction. Due to the strong polarization field, a thinner electron distribution profile of In$_{0.12}$Al$_{0.88}$N/AlN/GaN is observed as being comparable with the AlGaN/AlN/GaN structure. Therefore, improving the quality of interface roughness is essential for enhancing the transport property in these structures. A critical AlN thickness (~2.28 nm) was predicated when considering the 2DEG density in dependence on a lattice matched In$_{0.17}$Al$_{0.83}$N thickness. It is clear that the InAlN barrier predominately induces the 2DEG for a thinner AlN layer (1 nm), however, the AlN layer dominates the contribution to 2DEG formation when it grows beyond critical thickness.

We found the carrier transport for a generic InAlN/GaN-based HEMT under sub-threshold region can be well described even by the classical approach using Fermi-Dirac statistics. As the gate voltage increases, both the polarization charge and applied voltage give rise to potential well and the conduction band has been splitted into discrete sub-bands. A significant quantization effect also arises and must be carefully treated to achieve a reliable description of those small-geometry devices.

Although some previous C-V measurements indicate that many-body effects such as exchange and correlation energies are of importance [24,26,27], these effects have been neglected for the sake of simplicity. Nevertheless, since these effects can actually affect device performance, more work should be dedicated to these effects in future so as to find an easy way to incorporate them into Schrödinger's equation. The application of the presented results given in this work server as a guide for optimization of the epilayer structure in engineering, and would always be a substantial improvement with respect to the widely adopted semi-classical model.

Author Contributions: Conceptualization, J.Q.; Investigation & Methodology, J.Q.; Validation, J.Q. and B.y.L.; Formal analysis, J.Q. and Q.b.Z.; Writing—original draft preparation, J.Q. and Q.b.Z.; Writing—review and editing, H.W.; Supervision, H.W.

Funding: This research was funded by Science and Technologies plan Projects of Guangdong Province (Nos. 2017B010112003, 2017A050506013), and by Applied Technologies Research and Development Projects of Guangdong Province (Nos. 2015B010127013, 2016B010123004), and by Science and Technologies plan Projects of Guangzhou City (Nos. 201504291502518, 201605030014, 201604046021, 201704030139), and by Science and Technology project of Guangzhou Education Municipality (No. 201630328), and by Science and Technology Development Special Fund Projects of Zhongshan City (Nos. 2017F2FC0002, 2017A1009).

Conflicts of Interest: The authors declare no conflict of interest. The funders had no role in the design of the study.

References

1. Shen, L.; Heikman, S.; Moran, B.; Coffie, R. AlGaN/AlN/GaN high-power microwave HEMT. *IEEE Electron. Device Lett.* **2001**, *22*, 457–459. [CrossRef]

2. Palacios, T.; Chakraborty, A.; Heikman, S.; Keller, S.; DenBaars, S.P.; Mishra, U.K. AlGaN/GaN high electron mobility transistors with InGaN back-barriers. *IEEE Electron. Device Lett.* **2006**, *27*, 13–15. [CrossRef]

3. Menozzi, R.; Umana-Membreno, G.A.; Nener, B.D.; Parish, G.; Sozzi, G.; Faraone, L.; Mishra, U.K. Temperature-Dependent Characterization of AlGaN/GaN HEMTs: Thermal and Source/Drain Resistances. *IEEE Trans. Device Mater. Reliab.* **2008**, *8*, 255–264. [CrossRef]

4. Hao, Y.; Yang, L.; Ma, X.; Ma, J.; Cao, M.; Pan, C.; Wang, C.; Zhang, J. High-Performance Microwave Gate-Recessed AlGaN/AlN/GaN MOS-HEMT With 73% Power-Added Efficiency. *IEEE Electron. Device Lett.* **2011**, *32*, 626–628. [CrossRef]

5. Shinohara, K.; Regan, D.C.; Tang, Y.; Corrion, A.L.; Brown, D.F.; Wong, J.C.; Robinson, J.F.; Fung, H.H.; Schmitz, A.; Oh, T.C.; et al. Scaling of GaN HEMTs and Schottky Diodes for Submillimeter-Wave MMIC Applications. *IEEE Trans. Electron. Devices* **2013**, *60*, 2982–2996. [CrossRef]

6. Yu, T.H.; Brennan, K.F. Theoretical study of the two-dimensional electron mobility in strained III-nitride heterostructures. *J. Appl. Phys.* **2001**, *89*, 3827–3834. [CrossRef]

7. Yarar, Z.; Ozdemir, B.; Ozdemir, M. Electron mobility in a modulation doped AlGaN/GaN quantum well. *Eur. Phys. J. B* **2006**, *49*, 407–414. [CrossRef]

8. Meng, F.; Zhang, J.; Zhou, H.; Ma, J.; Xue, J.; Dang, L.; Zhang, L.; Lu, M.; Ai, S.; Li, X.; et al. Transport characteristics of AlGaN/GaN/AlGaN double heterostructures with high electron mobility. *J. Appl. Phys.* **2012**, *112*. [CrossRef]

9. Gonschorek, M.; Carlin, J.-F.; Feltin, E.; Py, M.A.; Grandjean, N. High electron mobility lattice-matched AlInN/GaN field-effect transistor heterostructures. *Appl. Phys. Lett.* **2006**, *89*, 062106. [CrossRef]

10. Jeganathan, K.; Shimizu, M.; Okumura, H.; Yano, Y.; Akutsu, N. Lattice-matched InAlN/GaN two-dimensional electron gas with high mobility and sheet carrier density by plasma-assisted molecular beam epitaxy. *J. Cryst. Growth* **2007**, *304*, 342–345. [CrossRef]

11. Pietzka, C.; Denisenko, A.; Alomari, M.; Medjdoub, F.; Carlin, J.-F.; Feltin, E.; Grandjean, N.; Kohn, E. Effect of Anodic Oxidation on the Characteristics of Lattice-Matched AlInN/GaN Heterostructures. *J. Electron. Mater.* **2008**, *37*, 616–623. [CrossRef]

12. Dasgupta, S.; Lu, J.; Speck, J.S.; Mishra, U.K. Self-Aligned N-Polar GaN/InAlN MIS-HEMTs With Record Extrinsic Transconductance of 1105 mS/mm. *IEEE Electron. Device Lett.* **2012**, *33*, 794–796. [CrossRef]

13. Smith, M.D.; O'Mahony, D.; Conroy, M.; Schmidt, M.; Parbrook, P.J. InAlN high electron mobility transistor Ti/Al/Ni/Au Ohmic contact optimisation assisted by in-situ high temperature transmission electron microscopy. *Appl. Phys. Lett.* **2015**, *107*, 4. [CrossRef]

14. Tsou, C.W.; Lin, C.Y.; Lian, Y.W.; Hsu, S.S.H. 101-GHz InAlN/GaN HEMTs on Silicon with High Johnson's Figure-of-Merit. *IEEE Trans. Electron. Devices* **2015**, *62*, 2675–2678. [CrossRef]

15. Miao, Z.L.; Tang, N.; Xu, F.J.; Cen, L.B.; Han, K.; Song, J.; Huang, C.C.; Yu, T.J.; Yang, Z.J.; Wang, X.Q.; et al. Magnetotransport properties of lattice-matched In0.18Al0.82N/AlN/GaN heterostructures. *J. Appl. Phys.* **2011**, *109*, 016102. [CrossRef]

16. Xue, J.; Zhang, J.; Hao, Y. Demonstration of InAlN/AlGaN high electron mobility transistors with an enhanced breakdown voltage by pulsed metal organic chemical vapor deposition. *Appl. Phys. Lett.* **2016**, *108*, 013508. [CrossRef]

17. Zhou, H.; Lou, X.; Conrad, N.J.; Si, M.; Wu, H.; Alghamdi, S.; Guo, S.; Gordon, R.G.; Ye, P.D. High-Performance InAlN/GaN MOSHEMTs Enabled by Atomic Layer Epitaxy MgCaO as Gate Dielectric. *IEEE Electron. Device Lett.* **2016**, *37*, 556–559. [CrossRef]

18. Chen, P.G.; Tang, M.; Liao, M.H.; Lee, M.H. In$_{0.18}$Al$_{0.82}$N/AlN/GaN MIS-HEMT on Si with Schottky-drain contact. *Solid-State Electron.* **2017**, *129*, 206–209. [CrossRef]

19. Xing, W.; Liu, Z.; Qiu, H.; Ranjan, K.; Gao, Y.; Ng, G.I.; Palacios, T. InAlN/GaN HEMTs on Si With High f$_T$ of 250 GHz. *IEEE Electron. Device Lett.* **2018**, *39*, 75–78. [CrossRef]

20. Medjdoub, F.; Carlin, J.; Gonschorek, M.; Feltin, E.; Py, M.A.; Ducatteau, D.; Gaquiere, C.; Grandjean, N.; Kohn, E. Can InAlN/GaN be an alternative to high power/high temperature AlGaN/GaN devices? In Proceedings of the 2006 International Electron Devices Meeting, San Francisco, CA, USA, 11–13 December 2006; pp. 1–4.

21. Gonschorek, M.; Carlin, J.-F.; Feltin, E.; Py, M.A.; Grandjean, N.; Darakchieva, V.; Monemar, B.; Lorenz, M.; Ramm, G. Two-dimensional electron gas density in Al1−xInxN/AlN/GaN heterostructures (0.03 ≤ x ≤ 0.23). *J. Appl. Phys.* **2008**, *103*, 093714. [CrossRef]

22. Kong, Y.C.; Zheng, Y.D.; Zhou, C.H.; Gu, S.L.; Zhang, R.; Han, P.; Shi, Y.; Jiang, R.L. Two-dimensional electron gas densities in AlGaN/AlN/GaN heterostructures. *Appl. Phys. A* **2006**, *84*, 95–98. [CrossRef]

23. Lenka, T.R.; Panda, A.K. Characteristics study of 2DEG transport properties of AlGaN/GaN and AlGaAs/GaAs-based HEMT. *Semiconductors* **2011**, *45*, 650–656. [CrossRef]

24. Dianat, P.; Persano, A.; Quaranta, F.; Cola, A.; Nabet, B. Anomalous Capacitance Enhancement Triggered by Light. *IEEE J. Sel. Top. Quantum Electron.* **2015**, *21*, 1–5. [CrossRef]

25. Skinner, B.; Shklovskii, B. Anomalously large capacitance of a plane capacitor with a two-dimensional electron gas. *Phys. Rev. B* **2010**, *82*, 155111. [CrossRef]

26. Li, L.; Richter, C.; Paetel, S.; Kopp, T.; Mannhart, J.; Ashoori, R.C. Very Large Capacitance Enhancement in a Two-Dimensional Electron System. *Science* **2011**, *332*, 825–828. [CrossRef] [PubMed]

27. Dianat, P.; Prusak, R.; Persano, A.; Cola, A.; Quaranta, F.; Nabet, B. An Unconventional Hybrid Variable Capacitor with a 2-D Electron Gas. *IEEE Trans. Electron. Devices* **2014**, *61*, 445–451. [CrossRef]

28. Ando, T.; Fowler, A.B.; Stern, F. Electronic properties of two-dimmensional systems. *Rev. Mod. Phys.* **1982**, *54*, 437–672. [CrossRef]

29. Ambacher, O.; Smart, J.; Shealy, J.R.; Weimann, N.G.; Chu, K.; Murphy, M.; Schaff, W.J.; Eastman, L.F. Two-dimensional electron gases induced by spontaneous and piezoelectric polarization charges in N- and Ga-face AlGaN/GaN heterostructures. *J. Appl. Phys.* **1999**, *85*, 3222–3233. [CrossRef]

30. Ambacher, O.; Foutz, B.; Smart, J.; Shealy, J.R.; Weimann, N.G.; Chu, K.; Murphy, M.; Sierakowski, A.J.; Schaff, W.J.; Eastman, L.F.; et al. Two dimensional electron gases induced by spontaneous and piezoelectric polarization in undoped and doped AlGaN/GaN heterostructures. *J. Appl. Phys.* **2000**, *87*, 334–344. [CrossRef]

31. Vurgaftman, I.; Meyer, J.R. Band parameters for nitrogen-containing semiconductors. *J. Appl. Phys.* **2003**, *94*, 3675–3696. [CrossRef]

32. Yu, E.T.; Sullivan, G.J.; Asbeck, P.M.; Wang, C.D.; Qiao, D.; Lau, S.S. Measurement of piezoelectrically induced charge in GaN/AlGaN heterostructure field-effect transistors. *Appl. Phys. Lett.* **1997**, *71*, 2794–2796. [CrossRef]

33. Zhang, J.; Syamal, B.; Zhou, X.; Arulkumaran, S.; Ng, G.I. A Compact Model for Generic MIS-HEMTs Based on the Unified 2DEG Density Expression. *IEEE Trans. Electron. Devices* **2014**, *61*, 314–323. [CrossRef]

34. Delagebeaudeuf, D.; Linh, N.T. Metal-(n) AlGaAs-GaAs two-dimensional electron gas FET. *IEEE Trans. Electron. Devices* **1982**, *29*, 955–960. [CrossRef]

35. Tan, I.H.; Snider, G.L.; Chang, L.D.; Hu, E.L. A self-consistent solution of Schrödinger–Poisson equations using a nonuniform mesh. *J. Appl. Phys.* **1990**, *68*, 4071–4076. [CrossRef]

36. Sadi, T.; Kelsall, R.W.; Pilgrim, N.J. Investigation of self-heating effects in submicrometer GaN/AlGaN HEMTs using an electrothermal Monte Carlo method. *IEEE Trans. Electron. Devices* **2006**, *53*, 2892–2900. [CrossRef]

37. Medjdoub, F.; Alomari, M.; Carlin, J.F.; Gonschorek, M.; Feltin, E.; Py, M.A.; Grandjean, N.; Kohn, E. Barrier-layer scaling of InAlN/GaN HEMTs. *IEEE Electron. Device Lett.* **2008**, *29*, 422–425. [CrossRef]

38. Chen, C.Y.; Wu, Y.R. Studying the short channel effect in the scaling of the AlGaN/GaN nanowire transistors. *J. Appl. Phys.* **2013**, *113*. [CrossRef]

© 2018 by the authors. Licensee MDPI, Basel, Switzerland. This article is an open access article distributed under the terms and conditions of the Creative Commons Attribution (CC BY) license (http://creativecommons.org/licenses/by/4.0/).

![electronics logo] *electronics*

MDPI

Article

Nano-Particle VO$_2$ Insulator-Metal Transition Field-Effect Switch with 42 mV/decade Sub-Threshold Slope

Massood Tabib-Azar * and Rugved Likhite

Electrical and Computer Engineering Department, University of Utah, Salt Lake City, UT 84112, USA;
rugved.likhite@utah.edu
* Correspondence: azar.m@utah.edu; Tel.: +1-801-581-8775

Received: 25 October 2018; Accepted: 17 January 2019; Published: 1 February 2019

Abstract: The possibility of controlling the insulator-to-metal transition (IMT) in nano-particle VO$_2$ (NP-VO$_2$) using the electric field effect in a metal-oxide-VO$_2$ field-effect transistor (MOVFET) at room temperature was investigated for the first time. The IMT induced by current in NP-VO$_2$ is a function of nano-particle size and was studied first using the conducting atomic force microscope (cAFM) current-voltage (I-V) measurements. NP-VO$_2$ switching threshold voltage (V_T), leakage current ($I_{leakage}$), and the sub-threshold slope of their conductivity (S_c) were all determined. The cAFM data had a large scatter. However, V_T increased as a function of particle height (h) approximately as $V_T(V) = 0.034$ h, while $I_{leakage}$ decreased as a function of h approximately as $I_{leakage}$ $(A) = 3.4 \times 10^{-8} e^{-h/9.1}$. Thus, an asymptotic leakage current of 34 nA at zero particle size and a tunneling (carrier) decay constant of ~9.1 nm were determined. S_c increased as a function of h approximately as S_c $(mV/decade) = 2.1 \times 10^{-3} e^{h/6}$ and was around 0.6 mV/decade at h~34 nm. MOVFETs composed of Pt drain, source and gate electrodes, HfO$_2$ gate oxide, and NP-VO$_2$ channels were then fabricated and showed gate voltage dependent drain-source switching voltage and current (I_{DS}). The subthreshold slope (S_t) of drain-source current (I_{DS}) varied from 42 mV/decade at V$_G$ = -5 V to 54 mV/decade at V$_G$ = +5 V.

Keywords: insulator–metal transition (IMT); charge injection; Mott transition; conductive atomic force microscopy (cAFM); gate field effect; atomic layer deposition (ALD)

---

1. Introduction

Vanadium dioxide (VO$_2$) is a model insulator–metal transition (IMT) material that displays a first-order transition from a monoclinic insulating phase to a tetragonal metallic phase at a critical temperature (T_C) of 341 K in its bulk form [1]. This transition in addition to the semiconductor field-effect in VO$_2$ provides an opportunity to potentially realize switches with steep subthreshold slope of better than 60 mV/decade of silicon for energy efficient devices and applications.

In addition to heat (temperature), the IMT in VO$_2$ can be induced by charge injection, light, mechanical stress, terahertz signals, and many other stimuli. In early reports, the charge-induced IMT was ascribed to local heating of the VO$_2$ material over the critical temperature (T$_{IMT}$) by the current flowing through the device [2]. Recently, however, other non-thermal mechanisms of the voltage-triggered IMT were proposed [3–5]. Specifically, the electric field, rather than local dissipated power due to Joule heating, was suggested to be the origin of the IMT in some cases [2,4–6]. Charge injection induced IMT in VO$_2$ was reported in the past and was used to realize negative differential gate capacitance to improve the subthreshold current slope (S$_t$) [4]. Additionally, the investigation of electric field effect on the IMT has gained importance as a field-controlled IMT effect in fast and reliable electronic devices.

Our main motivation in this study is to develop field-effect transistor (FET) devices with very steep sub-threshold current slopes for energy efficient switches. In crystalline VO_2 the insulating phase carrier concentration is quite high on the order of 10^{21} cm^{-3}. Therefore, the effect of gate electric field in modulating the I_{DS} is small since the channel has large conductivity to begin with. In VO_2 nano-particles and polycrystalline films, small crystalline regions are separated by grain boundary regions that have lower conductivity (n~ 10^{17}–10^{18} cm^{-3}) [7]. In these materials, the electric field-effect in the boundary regions appears to control the channel turn-on voltage and the sub-threshold current slope as discussed here. The electric field effect and the transition temperature can be related to each other. We have recently shown that the transition temperature in VO_2 reduces at higher applied gate voltages, and the relationship is approximately given by: $T_{transition}$ = 1.2 V_G + 63.8 °C where V_G < 0 [7].

Recently, there has been an increased interest in developing high energy efficiency electronic switches using transition metal dichalcogenides [8] and IMT materials [9,10]. The grain size in polycrystalline VO_2 affects the IMT temperature as discussed in [11]. We note that to realize efficient transistors, IMT materials are great candidates, since in addition to providing the usual semiconducting carrier control mechanism, they also provide very steep IMT transitions. VO_2 has a transition temperature of 63 °C and is used here to demonstrate the feasibility of metal-oxide-VO_2 field-effect transistor (MOVFET). A better channel material is Cr-doped V_2O_3 with possible transition temperature of above 150 °C [12].

2. Nano-Particle Studies

VO_2 nano-particles, obtained from Strem Chemicals (Newburyport, MA, USA) [13], were deposited on a gold covered oxidized silicon chip for atomic force microscope (cAFM) studies. To fabricate the chips, we started with a p-type silicon wafer that was cleaned using piranha and buffered oxide etches. Subsequently, the silicon was oxidized (~90 nm) using wet thermal oxidation and then it was coated with 100 nm of Au on 20 nm of Ti adhesion layer. Finally, the wafer was diced into 1 cm^2 square chips suitable for cAFM scans and measurements (Figure 1). The VO_2 powder was mixed with ethanol and ultra-sonicated for 3 min to prevent agglomeration. The mixture was poured over the central region of the gold-covered silicon chip (Figure 1b) and allowed to dry to obtain evenly spread VO_2 particles over the sample. Conductive atomic force microscopy (cAFM) measurements were then carried out on a Multimode AFM using a Pt conducting tip connected to a semiconductor parameter analyzer for I-V measurements (Figure 1a). To measure the thickness of each particle, we used the metrology capability of the AFM system that provided a numerical value for particle height. We assumed that the particles are spherical and used the particle height as its diameter. The I-V measurements were carried out by locating a VO_2 particle using the cAFM probe, which also acted as the top contact to the particle with the gold substrate as the bottom contact. All experiments were done at room temperature.

Figure 1c shows a representative AFM scan performed over the Au substrate containing VO_2 particles. A wide distribution of particle heights was obtained during a single scan. Particle heights were determined using AFM line scans shown by dashed lines in Figure 1c and displayed with numerical values by the AFM software in Figure 1d.

Figure 2 shows I-V measurement results obtained for 48 nm and 35 nm particles. Sharp transitions in I-Vs were observed in the NP-VO_2 ranging in size from 13 nm to 53 nm with turn-on voltages (V_T) ranging from 0.5–2.0 V. The leakage current and the sub-threshold slopes were calculated for each nano-particle at V = $V_T/2$ and V = V_T, respectively, as shown in Figure 2b,d.

Figure 3a shows V_T (defined in Figure 2) as a function of the particle size. Assuming that V_T = 0 V for h = 0 nm, the scattered data tentatively fits a line given by V_T (V) = 0.034 h with "h" in nm. Therefore, the critical transition field (E_c = V_T/h) is E_c ~ 3.44 × 10^5 V/cm. The leakage current as a function of particle size is shown in Figure 3b and has an approximate dependence given by $I_{leakage}$(A) = 3.4 × $10^{-8}e^{-h/9.1}$. This indicates an asymptotic leakage current of 34 nA at zero particle size and tunneling (carrier) decay constant of 9.1 nm inside the NP-VO_2.

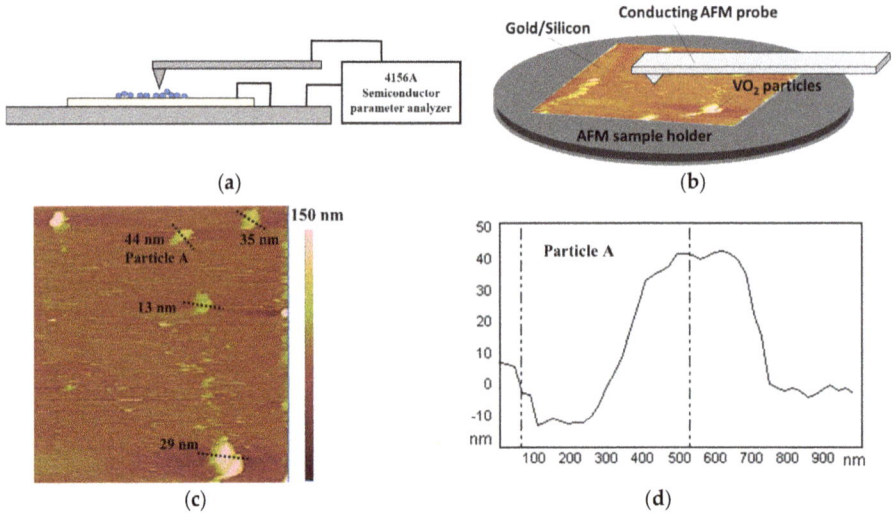

(a)

(b)

(c)

(d)

Figure 1. Schematic of AFM setup for scanning NP-VO$_2$ over gold covered silicon substrate. (**a**) AFM setup modified for imaging as well as current-voltage (I-V) measurement using Pt tip. (**b**) AFM scan showing some of the NP-VO$_2$ where I-V relationships were calculated. (**c**) Different particle heights ranging from 29 nm–44 nm are seen in the above image. (**d**) Section scan showing the lateral profile of NP-VO$_2$ with a height of 44nm.

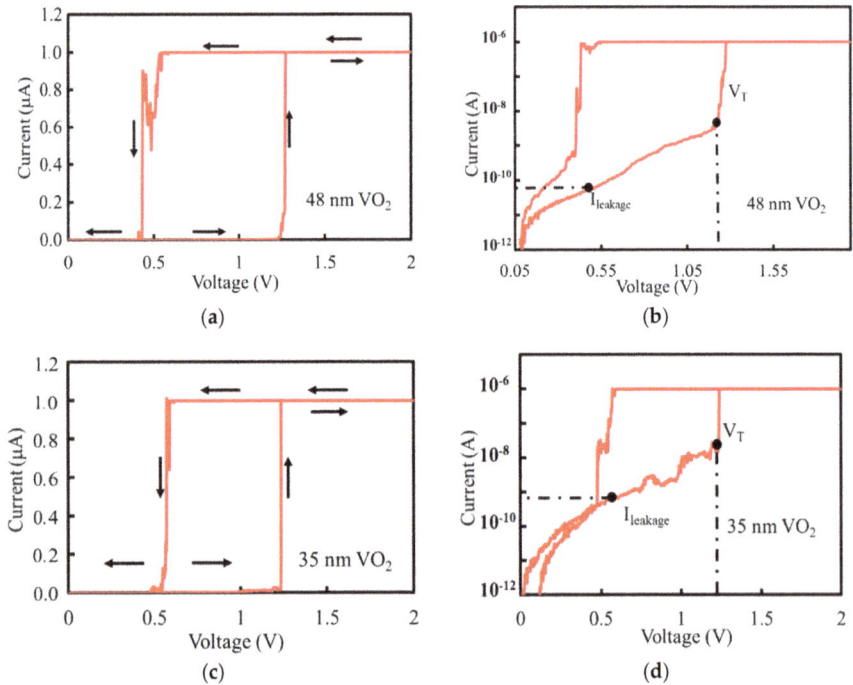

(a)

(b)

(c)

(d)

Figure 2. Representative current versus voltage (I-V) measurement results between the Pt AFM probe tip and NP-VO$_2$ on gold. (**a**) I-V of a 48 nm VO$_2$ particle and (**b**) the same I-V in log scale clearly showing V$_T$ and I$_{leakage}$ at V$_T$/2. (**c**) I-V of a 35 nm VO$_2$ particle and (**d**) its log scale.

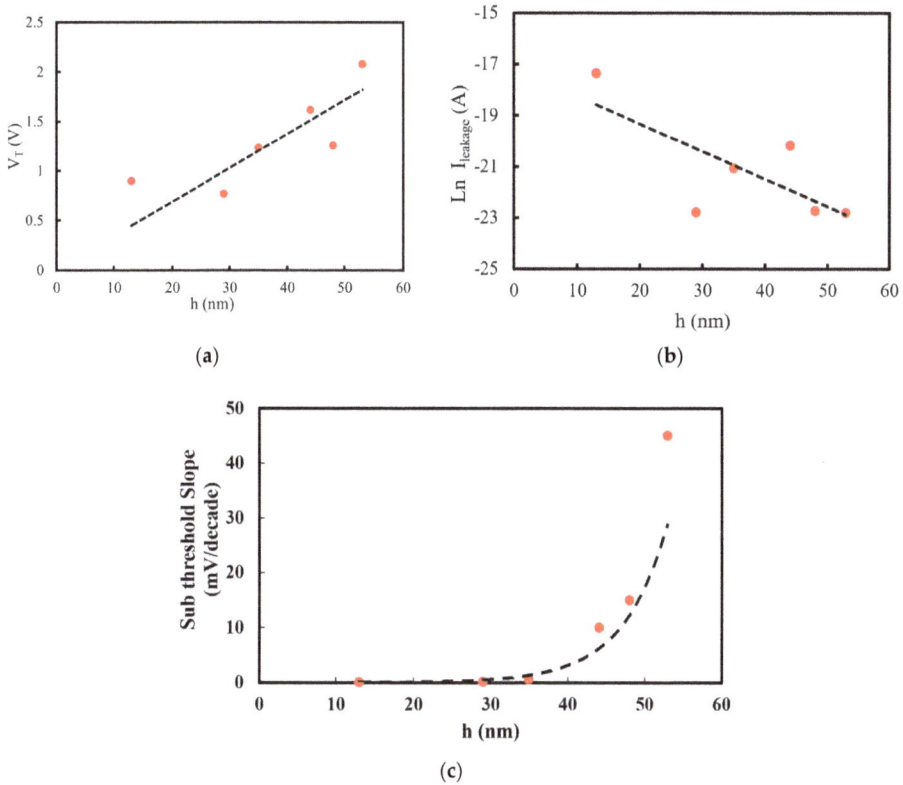

Figure 3. (a) V_T, (b) $I_{leakage}$ and (c) switching current slope (S_c) as a function of VO$_2$ nanoparticle size. The values of V_T and leakage currents were taken from the I-V curves of Figure 2.

Next, we examined the slope of the I-V at $V_T/2$ (see Figure 2b). This slope can be viewed as the switching current slope denoted by S_c. S_c increased as a function of particle size and was approximately given by S_c (mV/decade) = $2.1 \times 10^{-3}e^{h/6}$ as shown in Figure 3c and was around 0.6 mV/decade for particle size of 34 nm. The slope of I-V curve indicated by S_c is different than the slope of the MOVFET I_{DS} versus V_G curve indicated by S_t. The exponential dependence of S_c on particle size indicates that as the particle size becomes larger, the conduction tunnel paths and energies through the particle become more numerous leading to shallower I-V switching curve.

3. Device Studies

We next incorporated the NP-VO$_2$ in the channel of an FET that were fabricated on a glass substrate with a 100 nm Pt gate covered by a 50 nm atomic layer deposited (ALD) HfO$_2$ dielectric and a 100 nm Pt drain and source electrodes, as shown in Figure 4a,b. The fabrication process is discussed in [14] and was started with etching a 4″ glass substrate by immersing it in buffered oxide etch (BOE) for 1 min to create 100 nm deep trenches for the gate metallization regions. After 100 nm Pt gate metal deposition and patterning, 50 nm-thick HfO$_2$ was deposited using atomic layer deposition technique as gate oxide. 100 nm of Pt was then sputter deposited and patterned to create the source-drain regions. Figure 4b,c show the SEM image of the fabricated open-channel FET device.

The open-channel device geometry enables any material to be deposited in the channel region of the FET. To form the NP-VO$_2$ channel, we mixed the VO$_2$ powder with n-butyl-acetate and a small amount of silver paint (1% to 10% in weight) and then ultrasonicated the mixture that was subsequently deposited using a fine brush in the MOVFET's channel region, as shown in Figure 4d.

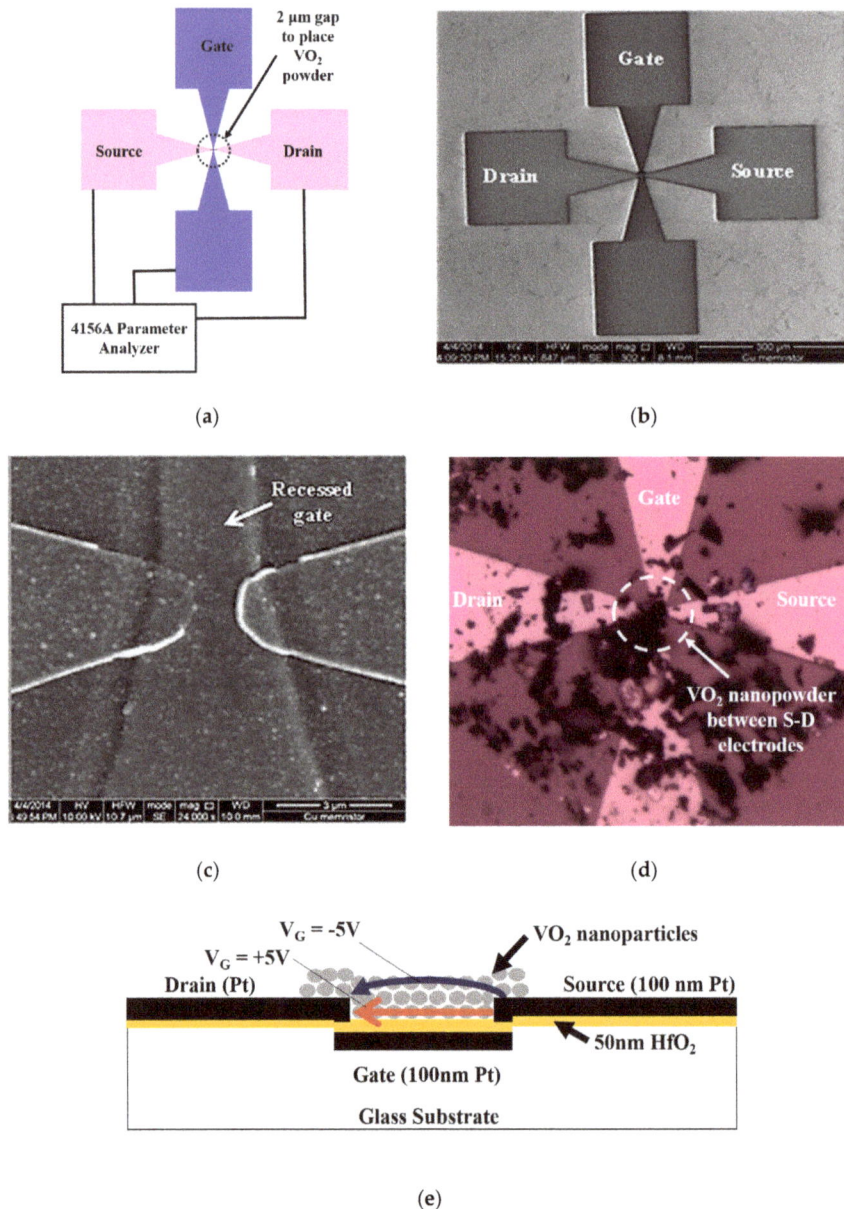

Figure 4. (a) VO$_2$ MOSFET was connected to a semiconductor parameter analyzer to obtain its electrical characteristics. (b) and (c) SEM images of the open-face MOSFET used in our experiments [14]. (d) Optical image of the open-face MOSFET with drop-cast NP-VO$_2$. Schematic cross-section view of the VO$_2$ MOSFET device architecture with gate embedded inside a glass wafer. (e) Schematic of the cross section of the open-face MOSFET with VO$_2$ channel showing the control of the conduction path using the gate field effect.

Figure 5a shows a typical I_{DS}-V_{DS} characteristic of the MOVFET. The transition from insulator to metal results in the very sharp increase in the I_{DS} as a function of increasing V_{DS}. I_{DS} transitions for decreasing V_{DS} are also very sharp. In these experiments the I_{DS} was limited to 0.5 mA by the instrument to prevent device breakdown. The $I_{DS\text{-}ON}/I_{DS\text{-}OFF}$ in these devices, measured using pulsed voltages, were higher than 1000, but we limited the current in static measurements. Figure 5b shows the sub-threshold current as a function of V_G. The sub-threshold slope increased as a function of gate voltage. Figure 5c shows the I_{DS} at $V_T/2$ (= $I_{leakage}$) as a function of V_G, clearly indicating that the leakage current is lower at the negative gate voltages compared to zero and positive gate voltages. We also note that positive gate voltages increased the V_T (Figure 5d), while the negative gate voltage reduced it. The V_{DS} voltage step in these experiments was 100 mV.

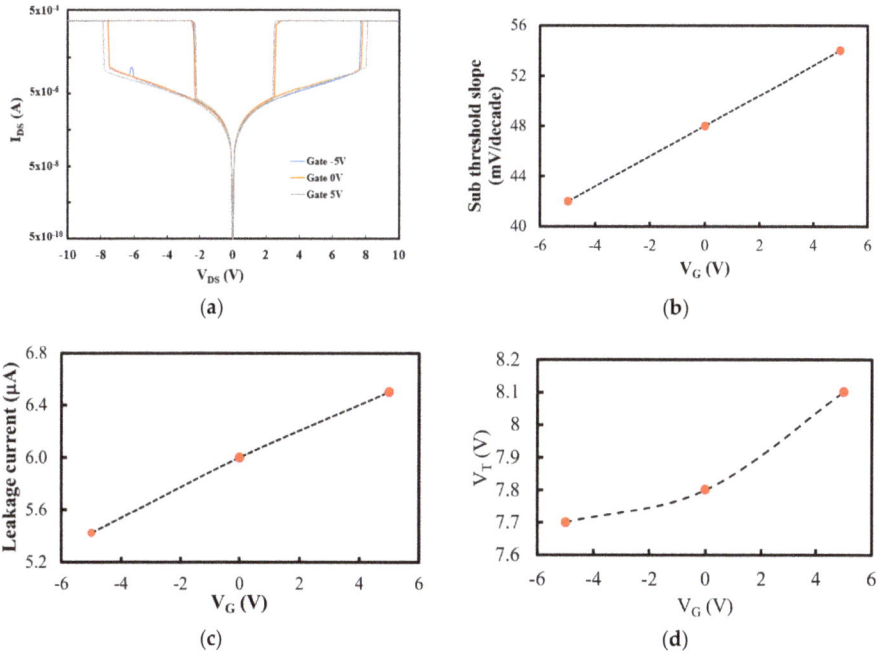

Figure 5. (a) NP-VO$_2$ MOSFET characteristics at room temperature. The current was limited to 0.5 mA, and the V_{DS} voltage steps was 100 mV. (b) Sub-threshold slope (S_t) of the I_{DS} versus V_G. (c) Leakage current as a function of V_G. (d) Threshold voltage as a function of gate voltage.

4. Discussion

The main charge carriers in VO$_2$ are believed to be electrons [2,15]. It is also assumed that the IMT is triggered by the onset of a critical density of electrons in the channel [2,16]. Thus, one expects the threshold voltage in the MOVFET to become smaller at more positive gate voltages. However, we note that just the opposite is observed in Figure 5c. We also note in Figure 3a that in nano-particles, the transition voltage became smaller in smaller particles. Putting these two observations together, we conclude that the gate electric field in our MOVFET changes the current path between the drain and source, as schematically shown by the white and gray (red) arrows in Figure 4e. When the gate voltage is positive, the conduction path is much wider than when the gate voltage is negative. A negative gate voltage "pushes" the conduction path away from the channel-gate interface region and confines it to a thinner layer at the top. Thus, the conduction occurs over the smaller channel cross section. The V_G dependence of all other parameters (S_t, S_c and $I_{leakage}$) agree with this observation.

We also note that other groups working with single crystal thin-film VO_2 reported difficulty demonstrating gate field effect [17]. This is attributed to the presence of high carrier (electron) concentration ~10^{18}–10^{21} cm^{-3} [7] in the insulator phase of the VO_2 that is difficult to modulate and requires very high breakdown gate insulator material. In our devices, we used NP-VO_2 that also show the same levels of electron concentrations as calculated from the leakage currents of Figure 2. However, nano-particles couple to each other through their outer boundary layers with lower electron concentrations. The relatively weak coupling between NP-VO_2 enables the gate field effect to modulate the current path that gives rise to the modulation of the V_T, $I_{leakage}$, and S_t.

While the exact origin of the field-effect induced IMT is still under investigation, it is believed to be non-thermal. Very fast switching (10 ns) in optimized two-terminal devices with thin-film VO_2 active regions is also reported [18] indicating the possibility of non-thermal switching mechanisms in these materials.

5. Conclusions

The current-induced IMT in NP-VO_2 was investigated using cAFM measurements. The transition voltage in NP-VO_2 decreased while the leakage current increased as a function of the particle size. These measurements were followed by construction of NP-VO_2 FET devices and the related I_{DS}-V_{DS}/V_G measurements. We used open-channel FET devices and drop-casted NP-VO_2 to form their channels and measured the resulting MOVFET characteristics that showed gate voltage dependent I_{DS}-V_{DS}.

Author Contributions: Investigation, R.L.; Original Idea and Supervision, M.T.-A.

Funding: This research was funded by the Utah Science and Technology and Research Program.

Conflicts of Interest: The authors declare no conflict of interest.

References

1. Tan, X.; Yao, T.; Long, R.; Sun, Z.; Feng, Y.; Cheng, H.; Wei, S. Unraveling metal-insulator transition mechanism of VO_2 triggered by tungsten doping. *Sci. Rep.* **2012**, *2*, 466. [CrossRef] [PubMed]
2. Stefanovich, G.; Pergament, A.; Stefanovich, D. Electrical switching and Mott transition in VO_2. *J. Phys. Condens. Matter* **2000**, *12*, 8837. [CrossRef]
3. Zylbersztejn, A.; Mott, N.F. Metal-insulator transition in vanadium dioxide. *Phys. Rev. B* **1975**, *11*, 4383. [CrossRef]
4. Lee, S.H.; Kim, M.K.; Lee, J.W.; Yang, Z.; Ramanathan, S.; Tiwari, S. Vanadium dioxide (VO_2) is also a ferroelectric: Properties from memory structures. In Proceedings of the 11th IEEE Conference on Nanotechnology (IEEE-NANO), Portland, OR, USA, 15–18 August 2011; pp. 735–739.
5. Nakano, M.; Shibuya, K.; Okuyama, D.; Hatano, T.; Ono, S.; Kawasaki, M.; Iwasa, Y.; Tokura, Y. Collective bulk carrier delocalization driven by electrostatic surface charge accumulation. *Nature* **2012**, *487*, 459–462. [CrossRef] [PubMed]
6. Wei, J.; Wang, Z.; Chen, W.; Cobden, D.H. New aspects of the metal–insulator transition in single-domain vanadium dioxide nanobeams. *Nat. Nanotechnol.* **2009**, *4*, 420–424. [CrossRef] [PubMed]
7. Jin, F.; Tabib-Azar, M. Optical Tweezer Assembled VO_2 Particles Aligned by Drain-Source Bowtie Antenna MOSFET with 10 mV/dec Sub-Threshold Slopes. *Electron Devices Lett.* **2019**. under review.
8. Oliva, N.; Casu, E.A.; Yan, C.; Krammer, A.; Magrez, A.; Schueler, A.; Martin, O.; Ionescu, M.A. MoS2/VO2 vdW Heterojunction Devices: Tunable Rectifiers, Photodiodes and Field Effect Transistors. Available online: https://infoscience.epfl.ch/record/253434 (accessed on 12 December 2018).
9. Chen, C.-K.; Lin, C.-Y.; Chen, P.-H.; Chang, T.-C.; Shih, C.-C.; Tseng, Y.-T.; Zheng, H.-X.; Chen, Y.-C.; Chang, Y.-F.; Lin, C.-C.; et al. The Demonstration of Increased Selectivity during Experimental Measurement in Filament-Type Vanadium Oxide-Based Selector. *IEEE Trans. Electron Devices* **2018**, *99*, 1–6. [CrossRef]
10. Vitale, A.; Casu, E.A.; Biswas, A.; Rosca, T.; Alper, C.; Krammer, A.; Luong, G.V.; Zhao, Q.-T.; Mantl, S.; Schuler, A.; et al. A Steep-Slope Transistor Combining Phase Change and Band-to-Band Tunneling to Achieve a sub-Unity Body Factor. *Sci. Rep.* **2017**, *7*, 355. [CrossRef] [PubMed]

11. Wang, C.Q.; Shao, J.; Liu, X.L.; Chen, Y.; Xiong, W.M.; Zhang, X.Y.; Zheng, Y. Phase Transition Characteristics in the Conductivity of VO_2(A) nanowires: Size and Surface Effects. *Phys. Chem. C* **2016**, *18*, 10262. [CrossRef] [PubMed]

12. Yethiraj, M.; Werner, S.A.; Yelon, W.B.; Honig, J.M. Phase transitions in pure and Cr-doped V_2O_3. *Phys. B+C* **1986**, *136*, 458–460. [CrossRef]

13. Strem Chemicals. Available online: http://www.strem.com/catalog/v/93-2309/80/vanadium_12036-21-4 (accessed on 12 December 2018).

14. Mou, N.I.; Zhang, Y.; Pai, P.; Tabib-Azar, M. Steep Sub-threshold Current Slope (~2mV/dec) Pt/Cu2S/Pt Gated Memristor with Ion/Ioff>100. *Solid-State Electron.* **2017**, *127*, 20–25. [CrossRef]

15. Duchene, J.; Terraillon, M.; Pailly, P.; Adam, G. Filamentary Conduction in VO2 Coplanar Thin-Film Devices. *Appl. Phys. Lett.* **1971**, *19*, 115. [CrossRef]

16. Okimura, K.; Ezreena, N.; Sasakawa, Y.; Sakai, J. Electric Field-Induced Multi-Step Resistance Switching Phenomena in a Planer VO2/c-Al2O3 Structure. *Jpn. J. Appl. Phys.* **2009**, *48*, 065003. [CrossRef]

17. Kim, H.T.; Chae, B.G.; Youn, D.H.; Kim, G.; Kang, K.Y.; Lee, S.J.; Kim, K.; Lim, Y.S. Raman study of electric-field-induced first-order metal-insulator transition in VO2-based devices. *Appl. Phys. Lett.* **2005**, *86*, 242101. [CrossRef]

18. Kim, H.T.; Lee, Y.W.; Kim, B.J.; Chae, B.G.; Yun, S.J.; Kang, K.Y.; Lim, Y.S. Monoclinic and correlated metal phase in VO_2 as evidence of the Mott transition: Coherent phonon analysis. *Phys. Rev. Lett.* **2006**, *97*, 266401. [CrossRef] [PubMed]

© 2019 by the authors. Licensee MDPI, Basel, Switzerland. This article is an open access article distributed under the terms and conditions of the Creative Commons Attribution (CC BY) license (http://creativecommons.org/licenses/by/4.0/).

electronics

MDPI

Article

Optimization of Line-Tunneling Type L-Shaped Tunnel Field-Effect-Transistor for Steep Subthreshold Slope

Faraz Najam and Yun Seop Yu *

Department of Electrical, Electronic and Control Engineering and IITC, Hankyong National University, Anseong 17579, Korea; faraznajam@hknu.ac.kr
* Correspondence: ysyu@hknu.ac.kr; Tel.: +81-31-670-5293

Received: 12 October 2018; Accepted: 22 October 2018; Published: 24 October 2018

Abstract: The L-shaped tunneling field-effect-transistor (LTFET) has been recently introduced to overcome the thermal subthreshold limit of conventional metal-oxide-semiconductor field-effect-transistors (MOSFET). In this work, the shortcomings of the LTFET was investigated. It was found that the corner effect present in the LTFET effectively degrades its subthreshold slope. To avoid the corner effect, a new type of device with dual material gates is presented. The new device, termed the dual-gate (DG) LTEFT (DG-LTFET), avoids the corner effect and results in a significantly improved subthreshold slope of less than 10 mV/dec, and an improved ON/OFF current ratio over the LTFET. The DG-LTFET was evaluated for different device parameters and bench-marked against the LTFET. This work presents the optimum configuration of the DG-LTFET in terms of device dimensions and doping levels to determine the best subthreshold, ON current, and ambipolar performance.

Keywords: band-to-band tunneling; L-shaped tunnel field-effect-transistor; double-gate tunnel field-effect-transistor; corner-effect

1. Introduction

Tunnel field-effect-transistors (TFETs) are being actively pursued as a potential replacement to conventional metal-oxide-semiconductor (MOS) technology [1]. TFETs offer a sub-thermal subthreshold slope (SS) but suffer from limited ON current I_{ON} performance [2]. To overcome the limit, different types of line tunneling type TFETs have been introduced, including L-shaped [3] (LTFETs), U-shaped [4] (UTFETs), and Z-shaped [5] TFETs (ZTFETs). Among them, only the LTFET has been experimentally demonstrated [3].

It was found using device simulations that the 2D corner effect [6] present in LTFETs degrades its subthreshold performance. In order to remove SS degradation due to the kink effect induced by the source corner, the fully depleted rounded corner with a gradual doping profile was used [6]. The LTFET still achieves a sub-thermal SS, but as shown in this work there is room for significant improvement in the subthreshold performance of LTFETs. To achieve this improvement, a new device based on the original LTFET is introduced. The new device uses a dual-gate (DG) structure and is termed the DG-LTFET. The two gates (gate1 and gate2) have different workfunctions and different heights. The DG-LTFET was thoroughly evaluated for different device parameters, including the source region height, gate1 and gate2 heights, gate1 and gate2 workfunctions, channel thickness, and drain doping levels. Optimum dimensions and drain doping level were determined for the DG-LTFET. Section 2 briefly discusses the corner-effect problem of the LTFET. Section 3 introduces the DG-LTFET and compares its results with the LTFET. Section 4 presents the conclusion.

2. The LTFET: The Corner Effect

Figure 1 shows a schematic for LTFET. The p$^+$ (10^{20} cm^{-3}) doped source region overlaps the gate with the n$^-$ (10^{12} cm^{-3}) channel sandwiched in between them. This sandwiched channel region is termed as $R_{nonoffset}$. There is also a part of the channel termed R_{offset} in which there is an offset present between the source and the gate, as indicted in Figure 1. The following parameters were used for all devices considered in this work unless otherwise specified: source height (H_s) = 40 nm, oxide thickness (t_{ox}) = 2 nm, length of $R_{nonoffset}$ (T_j) = 5 nm, channel length (L_{ch}) = 50 nm, height of R_{offset} (H_{offset}) = 10 nm, height of $R_{nonoffset}$ ($H_{nonoffset}$) = H_s, gate height (H_{g1}) = H_s + (H_{offset} − t_{ox}) = 48 nm, dielectric permittivity ε_{ox} = 25, metal gate workfunction W_{rk_LTFET} = 4.72 eV, and drain doping (N_d) = 10^{20} cm^{-3}.

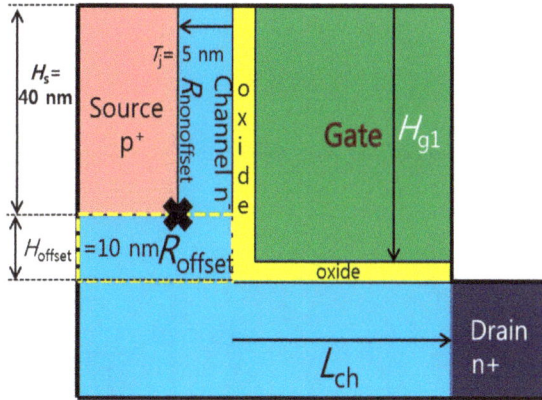

Figure 1. Schematic of the L-shaped tunneling field-effect-transistor (LTFET).

Sentaurus technology-computer-aided-design tool (TCAD) was used as the simulator [7]. The following models were used in the simulation: the dynamic nonlocal band-to-band-tunneling (BTBT) model, Fermi statistics, and the constant mobility model. The dynamic nonlocal BTBT model calculates BTBT in both lateral and 1D directions. Crystal orientation is assumed to be <100> in all devices. A constant electron effective tunneling mass of 0.19 m_0 was used in all simulations [8]. All simulations were performed at a drain source bias V_{ds} = 0.1 V unless otherwise specified.

For analysis to follow, drain-source current (I_{ds}) versus gate-source bias (V_{gs}) characteristics of the LTFET are shown in Figure 2a. There is a direct overlap between gate and source in $R_{nonoffset}$, and the electric field in $R_{nonoffset}$ is in the 1D direction. In R_{offset}, however, the electric field from the gate converges around the sharp source corner marked by an **X** in Figure 1. This increases the potential in R_{offset} as compared to $R_{nonoffset}$ for any given bias (until potential saturates due to electron inversion). Figure 2b shows the surface potential at V_{gs} = 0 V. It can be seen that, because the electric field converges around the sharp source corner [6], the potential in R_{offset} has increased. Since the potential is higher in R_{offset} as compared to $R_{nonoffset}$, the threshold voltage for BTBT in R_{offset} ($V_{th\_Roffset}$) is lower than the threshold voltage for BTBT in $R_{nonoffset}$ ($V_{th\_Rnonoffset}$).

Figure 3a,b show the tunneling rate (G_{tun}) contour plot and G_{tun}, respectively, at V_{gs} = 0.21 V which is the bias needed to generate I_{ds} = 10^{-13} A (from Figure 2a). It is obvious from Figure 3 that the BTBT only takes place in R_{offset}, whereas $R_{nonoffset}$ is completely switched off. Figure 4a shows G_{tun} at several V_{gs} values. From Figure 4a, $V_{th\_Roffset}$ and $V_{th\_Rnonoffset}$ can be found to be around V_{gs} = 0.17 V and 0.24 V, respectively. Figure 4b shows the G_{tun} contour plot at V_{gs} = $V_{th\_Rnonoffset}$ = 0.24 V. Figure 4a shows that G_{tun} in $R_{nonoffset}$ just after it turns on, is always higher and has a much larger BTBT area (in the *y* direction) as compared to R_{offset}. Thus, whenever $R_{nonoffset}$ turns on, it dominates over R_{offset}. The reason why G_{tun} is higher in $R_{nonoffset}$ is simply because the BTBT paths in R_{offset} are laterally oriented or 2D from source to the surface in R_{offset}, whereas the BTBT paths in $R_{nonoffset}$ are 1D. The 2D BTBT paths being naturally longer than the 1D paths result in a lower G_{tun} in R_{offset}.

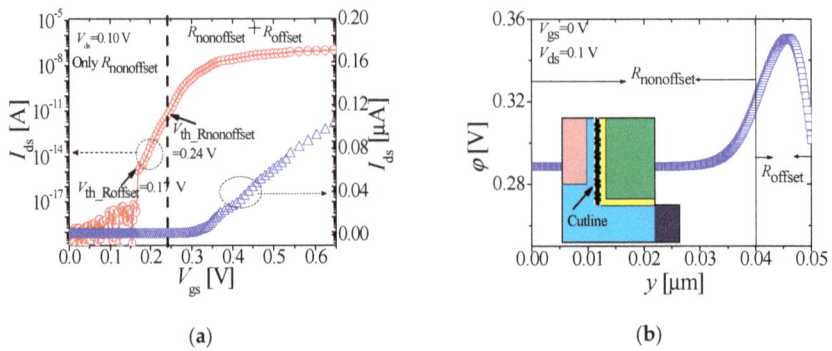

(a)   (b)

Figure 2. (a) I_{ds}-V_{gs} transfer characteristics of the LTFET. $V_{th\_Rnonoffset}$ = 0.24 V and $V_{th\_Roffset}$ = 0.17 V. (b) Potential along the cutline shown in the inset at V_{gs} = 0 V. Potential is higher in R_{offset}.

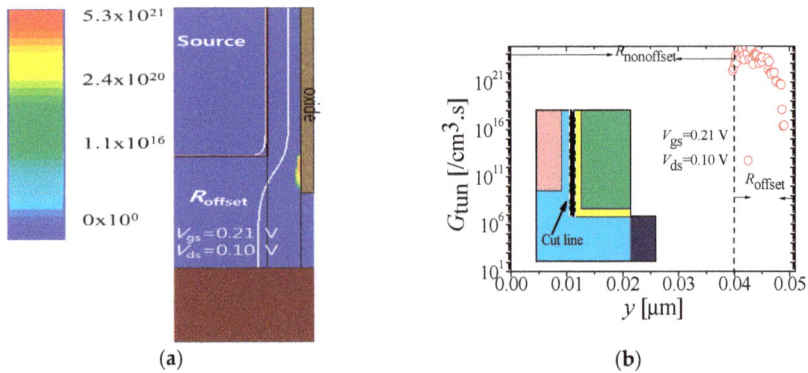

(a)   (b)

Figure 3. (a) G_{tun} contour plot at V_{gs} = 0.21 V, which is the bias needed to generate I_{ds} = 10^{-13} A and (b) G_{tun} extracted from (a).

From Figure 4a, it can be observed that, for a large part of the subthreshold region (V_{gs} < 0.24 V), only R_{offset} with the longer 2D BTBT paths and lower G_{tun} is contributing to the BTBT current and the more efficient $R_{nonoffset}$ makes no contribution to the current. In other words, the LTFET underperforms in the subthreshold region. If $R_{nonoffset}$ could be forced to turn on at a lower bias than R_{offset}, which is the condition $V_{th\_Rnonoffset}$ < $V_{th\_Roffset}$, $R_{nonoffset}$ will turn on in the subthreshold region, and with the condition G_{tun} in $R_{nonoffset}$ > G_{tun} in R_{offset}, demonstrated in Figure 4a, a significant improvement in SS could be expected.

In other words, the $R_{nonoffset}$ could be regarded as a parasitic region with a parasitic, fringing capacitance originating from the bottom of the gate to the sharp source corner. Since the potential is different in this area (Figure 2b), the capacitance associated with this region is different from the $R_{nonoffset}$ region. If $V_{th\_Rnonoffset}$ < $V_{th\_Roffset}$ could be achieved, as is demonstrated below, the effect of this parasitic capacitance could be practically eliminated, and this is the purpose of the device proposed below. Since drain is not in close proximity to R_{offset}/$R_{nonoffset}$, where the BTBT current is generated, gate-drain capacitance fringing capacitance is not expected to influence the potential and BTBT significantly at high frequency.

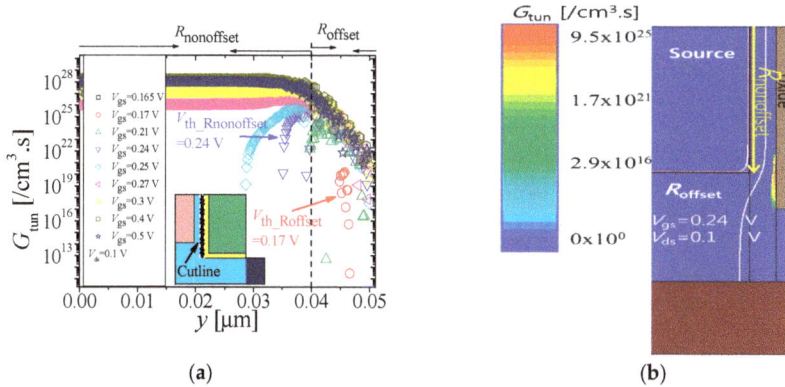

Figure 4. (a) G_{tun} at different V_{gs}. (a) $V_{th\_Rnonoffset} = 0.24$ V and $V_{th\_Roffset} = 0.17$ V. (b) G_{tun} contour plot at $V_{gs} = V_{th\_Rnonoverlap} = 0.24$ V. In (**b**), yellow arrow indicates the height of $R_{nonoffset}$.

3. DG-LTFET

3.1. The DG-LTFET: Basic Device Physics

In order to achieve the condition $V_{th\_Rnonoffset} < V_{th\_Roffset}$, the DG-LTFET is presented in Figure 5a. DG-LTFET uses dual material gates denoted by gate1 and gate2, each with a different workfunction ($W_{rk\_gate1/2}$) and height ($H_{g1/2}$). $H_{g1} = H_{nonoffset} = H_s = 40$ nm, $H_{offset} = 10$ nm, $H_{g2} = H_{nonoffset} - H_{g1} + (H_{offset} - t_{ox}) = 8$ nm, and $T_j = 5$ nm. W_{rk_gate1} is always lower than W_{rk_gate2}. W_{rk_gate2} is fixed at $W_{rk\_LTFET} = 4.72$ eV for all DG-LTFET considered in this work. The DG-LTFET process-flow is indicated in Figure 5a. The process-flow is based on the LTFET process-flow [3]. The DG-LTFET process-flow follows the LTFET process-flow until the chemical vapor deposition (CVD) of gate2 (similar to the gate deposition in the LTFET). After this, two additional steps are required. The device is masked to protect the gate oxide and channel areas, and gate2 is selectively etched according to the desired height. Gate1 is then deposited in the recess created by gate2-etching by a low-temperature atomic layer deposition process. Similar dual-material gate structures have been extensively reported in the literature including [9–11].

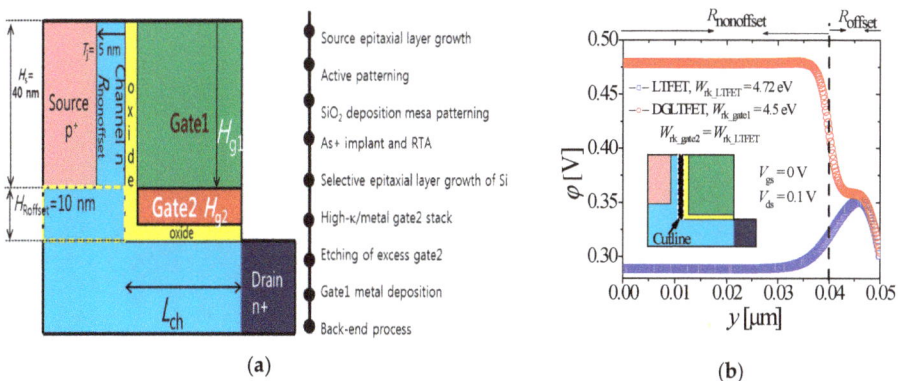

Figure 5. (**a**) Schematic of DG-LTFET with process-flow indicated alongside and (**b**) V_{fb} of DG-LTFET (red symbols) compared with that of the LTFET (blue symbols). In the DG-LTFET, $W_{rk\_gate1} = 4.5$ eV and $W_{rk\_gate2} = W_{rk\_LTFET} = 4.72$ eV were used.

Lower W_{rk_gate1} results in an increased flatband voltage [12] (V_{fb}) in $R_{nonoffset}$ as compared to R_{offset}. Figure 5b shows V_{fb} of DG-LTFET (red symbols) with W_{rk_gate1} = 4.5 eV and W_{rk_gate2} = W_{rk_LTFET}. Also shown for the reference is V_{fb} of the LTFET (blue symbols). Expectedly, the DG-LTFET potential increases in $R_{nonoffset}$. The potential does not change abruptly from gate1 to gate2 because of the presence of 2D effects around the source corner. Electric field from the bottom of gate2 converges around the source corner. Around the middle of R_{offset}, equilibrium is established between the two gates and DG-LTFET potential overlaps LTFET potential since W_{rk_gate2} = W_{rk_LTFET}. With W_{rk_gate1} < W_{rk_gate2}, the increased potential in $R_{nonoffset}$ reduces $V_{th\_Rnonoffset}$. If $W_{rk\_gate1/2}$ are appropriately tuned with W_{rk_gate1} < W_{rk_gate2}, the condition $V_{th\_Rnonoffset}$ < $V_{th\_Roffset}$ = 0.17 V can be achieved. Because W_{rk_gate2} = W_{rk_LTFET} = 4.72 eV, $V_{th\_Roffset}$ (in the DG-LTFET) is equal to $V_{th\_Roffset}$ (in the LTFET).

Figure 6a–c show I_{ds}-V_{gs} characteristics at different W_{rk_gate1}, SS, and I_{ON}/I_{OFF} of the DG-LTFET with constant W_{rk_gate2} = W_{rk_LTFET} = 4.72 eV for all DG-LTFET, respectively. Also shown for the reference is the I_{ds}-V_{gs} characteristics of the LTFET (black squares). I_{ON} is extracted at V_{gs} = 0.7 V, and I_{OFF} is defined as I_{ds} = 10^{-17} A. With W_{rk_gate1} = 4.675 eV (red circles), the $V_{th\_Rnonoffset}$ is reduced to 0.189 V. Compared with the LTFET, $R_{nonoffset}$ now turns on earlier in the subthreshold region, along with R_{offset}. Since the BTBT is more efficient in $R_{nonoffset}$ (Figure 4a) as compared to R_{offset}, I_{ds} increases more rapidly within the subthreshold region.

Hence, just at the transition point, where $R_{nonoffset}$ turns on (V_{gs} ~0.189), a kink appears in the I_{ds}-V_{gs} curve. With W_{rk_gate1} = 4.65 eV (green triangles), $V_{th\_Rnonoffset}$ is reduced to V_{gs} = 0.167 V and the condition $V_{th\_Rnonoffset}$ < $V_{th\_Roffset}$ is achieved, and DG-LTFET exhibits a remarkable SS with values less than 10 mV/dec as seen in Figure 6b. With W_{rk_gate1} = 4.625 eV (blue stars), $V_{th\_Rnonoffset}$ reduces further to 0.1448 V, which is < $V_{th\_Roffset}$. If $V_{th\_Rnonoffset}$ < $V_{th\_Roffset}$ is established, then any increase in $V_{th\_Roffset}$ − $V_{th\_Rnonoffset}$ simply shifts the I_{ds}-V_{gs} to the left without any change in SS as shown by the blue stars (W_{rk_gate1} = 4.625 eV) and orange diamonds (W_{rk_gate1} = 4.5 eV) in Figure 6a,b, respectively. An improvement of ~16% is observed in the I_{ON}/I_{OFF} of the DG-LTFET (with W_{rk_gate1} = 4.625 eV) over the LTFET.

Figure 7a shows the G_{tun} contour plot of DG-LTFET at a V_{gs} (= 0.172 V) bias needed to achieve an equivalent I_{ds} of 10^{-13} A in DG-LTFET with W_{rk_gate1} = 4.65 eV. Figure 7b shows the contour plot extracted from Figure 7a. For reference, Figure 7b also shows that G_{tun} needed to generate an equivalent amount of I_{ds} in the LTFET (at a V_{gs} bias of 0.21 V, Figure 3b). As can be seen in Figure 7b, the LTFET needs contribution only from R_{offset}, but generating the same amount of I_{ds} DG-LTFET depends heavily on $R_{nonoffset}$ with some contribution from R_{offset}. Because G_{tun} in $R_{nonoffset}$ is more efficient (Figure 4a), as the V_{gs} bias increases, G_{tun} increases exponentially in a much larger area in $R_{nonoffset}$, which results in the DG-LTFET exhibiting a much steeper subthreshold swing, while the LTFET continues to depend only on the inefficient BTBT in R_{offset} until around $V_{th\_Rnonoffset}$ = 0.24 V.

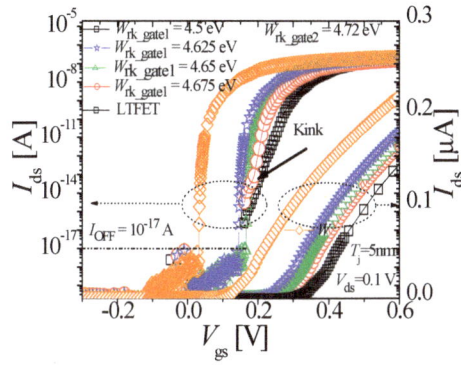

(a)

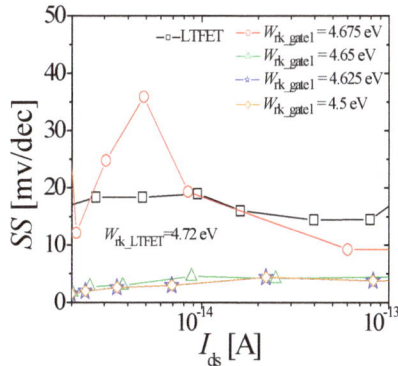

(b)

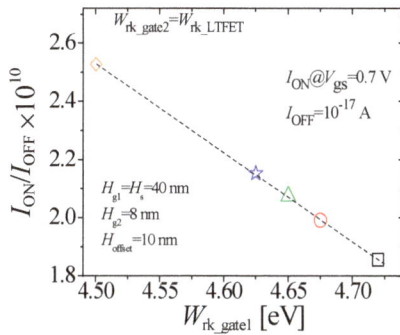

(c)

Figure 6. (a) I_{ds}-V_{gs} characteristics of DG-LTFET with different W_{rk_gate1}s and fixed $W_{rk\_gate2} = W_{rk\_LTFET}$. Also shown are I_{ds}-V_{gs} characteristics of the LTFET (black squares). (b) SS extracted from I_{ds}-V_{gs} characteristics in Figure 8a. (c) I_{ON}/I_{OFF} ratio extracted from I_{ds}-V_{gs} characteristics in Figure 8a. Red circles: W_{rk_gate1} = 4.675 eV; green triangles: W_{rk_gate1} = 4.65 eV; blue stars: W_{rk_gate1} = 4.625 eV; orange diamonds: W_{rk_gate1} = 4.5 eV.

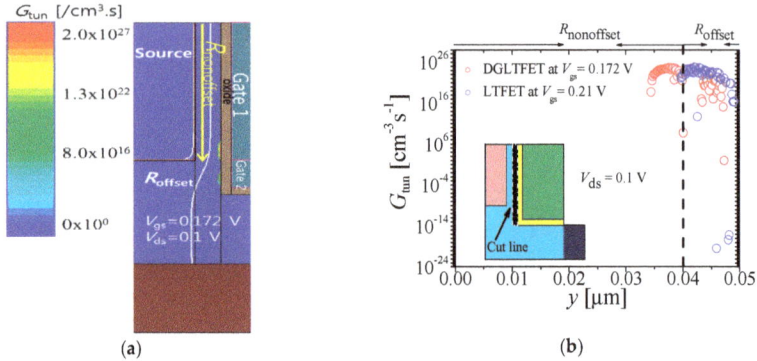

Figure 7. (a) G_{tun} contour plot of DG-LTFET at $V_{gs} = 0.172$ V, which is needed to generate $I_{ds} = 10^{-13}$ A and (b) G_{tun} extracted from (a) (red symbols). Also shown for reference is G_{tun} (blue symbols) of the LTFET at a V_{gs} bias needed to generate $I_{ds} = 10^{-13}$ A. In (a), yellow arrow indicates the height of $R_{nonoffset}$.

3.2. Device Optimization

To optimize device performance, the impact of variations in key parameters including $H_{g1/2}$, H_s/T_j, and N_d was investigated. To investigate the impact of $H_{g1/2}$ values, I_{ds}-V_{gs} characteristics for the DG-LTFET at different H_{g1} and $H_{g2} = H_{nonoffset} - H_{g1} + (H_{offset} - t_{ox})$ with fixed $W_{rk\_gate1} = 4.5$ eV and $W_{rk\_gate2} = W_{rk\_LTFET}$, $H_s = H_{nonoffset} = 40$ nm, $H_{offset} = 10$ nm, and $T_j = 5$ nm is presented in Figure 8. It can be seen that I_{ds} is independent of $H_{g1/2}$.

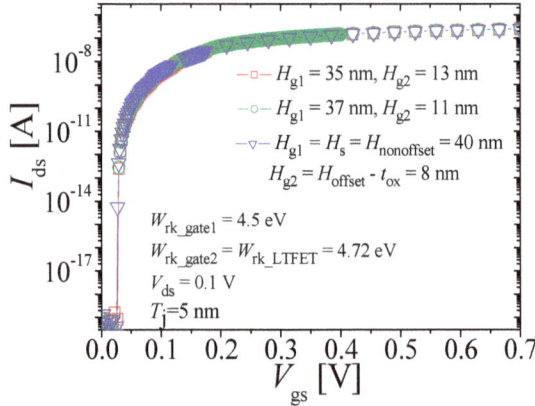

Figure 8. I_{ds}-V_{gs} characteristics for several $H_{g1/2}$s with $W_{rk\_gate1/2} = 4.5$ eV and $W_{rk\_gate2} = W_{rk\_LTFET}$. Red squares, green circles, and blue triangles: $H_{g1} = 35$, 37, and 40 nm, respectively.

Next, to investigate the effect of T_j on device performance, I_{ds}-V_{gs} characteristics, SS, and I_{ON}/I_{OFF} of DG-LTFET are presented for different T_j with fixed $W_{rk\_gate1} = 4.5$ eV and $W_{rk\_gate2} = W_{rk\_LTFET}$, $H_{g1} = H_{nonoffset} = 40$ nm, $H_{offset} = 10$ nm, and $H_{g2} = H_{nonoffset} - H_{g1} + (H_{offset} - t_{ox}) = 8$ nm in Figure 9a–c, respectively. It was found that the increasing T_j results in a degradation of the I_{ON}/I_{OFF} ratio. It is simply because of the increase in BTBT path length with the increase in T_j. The T_j of 5 nm was found to be optimum in this work as any further reduction will bring significant quantum confinement effect into play, which is well known to degrade device performance [4,5,13–15].

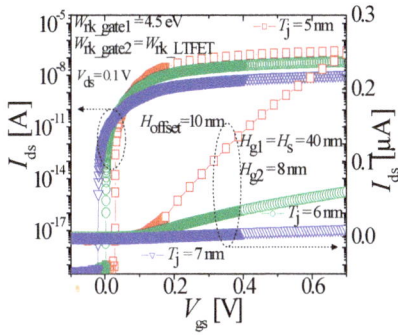

(a)

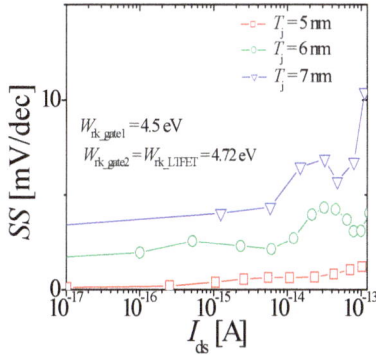

(b)

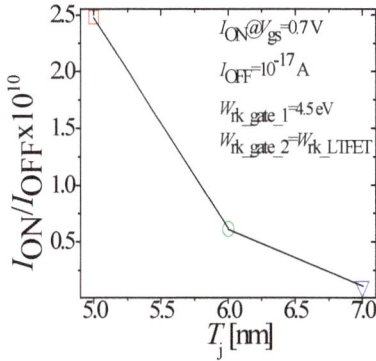

(c)

Figure 9. (a) I_{ds}-V_{gs} characteristics of the DG-LTFET with different T_j and fixed W_{rk_gate1} = 4.5 eV, W_{rk_gate2} = W_{rk_LTFET}, and H_{g1} = H_s = $H_{nonoffset}$ = 40 nm, H_{g2} = H_{offset} (10 nm) − t_{ox} = 8 nm. (b) The SS of I_{ds}-V_{gs} shown in Figure 8a. (c) I_{ON}/I_{OFF} ratio of I_{ds}-V_{gs} characteristics shown in Figure 8a. Red squares, green circles, and blue triangles: T_j = 5, 6 and 7 nm, respectively.

Next, the impact of varying H_s was investigated. I_{ds}-V_{gs} characteristics of the DG-LTFET for several H_s with fixed W_{rk_gate1} = 4.5 eV and W_{rk_gate2} = W_{rk_LTFET}, H_{g1} = H_s = $H_{nonoffset}$, H_{g2} = $H_{nonoffset}$ − H_{g1} + (H_{offset} − t_{ox}) = 8 nm, and T_j = 5 nm is presented in Figure 10. By maintaining H_{g1} = H_s, H_{offset} = 10 nm, and H_{g2} = 8 nm, the electric field vector distribution within the DG-LTFET remains the same as H_s is varied, and the BTBT area simply scales with H_s. An increase (decrease) in the BTBT area with H_s simply results in an increased (decreased) I_{ON}/I_{OFF} ratio as shown in Figure 10b with no change in SS, as evident from Figure 10a.

Finally, the ambipolar current of the DG-LTFET is discussed. Ambipolar I_{ds} of TFET depends on the drain-channel junction. In the DG-LTFET, the drain-channel junction is controlled by gate2 with W_{rk_gate2} = W_{rk_LTFET}. With the same workfunction, the electrostatics of the drain-channel junction in the DG-LTFET is exactly the same as that in the LTFET. Figure 11a shows ambipolar I_{ds} of the DG-LTFET compared with the LTFET. Any change in W_{rk_gate1} in the DG-LTFET does not affect the drain-channel junction. The same argument applies for any other design parameter variation in DG-LTFET including H_s, $H_{g1/2}$, and T_j; that is, as long as the electrostatics of the drain-channel junction remains unaffected, the DG-LTFET will exhibit an equivalent ambipolar I_{ds} as the LTFET. Further, the impact of N_d on ambipolar I_{ds} was considered. Different N_d values were considered for a DG-LTFET with W_{rk_gate1} = 4.5 eV and W_{rk_gate2} = W_{rk_LTFET}, H_{g1} = $H_{nonoffset}$ = 40 nm, H_{g2} = H_{offset} − t_{ox} = 8 nm, and T_j = 5 nm, and the results are shown in Figure 11b. A drain doping level of 10^{18} cm^{-3} was found to suppress ambipolar I_{ds} appreciably without affecting the I_{ON}.

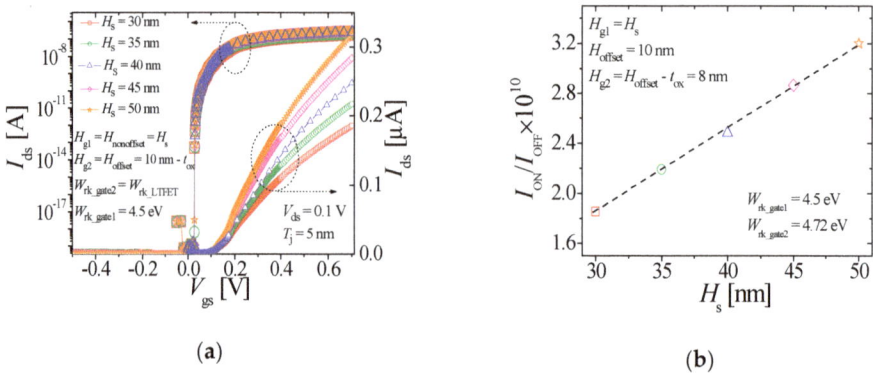

Figure 10. I_{ds}-V_{gs} characteristics of DG-LTFET with different H_s, fixed W_{rk_gate1} = 4.5 eV, W_{rk_gate2} = W_{rk_LTFET}, and H_{g1} = H_s = $H_{nonoffset}$, H_{g2} = H_{offset} (=10 nm) − t_{ox} = 8 nm. (**b**) An I_{ON}/I_{OFF} ratio of I_{ds}-V_{gs} characteristics shown in (**a**). Red squares, green circles, blue triangles, magenta diamonds, and orange stars: H_s = 30, 35, 40, 45, and 50 nm, respectively.

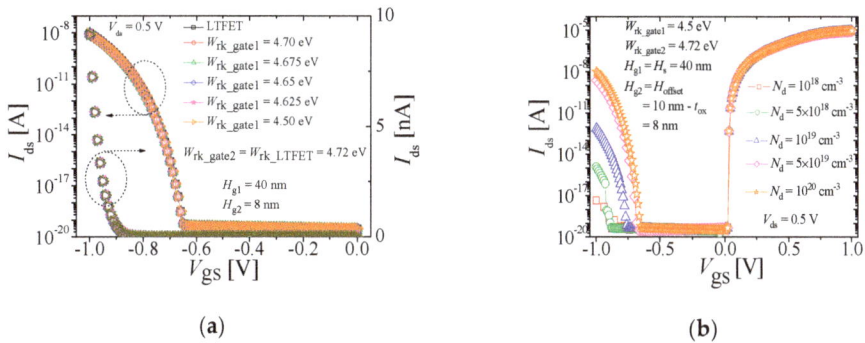

(a)   (b)

Figure 11. (a) I_{ds}-V_{gs} characteristics of DG-LTFET at V_{ds} = 0.5 V with different W_{rk_gate1} and W_{rk_gate2} = W_{rk_LTFET}, H_{g1} = H_{offset} = 10 nm, H_{g2} = 8 nm, T_j = 5 nm and N_d = 10^{20} cm^{-3}. Red circles, green triangles, blue diamonds, magenta stars, and orange right triangles: W_{rk_gate1} = 4.7, 4.675, 4.65, 4.625, and 4.5 eV. (b) DG-LTFET I_{ds} with different N_d. N_d = 10^{18} cm^{-3} demonstrates almost negligible ambipolar I_{ds}. Red squares, green circles, blue triangles, magenta diamonds, and orange stars: N_d = 10^{18}, 5×10^{18}, 10^{19}, 5×10^{19}, and 10^{20} cm^{-3}.

4. Conclusions

The device physics of the LTFET was investigated. It was found that a large part of the subthreshold region is dominated by the parasitic, lateral, 2D BTBT from the source to R_{offset} with a lower G_{tun}. The more efficient 1D BTBT from the source to $R_{nonoffset}$, with a higher G_{tun} takes place at a higher bias in the subthreshold region. In other words, the condition, that is, $V_{th\_Rnonoffset} > V_{th\_Roffset}$, exists in the LTFET. With $R_{nonoffset}$ not conducting the device does not utilize its channel fully during the subthreshold region. A new type of device based on the LTFET was introduced in this work. The device uses a dual gate structure with $W_{rk\_gate1} < W_{rk\_gate2}$. This increases the potential in $R_{nonoffset}$ and lowers $V_{th\_Rnonoffset}$. The DG-LTFET reverses the threshold condition of the LTFET, that is, it lowers $V_{th\_Rnonoffset}$ and makes it $<V_{th\_Roffset}$. $R_{nonoffset}$ with higher G_{tun} turns on earlier than R_{offset} in the subthreshold region in the DG-LTFET and the device exhibits an SS of less than 10 mV/dec. It was found that W_{rk_gate1} in the DG-LTFET needs to be sufficiently less than W_{rk_gate2} to achieve the sub 10 mv/dec SS. It was found that I_{ds} and SS are independent of $H_{g1/2}$. The DG-LTFET was further evaluated for different device dimensions including T_j and H_s while maintaining the electric field vector distribution equivalent. I_{ds} decreases with an increase in T_j and scales with H_s. The N_d value of 10^{18} cm^{-3} was found to appreciably reduce ambipolar I_{ds}. With the results presented in this work, the DG-LTFET could be considered as a viable potential replacement to conventional MOSFET and 3D integrations [16].

Author Contributions: Conceptualization, F.N. and Y.S.Y.; methodology, F.N. and Y.S.Y.; investigation, F.N. and Y.S.Y.; data curation, F.N.; writing—original draft preparation, F.N.; writing—review and editing, F.N., and Y.S.Y.; supervision, Y.S.Y.; project administration, Y.S.Y.; funding acquisition, Y.S.Y.

Funding: This research was funded by Ministry of Trade, Industry & Energy (MOTIE), project number 10054888 and Korea Semiconductor Research Consortium (KSRC) support program for the development of future semiconductor devices.

Acknowledgments: This work was supported by IDEC (EDA tool).

Conflicts of Interest: The authors declare no conflict of interest. The funders had no role in the design of the study; in the collection, analyses, or interpretation of data; in the writing of the manuscript; or in the decision to publish the results.

References

1. Avci, U.E.; Morris, D.H. Tunnel field-effect transistors. *IEEE J. Electron. Devices Soc.* **2015**, *3*, 88–95. [CrossRef]
2. Ionescu, A.M.; Riel, H. Tunnel field-effect transistors as energy efficient electronic switches. *Nature* **2011**, *479*, 329–337. [CrossRef] [PubMed]
3. Kim, S.W.; Kim, J.H.; Liu, T.K.; Choi, W.Y.; Park, B. Demonstration of L-shaped tunnel field-effect transistors. *IEEE Trans. Electron. Devices* **2016**, *63*, 1774–1778. [CrossRef]
4. Yang, Z. Tunnel field-effect transistor with an L-shaped gate. *IEEE Electron. Device Lett.* **2016**, *4*, 839–842. [CrossRef]
5. Imenabadi, R.M.; Saremi, M.; Vandenberghe, W.G. A novel PNPN-like Z-shaped tunnel field-effect transistor with improved ambipolar behavior and RF performance. *IEEE. Trans. Electron. Devices* **2017**, *64*, 4752–4758. [CrossRef]
6. Kim, S.W.; Choi, W.Y.; Sun, M.C.; Park, B.G. Investigation on the corner effect of L-shaped tunneling field-effect transistors and their fabrication method. *J. Nanosci. Nanotechnol.* **2016**, *9*, 6376–6381. [CrossRef]
7. *Sentaurus User Manual*; Version L-2016.03 March; Synopsys, Inc.: Mountain View, CA, USA, 2016.
8. Kao, K.H.; Verhulst, A.S.; Vandenberghe, W.G.; Soree, B.; Groeseneken, G.; Meyer, K.D. Direct and indirect band-to-band-tunneling in germanium-based TFETs. *IEEE Trans. Electron. Devices* **2012**, *59*, 292–301. [CrossRef]
9. Saxena, R.S.; Kumar, M.J. Dual-material gate technique for enhanced transconductance and breakdown voltage of trench power MOSFETs. *IEEE. Trans. Electron. Devices* **2009**, *56*, 517–522. [CrossRef]
10. Long, W.; Ou, H.; Kuo, J.-M.; Chin, K.K. Dual-material gate (DMG) Field Effect Transistor. *IEEE. Trans. Electron. Devices* **1999**, *46*, 865–870. [CrossRef]
11. Polishchuk, I.; Ranade, P.; King, T.-J.; Hu, C. Dual work function metal gate CMOS technology using metal interdiffusion. *IEEE Electron. Device Lett.* **2001**, *9*, 444–446. [CrossRef]
12. Sze, S.M.; Kwok, K.N. *Physics of Semiconductor Devices*, 3rd ed.; John Wiley & Sons: Hoboken, NJ, USA, 2006; ISBN 9780471143239.
13. Walke, A.M.; Verhulst, A.S.; Vandooren, A.; Verreck, D.; Simeon, E.; Rao, V.R.; Groeseneken, G.; Collaert, N.; Thean, A.V.Y. Part I: Impact of field-induced quantum confinement on subthreshold swing behavior of line TFETs. *IEEE Trans. Electron. Devices* **2013**, *60*, 4057–4064. [CrossRef]
14. Padilla, J.L.; Gamiz, F.; Godoy, A. A simple approach to quantum confinement in tunneling field-effect transistors. *IEEE Electron. Device Lett.* **2012**, *33*, 1342–1344. [CrossRef]
15. Padilla, J.L.; Alper, C.; Gamiz, F.; Ionescu, A.M. Assessment of field-induced quantum confinement in heterogate germanium electron-hole bilayer tunnel-field transistor. *Appl. Phys. Lett.* **2014**, *105*, 082108. [CrossRef]
16. Lim, S.K. Bringing 3D ICs to Aerospace: Needs for Design Tools and Methodologies. *J. Inf. Commun. Converg. Eng.* **2017**, *15*, 117–122. [CrossRef]

© 2018 by the authors. Licensee MDPI, Basel, Switzerland. This article is an open access article distributed under the terms and conditions of the Creative Commons Attribution (CC BY) license (http://creativecommons.org/licenses/by/4.0/).

electronics

MDPI

Article

Simulation Analysis in Sub-0.1 μm for Partial Isolation Field-Effect Transistors

Young Kwon Kim [1], Jin Sung Lee [1], Geon Kim [1], Taesik Park [2,*], HuiJung Kim [3], Young Pyo Cho [4], Young June Park [3] and Myoung Jin Lee [1,*]

[1] School of Electronics and Computer Engineering, Chonnam National University, Gwangju 500-757, Korea; yyong13@naver.com (Y.K.K.); lsc5176@naver.co (J.S.L.); rjsdlchd@naver.com (G.K.)
[2] Department of Electrical and Control Engineering, Mokpo National University, Jeollanam-do 534-729, Korea
[3] School of Electrical Engineering, Seoul National University, Seoul 151-742, Korea; khj95@snu.ac.kr (H.J.K.); ypark@snu.ac.kr (Y.J.P.)
[4] The KEPCO Research Institute, Daejeon 305-760, Korea; yp.zo@kepco.co.kr
* Correspondence: tspark@mokpo.ac.kr (T.P.); mjlee@jnu.ac.kr (M.J.L.); Tel.: +82-62-530-1810 (M.J.L.)

Received: 20 September 2018; Accepted: 28 September 2018; Published: 2 October 2018

Abstract: In this paper, we extensively analyzed the drain-induced barrier lowering (DIBL) and leakage current characteristics of the proposed partial isolation field-effect transistor (PiFET) structure. We then compared the PiFET with the conventional planar metal-oxide semiconductor field-effect transistor (MOSFET) and silicon on insulator (SOI) structures, even though they have the same doping profile. Two major features of the PiFET are potential condensation and potential modulation by a buried insulator. The potential modulation near the drain region can control the electric field in the overlapped region of the drain and gate, because it causes a high gate-fringing field. Therefore, we suggest guidelines with respect to the optimal PiFET structure.

Keywords: drain-induced barrier lowering (DIBL); gate-induced drain leakage (GIDL); silicon on insulator (SOI)

1. Introduction

As the design rule of the dynamic random-access memory (DRAM) cell shrinks, it has become very difficult to obtain sufficient data retention times, because of the short channel effect and leakage current [1]. One of the solutions for these issues is the silicon on insulator (SOI) metal-oxide semiconductor field-effect transistor (MOSFET) [2–4]. However, it suffers from a critical low threshold voltage, a back-gate interface issue, a floating body effect, and a high price, even though it shows low power consumption, a self-limited shallow junction, and an improved drain-induced barrier lowering (DIBL) [5–9]. Therefore, a partial isolation field-effect transistor (PiFET) structure has been proposed. The PiFET is a type of transistor in which, unlike in the SOI MOSFET, the buried insulator is penetrated to a certain depth in the channel direction under the drain doping region, and does not deplete the entire channel [10,11]. Therefore, the kink effect due to the floating body effect, which is one of the most significant weaknesses of the SOI MOSFET, is structurally completely blocked. In this paper, we analyzed the various types of PiFET structures according to the slopes, dielectric constants, and silicon film thicknesses. We also defined two effects that determine the performance of the PiFET via the potential contour map near the drain. We considered the following factors: off-current, short-channel effect (SCE), on-current, threshold voltage, and $G_{m.max}$, as they are the most important determinants of DRAM cell performance. The mechanism of improving DIBL and the gate-induced drain leakage (GIDL) characteristics is discussed in the subsequent sections, based on a technology computer-aided design (TCAD) device simulation. The simulator is well tuned to predict drain leakage current, such as the GIDL component, by applying the Hurkx band-to-band tunneling model [12,13]. From the result, we propose a PiFET structure to achieve better DIBL characteristics, V_{TH} controllability, and a low off-current for DRAM cell operation.

2. Experimental Methods

In this study, the PiFET structure has a channel length of 0.1 µm and an oxide thickness of 6 nm, according to the buried insulator material, with a relative dielectric constant of 3.9 (SiO$_2$), 7.5 (Si$_3$N$_4$), and 25 (HfO$_2$). The S/D peak doping and uniform body doping concentration of the PiFETs, planar MOSFET, and SOI MOSFET are 1.1×10^{19} cm^{-3} and 7×10^{17} cm^{-3}, respectively. To analyze the electrical characteristics of the PiFET, a sentaurus TCAD device simulator was used. Figure 1 shows the I$_{DS}$-V$_{GS}$ characteristics for the V$_{DS}$ of 0.1 [V] and 1.6 [V] in the planar MOSFET and the PiFET. The simulated drain current and the measured drain current of the PiFET structures, which are fabricated with 0.1 µm DRAM technology, are well fitted [10].

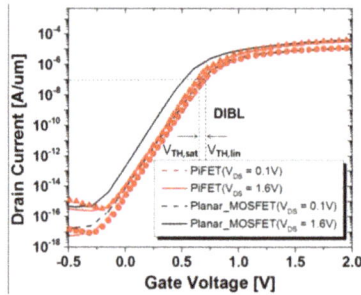

Figure 1. I$_{DS}$-V$_{GS}$ characteristics for a V$_{DS}$ of 0.1 [V] and 1.6 [V], and the definition of DIBL. The red symbol shows the measured drain current of the fabricated PiFET structure with 0.1 µm DRAM technology.

The threshold voltage is estimated based on the constant current (10^{-7} A/µm) method. The V_{TH} in the PiFET, according to V$_{DS}$, is higher than that in the planar MOSFET, meaning the PiFET has the better DIBL characteristics. We also found that the PiFET has a lower off-current than that in the planar MOSFET when the V$_{GS}$ = 0 V or less.

3. Results and Discussion

3.1. Five Slope

The thickness of the buried insulator of an SOI MOSFET is limited for improved electrical properties, but the buried insulator thickness of the PiFET can be easily controlled for better electrical properties [8,9]. As shown in Figure 2, buried insulator types for PiFET were divided into five slopes, in order to investigate the electrical characteristics according to the lateral encroachment levels in the channel direction. The lateral encroachment of the buried insulator of slope 5 is shorter in the group than in the other slopes. Figure 3 shows the DIBL and the threshold voltage according to the PiFET structures having various types of slopes compared to the planar MOSFET, and the SOI MOSFET at V$_{GS}$ = 0 V and V$_{DS}$ = 1.6 V. The PiFET having slope 5 exhibits better DIBL and a higher threshold voltage than the conventional MOSFETs and the PiFETs, which have gentle slopes 1–4, since the electric field near the source region is smaller. These results are also consistent with those of the planar MOSFETs and SOI MOSFETs. The smaller electric field is formed near the source region in the PiFET having slope 5, mainly because most of the electric field between the drain and the source condenses into the buried insulator, at the bottom of the drain region. This potential condensation phenomenon causes a large threshold voltage, even under low drain voltage conditions. Thus, to maintain a high V$_{TH}$ and DIBL, we found that the buried insulator penetrating into the drain region should be minimized below the channel region. Also, the off-currents, according to slope type of the PiFET, remained almost constant, because the electric field near the overlapped region of the gate and the drain region remains almost constant regardless of the slope type.

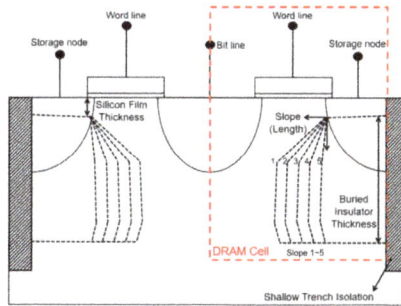

Figure 2. Several types of PiFET structures, according to the silicon film thickness and the slope.

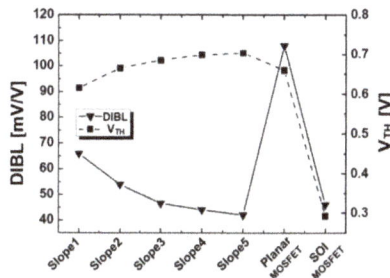

Figure 3. Simulation results for DIBL and threshold voltage for the planar MOSFET, the SOI MOSFET, and PiFETs with a silicon film thickness of 15 nm and the buried insulators (slopes 1–5) of the SiO_2 material.

3.2. Various Dielectric Constants

Figure 4 shows the simulation results of the potential contour near the drain region for the PiFETs and the planar MOSFET, when $V_{GS} = 0$ V and $V_{DS} = 1.6$ V. The electrical characteristics, according to the dielectric constants of the buried insulator, could be explained by potential condensation and potential modulation phenomenon occurring near the drain region [14]. In the potential contours shown in Figure 4b–d, the potential modulation means that the starting point of the potential drop is shifted from an n^- doped region to an n^+ doped region, near the drain region. The potential modulation occurs near the drain region because the penetration of the buried insulator induces a potential drop at a high-doped drain region. Namely, since the potential drop in the buried insulator of the PiFETs remains relatively constant, it results in a strong voltage change near the contact above the buried insulator. Therefore, the depletion region is formed, even in the n^+ doped region, where it is difficult for the voltage change to occur in the planar MOSFET. As shown in Figure 4b–d, as the dielectric constant increases, a high drain potential edge is also pushed forward and moves toward the n^+ drain region, due to an interaction increase of the boundary surface between the silicon and buried insulator. In other words, the high dielectric constant of the buried insulator deepens the potential modulation. It is found that the potential modulation results in an expanding depletion region near the n^+ doped region by a high gate-fringing field. Although a high gate fringing field can cause unintended fringing field-induced barrier-lowering (FIBL) effects, reducing the channel-to-drain barrier, it can reduce the electric field across the gate oxide between the gate and drain regions [15,16]. As shown in

Figure 5a, the peak of the vertical field of the overlapped region between the drain and gate of the PiFETs decreases, and is shifted from an n^- doped region to an n^+ doped region, due to the increased gate-fringing field. This means that the band-to-band tunneling (BTBT) in the overlapped region between the gate and the drain is reduced [17–20]. Therefore, the PiFET structure can effectively reduce the off-current by GIDL. Meanwhile, Figure 5b shows the potential drop of the PiFETs and planar MOSFET at the channel surface by potential modulation. A deep potential modulation phenomenon in the drain region can consequently contribute to improving the DIBL characteristics, by effectively blocking the electric field transmitted to the source.

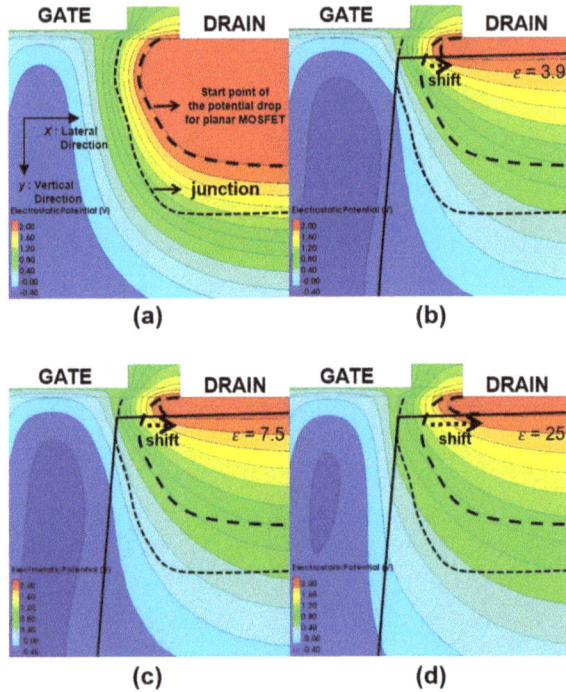

Figure 4. Two-dimensional potential contour profiles (at $V_{GS} = 0.0$ V, $V_{DS} = 1.6$ V) of 0.2 steps near the drain region in (**a**) planar MOSFET, (**b**–**d**) PiFETs with slope 5 and silicon film thickness of 15 nm. (**b**) k = 3.9, (**c**) k = 7.5, (**d**) k = 25.

Figure 6 shows the DIBL and the threshold voltage dependence on the dielectric constant of the buried insulator with slope 5. The PiFET, with a high dielectric constant of 25, has a relatively small electric field inside the buried insulator, thereby reducing the potential drop occurring near the buried insulator formed under the drain. This means that a large electric field penetrates the channel region. That is, as the dielectric constant of the buried insulator with a constant slope increases, the potential condensation phenomenon is weakened. Although the potential modulation phenomenon is strengthened in the PiFET structure of a buried insulator having a higher dielectric constant, the potential condensation phenomenon is the most crucial factor for improving the DIBL in the PiFET structure; it is even more important in PiFET structures with low dielectric constants. Therefore, the PiFET with a high dielectric constant exhibits DIBL and V_{TH} characteristics that are less improved than those of the PiFET having a low dielectric constant; however, the PiFETs exhibit DIBL and a high V_{TH} superior to conventional MOSFETs. Figure 7 shows the drain current in the subthreshold region of the PiFETs and the planar MOSFET. Tunneling of the electrons from gate to drain occurs with difficulty, due to the high gate-fringing field, as shown in Figure 5a. The PiFET having a dielectric constant of 25 exhibits a small drain current compared with the PiFETs with low dielectric constant and planar MOSFET in the region where $V_{GS} = 0$ V or less. The results in Figures 6 and 7 show that PiFET structures can provide an improved leakage current characterization mechanism for DRAM cell operation. This is because the leakage current characteristics, such as DIBL and GIDL, are likely to be degraded due to the DRAM cell operation when the storage node of the DRAM cell is in a high voltage state.

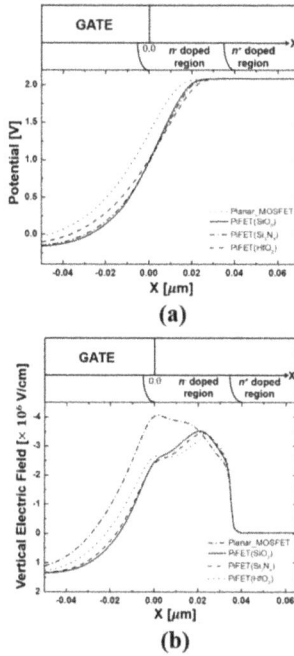

Figure 5. (a) Vertical electric field distribution and (b) lateral potential distribution on the silicon surface near the drain region for the planar MOSFET and the PiFETs when $V_{GS} = -0.5$ V and $V_{DS} = 1.6$ V.

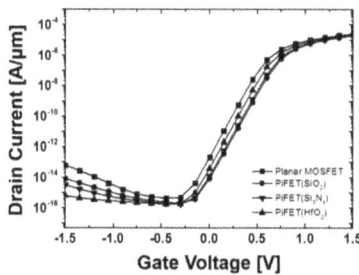

Figure 6. I_{DS}-V_{GS} indicating off-current characteristics for the planar MOSFET, the PiFET with a silicon film thickness of 15 nm, and the buried insulators of the SiO_2, Si_3N_4, and HfO_2 materials.

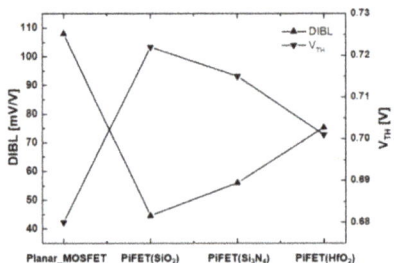

Figure 7. Simulation results for the DIBL and the threshold voltage for the planar MOSFET and PiFETs with a silicon film thickness of 15 nm and the buried insulators of SiO_2, Si_3N_4, and HfO_2 materials.

Figure 8 shows the V_{DS}-I_{DS} characteristics according to the dielectric constant of PiFETs. Even with the high voltage in the drain region, the potential modulation and the potential condensation phenomenon in the PiFET structure minimizes the reduction of the effective channel length. Consequently, the saturation current is kept constant compared to the planar MOSFET, and the PiFET having a low dielectric constant of 3.9 has the most constant saturation current. Figure 9 shows the DIBL characteristics, according to the gate channel lengths of PiFET, SOI MOSFET, and the planar MOSFET. As the gate channel length is reduced, the planar MOSFET and the PiFET with a dielectric constant of 25 show a dramatically increased slope with increasing DIBL, while the PiFET with a dielectric constant of 3.9 maintains a DIBL of less than 150.

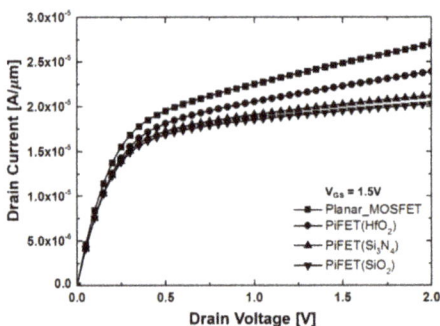

Figure 8. I_{DS}-V_{DS} characteristics for the planar MOSFET; the PiFET with a silicon film thickness of 15 nm; and the buried insulators of the SiO_2, Si_3N_4, and HfO_2 material.

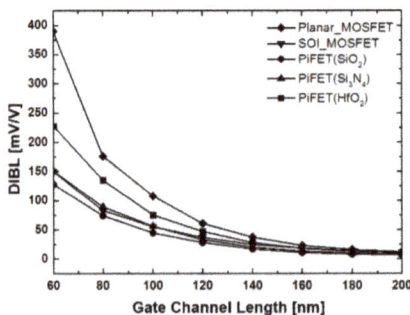

Figure 9. Simulation results of the DIBL characteristic of planar MOSFET, SOI MOSFET, and PiFETs, each with a silicon film thickness of 15 nm, as well as the buried insulators of the SiO_2, Si_3N_4, and HfO_2 materials.

3.3. Silicon Film Thickness

Figure 10 shows the DIBL and V_{TH} characteristics for the PiFET, having a buried insulator of slope 5 with a dielectric constant of 3.9, according to the silicon film thickness. While the SOI MOSFET exhibits improved DIBL characteristics, the threshold voltage is drastically reduced due to the thinner silicon film thickness, which reduces the fully depleted channel region area. However, the PiFET with a thin silicon film exhibits improved characteristics in both the DIBL and the threshold voltage without a fully depleting channel region as the silicon film thickness is reduced. As shown in Figure 11a,b, these improved characteristics also occur because the starting point of the potential drop shifts to the deep n^+ doped region when the silicon film thickness is 10 nm, compared to when it is 20 nm. These results indicate that the thin silicon film thickness of PiFET can also induce the potential modulation at the n^+ doped region immediately above the buried insulator. Therefore, a large potential drop occurs near the drain region, and the peak vertical electric field decreases near the overlapped region between the drain and gate region, resulting from the high gate-fringing field in the case of thin silicon film thickness, as shown in Figure 12a,b. We thus found that it is possible to enhance the potential modulation phenomenon without a buried insulator with a high dielectric constant, by maintaining the thin silicon film thickness. Figure 13 shows the off-current and $G_{m.max}$, according to the silicon film thickness of the PiFET. Actually, the on-current of the PiFETs is almost constant, because the PiFET structure is not related to the area of the source region, although a small drain region limited by the silicon film thickness can slightly reduce the on-current. However, as the silicon film thickness decreases, the off-current sharply decreases. We therefore do not need to seriously consider the $G_{m.max}$ reduction problem.

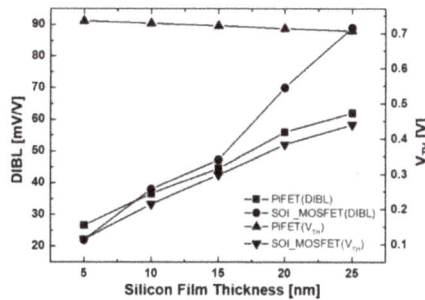

Figure 10. DIBL and threshold voltage dependence of silicon film thickness on the SOI MOSFET and the PiFET with slope 5.

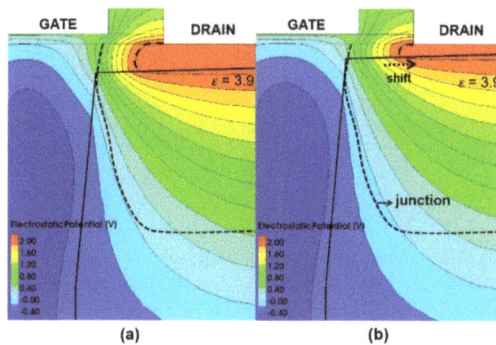

Figure 11. Two-dimensional potential contour profile (at V_{GS} = 0.0 V, V_{DS} = 1.6 V) of the 0.2 step near the drain region in (**a**) PiFET (silicon film thickness = 20 nm) and (**b**) PiFET (silicon film thickness = 10 nm) with slope 5.

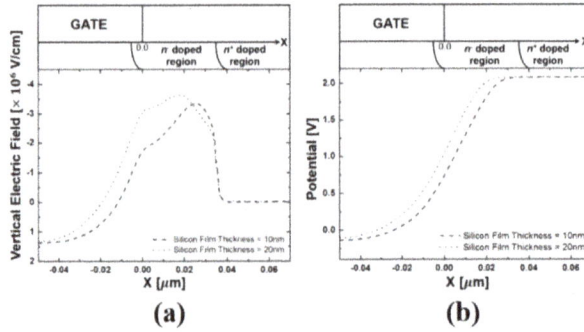

Figure 12. (**a**) Vertical electric field distribution and (**b**) lateral potential distribution on the silicon surface near the drain region for the PiFETs when $V_{GS} = -0.5$ V and $V_{DS} = 1.6$ V.

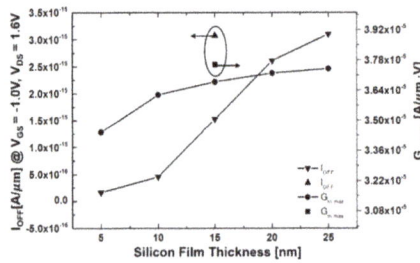

Figure 13. I_{OFF} and $G_{m.max}$ dependence of silicon film thickness on the PiFET with slope 5.

4. Conclusions

From the simulation analysis of partial isolation Field-Effect Transistor (PiFET), we found that the potential condensation and potential modulation phenomenon near the drain region caused by the buried insulator considerably improve the DIBL characteristics. In particular, we made the novel discovery that the gate-fringing field effect caused by the potential modulation phenomenon improved the GIDL characteristic by suppressing the BTBT component with a low vertical electric field. Therefore, the PiFET structure is a promising device for improving the properties in the memory cell when designing with precise consideration of the silicon film thicknesses, as well as the shape of the slope of the buried insulator. In addition, the optimized dielectric materials can be chosen as low dielectric constant materials.

Author Contributions: Conceptualization, Y.K.K.; Visualization, J.S.L. and G.K.; Resources, Y.P.C., H.J.K. and Y.J.P; Supervision, T.P. and M.J.L.

Funding: This research was supported in part by the Korea Electric Power Corporation (Grant number: R17XA05-78), in part by the National Research Foundation of Korea (NRF) grant funded by the Korea government (MSIT) (Gant number: 2018R1A2B6008216), in part by the R&D Special Zone Development Project (Technology Transfer Business Project: 17GJI003) funded by the Ministry of Science, and the ICT & INNOPOLIS Foundation.

Conflicts of Interest: The authors declare no conflict of interest.

References

1. Chang, L.; Choi, Y.-K.; Ha, D.; Ranade, P.; Xiong, S.; Bokor, J.; Hu, C.; King, T.-J. Extremely scaled silicon nano-CMOS device. *Proc. IEEE* **2003**, *91*, 1860–1873. [CrossRef]
2. Veeraraghavan, S.; Fossum, J.G. Short-Channel Effects in SOI MOSFET's. *IEEE Trans. Electron Devices* **1989**, *36*, 522–528. [CrossRef]

3. Risho, K. Buried Layer Engineering to Reduce the Drain-Induced Barrier Lowering of Sub-0.05 μm SOI-MOSFET. *Jpn. J. Appl. Phys.* **1999**, *38*, 2294–2299.
4. Wang, J.; Kistler, N.; Woo, J.; Viswanathan, C.R. Mobility-Field Behavior of Fully Depleted SO1 MOSFET's. *IEEE Electron Device Lett.* **1994**, *15*, 117–119. [CrossRef]
5. Veeraraghavan, S.; Fossum, J.G. A Physical Short-channel Model for the Thin-Film SOI MOSFET Applicable to Device and Circuit CAD. *IEEE Trans. Electron Devices* **1988**, *35*, 1866–1874. [CrossRef]
6. Koh, R.; Kato, H.; Matsumoto, H. Capacitance Network Model of the Short Channel Effect for 0.1 μm Fully Depleted SOI MOSFET. *Jpn. J. Appl. Phys.* **1996**, *35*, 996–1000. [CrossRef]
7. Kumar, M.J.; Chaudhry, A. Two-Dimensional Analytical Modeling of Fully Depleted DMG SOI MOSFET and Evidence for Diminished SCEs. *IEEE Trans. Electron Devices* **2004**, *51*, 569–574. [CrossRef]
8. Chaudhry, A.; Kumar, M.J. Investigation of the Novel Attributes of a Fully Depleted Dual-Material Gate SOI MOSFET. *IEEE Trans. Electron Devices* **2004**, *51*, 1463–1467. [CrossRef]
9. Joachim, H.O.; Yamaguchi, Y.; Ishikawa, K.; Inoue, Y.; Nishimura, T. Simulation and Two-Dimensional Analytical Modeling of Subthreshold Slope in Ultrathin-Film SOI MOSFET's Down to 0.1 um Gate Length. *IEEE Trans. Electron Devices* **1993**, *40*, 1812–1817. [CrossRef]
10. Lee, M.J.; Cho, J.H.; Lee, S.D.; Ahn, J.H.; Kim, J.W.; Park, S.W.; Park, Y.J.; Min, H.S. Partial SOI Type Isolation for Improvement of DRAM Cell Transistor Characteristics. *IEEE Trans. Electron Devices* **2005**, *26*, 332–334.
11. Kim, Y.K.; Lee, J.S.; Kim, G.; Park, T.; Kim, H.J.; Cho, Y.P.; Park, Y.J.; Lee, M.J. The optimized Partial Insulator Isolation MOSFET (PiFET). *J. Semicond. Technol. Sci.* **2017**, *17*, 729–732. [CrossRef]
12. Jin, S.; Lee, M.J.; Yi, J.H.; Choi, J.H.; Kang, D.G.; Chung, I.Y.; Park, Y.J.; Min, H.S. New Direct Evaluation Method to Obtain the Data Retention Time Distribution of DRAM. *IEEE Trans. Electron Devices* **1992**, *39*, 2244–2350. [CrossRef]
13. Hurkx, G.A.M.; Klaassen, D.B.M.; Knuvers, M.P.G. A New Recombination Model for Device Simulation Including Tunneling. *IEEE Trans. Electron Devices* **2006**, *53*, 331–338. [CrossRef]
14. Jang, E.; Shin, S.; Jung, J.W.; Kim, K.R. Gate induced drain leakage reduction with analysis of gate fringing field effect on high-κ/metal gate CMOS technology. *Jpn. J. Appl. Phys.* **2015**, *54*, 06FG10. [CrossRef]
15. Cheng, B.; Cao, M.; Rao, R.; Inani, A.; Voorde, P.V.; Greene, W.M.; Stork, J.M.; Yu, Z.; Zeitzoff, P.M.; Woo, J.C. The Impact of High-k Gate Dielectrics and Metal Gate Electrodes on Sub-100 nm MOSFET's. *IEEE Trans. Electron Devices* **1999**, *46*, 1537–1544. [CrossRef]
16. Ernst, T.; Tinella, C.; Raynaud, C.; Cristoloveanu, S. Fringing fields in sub-0.1 μm fully depleted SOI MOSFETs: Optimization of the device architecture. *Solid-State Electron.* **2002**, *46*, 373–378. [CrossRef]
17. Mizuno, T.; Kobori, T.; Saitoh, Y.; Sawada, S.; Tanaka, T. Gate-Fringing Field Effects on High Performance in High Dielectric LDD Spacer MOSFET's. *IEEE Trans. Electron Devices* **1992**, *39*, 982–989. [CrossRef]
18. Feng, W.S.; Chan, T.Y.; Hu, C. MOSFET Drain Breakdown Voltage. *IEEE Trans. Electron Devices* **1986**, *7*, 572–574.
19. Chen, J.; Chan, T.Y.; Chen, I.C.; Ko, P.K.; Hu, C. Subbreakdown Drain Leakage Current in MOSFET. *IEEE Trans. Electron Devices* **1987**, *8*, 449–450. [CrossRef]
20. Chen, J.; Assaderaghi, F.; Ko, P.K.; Hu, C. The Enhancement of Gate-Induced-Drain-Leakage (GIDL) Current in Short-Channel SO1 MOSFET and its Application in Measuring Lateral Bipolar Current Gain β. *IEEE Trans. Electron Devices* **1992**, *13*, 515–517.

© 2018 by the authors. Licensee MDPI, Basel, Switzerland. This article is an open access article distributed under the terms and conditions of the Creative Commons Attribution (CC BY) license (http://creativecommons.org/licenses/by/4.0/).

electronics

MDPI

Article

Partial Isolation Type Saddle-FinFET(Pi-FinFET) for Sub-30 nm DRAM Cell Transistors

Young Kwon Kim [1], Jin Sung Lee [1], Geon Kim [1], Taesik Park [2,*], Hui Jung Kim [3],
Young Pyo Cho [4], Young June Park [3] and Myoung Jin Lee [1,*]

[1]   School of Electronics and Computer Engineering, Chonnam National University, Gwangju 500-757, Korea;
      yyong13@naver.com (Y.K.K.); lsc5176@naver.com (J.S.L.); rjsdlchd@naver.com (G.K.)
[2]   Department of Electrical and Control Engineering, Mokpo National University, Jeollanam-do 534-729, Korea
[3]   School of Electrical Engineering, Seoul National University, Seoul 151-742, Korea; khj95@snu.ac.kr (H.J.K.);
      ypark@snu.ac.kr (Y.J.P.)
[4]   The KEPCO Research Institute, Daejeon 305-760, Korea; yp.zo@kepco.co.kr
[*]   Correspondence: tspark@mokpo.ac.kr (T.P.); mjlee@jnu.ac.kr (M.J.L.); Tel.: +82-62-530-1810 (M.J.L.)

Received: 1 November 2018; Accepted: 19 December 2018; Published: 21 December 2018

Abstract: In this paper, we proposed a novel saddle type FinFET (S-FinFET) to effectively solve problems occurring under the capacitor node of a dynamic random-access memory (DRAM) cell and showed how its structure was superior to conventional S-FinFETs in terms of short channel effect (SCE), subthreshold slope (SS), and gate-induced drain leakage (GIDL). The proposed FinFET exhibited four times lower I_{off} than modified S-FinFET, called RFinFET, with more improved drain-induced barrier lowering (DIBL) characteristics, while minimizing I_{on} reduction compared to RFinFET. Our results also confirmed that the proposed device showed improved drain-induced barrier lowering (DIBL) and I_{off} characteristics as gate channel length decreased.

Keywords: gate-induced drain leakage (GIDL); drain-induced barrier lowering (DIBL); recessed channel array transistor (RCAT); on-current (I_{on}); off-current (I_{off}); subthreshold slope (SS); threshold voltage (V_{TH}); saddle FinFET (S-FinFET); potential drop width (PDW); shallow trench isolation (STI); source/drain (S/D)

1. Introduction

With decreasing dynamic random-access memory (DRAM) cell size, a recessed channel array transistor (RCAT) has been proposed to overcome the short channel effect (SCE) of conventional MOSFETs with planar channels. Although the recessed channel of RCAT has improved short channel effect (SCE), RCAT suffers from low driving current and V_{TH} sensitivity due to the shape of the bottom corner of the recessed channel [1]. To solve these problems, a saddle FinFET (S-FinFET) has been proposed with a tri-gate that wraps both the recessed channel surface and the side surface [2–4]. S-FinFET not only exhibits excellent short channel effect characteristics, but also maintains excellent subthreshold swing (SS), high I_{on}, and nearly constant V_{TH}. However, S-FinFET has higher gate-induced drain leakage (GIDL) than RCAT because the overlap region between the gate and drain regions is wider in S-FinFET. Although modified S-FinFET, called RFinFET, has emerged to reduce leakage by GIDL, RFinFET still needs to operate in sub-30 nm cell size [5,6]. Minimizing I_{off} in DRAM applications is a critical issue to achieving long refresh time. If this problem is not addressed, conventional S-FinFETs, including S-FinFET and RFinFET, will experience significant drawbacks in the application of DRAM technology. Therefore, a new device structure with an improved I_{off} is required. In this paper, we proposed a new device with a partial isolation region under the storage node of conventional S-FinFET. This structure can be fabricated by using an isotropic dry etching technique for the buried insulator under the cell transistor [7,8]. We analyzed electrical characteristics of this proposed device and compared them with those of conventional S-FinFETs of the same size. We also showed the optimized parameters of the buried insulator using a three-dimensional (3D) device simulator in sub-30 nm cell size [9]. The device described in this paper has reliable source/drain (S/D) doping concentration with a Gaussian profile. The simulator is well tuned to predict DRAM cell transistor leakage distribution [10,11].

2. Device Structure

The partial isolation type S-FinFET (Pi-FinFET) is a structure with a buried insulator at a certain depth from the storage node of a conventional S-FinFET. Figure 1a shows a 3D schematic of a Pi-FinFET. Silicon film thickness, buried insulator thickness, and L_{in}, as shown in Figure 1a,b, are defined as the distance from the contact surface of the storage node to the buried insulator, the thickness of the buried insulator in the direction perpendicular to the channel, and the distance from the side gate oxide to the buried insulator, respectively. L_g, L_{side}, L_{ov_xj}, and L_{ov_side} represent gate channel length, the length of the side-gate in the direction parallel to the channel, the overlapped length of the side-gate and S/D doping region shown in Figure 1a, and the side-channel width shown in Figure 1c, respectively. When L_{side} and L_{ov_xj} are increased, the width of the overlap region of the gate and the S/D region will also increase. The n^+ poly gate with a gate work function of 4.2 eV was applied. L_g, L_{side}, L_{ov_side} and the recessed depth are 30 nm, 42 nm, 10 nm, and 100 nm, respectively. RFinFET is the modified S-FinFET with a structure in which the overlap region between the side-gate and the S/D region is removed [5]. Therefore, the Pi-FinFETs proposed in this paper can be divided into Pi-SFinFET and Pi-RFinFET depending on the presence or absence of the overlap region between the side-gate and the S/D region. Namely, in this paper, the L_{ov_xj} of Pi-RFinFET is 0 nm and that of Pi-SFinFET is 70 nm.

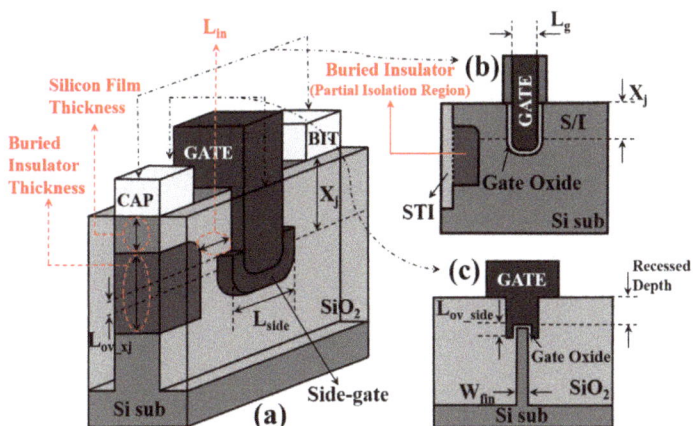

Figure 1. (a) 3D schematic view of Pi-FinFET; (b) Cross-sectional view across the gate; (c) Cross-sectional view of the thin body. The gate wraps three surfaces of the recessed channel, similar to a FinFET. The buried insulator material is used with SiO_2 below the storage node. The buried insulator is penetrated from the shallow trench isolation (STI) region. The X_j of the source/drain (S/D) is located 112 nm from the top surface of the S/D region with a Gaussian profile. The peak concentration of the S/D Gaussian doping profile is 1.5×10^{20} cm^{-3}, and the uniform body doping concentration is 5×10^{17} cm^{-3}.

3. Results and Discussion

Log and linear I_D-V_{GS} curves shown in Figure 2 are the result of comparing conventional S-FinFETs and RCAT with the Pi-FinFETs proposed in this paper. Silicon film thickness, buried insulator thickness, and L_{in} of Pi-FinFET was set at 20 nm, 100 nm, and 20 nm, respectively. As shown in Figure 2, the four saddle type FinFETs have significantly higher I_{on} than RCAT. Moreover, the dotted red ellipse in

Figure 2 show that Pi-SFinFET has about three times smaller I_{off} than S-FinFET, while Pi-RFinFET has about four times smaller I_{off} than RFinFET at V_{GS} of -0.5 V. To understand physical characteristics of I_{off} reduction for Pi-FinFET, device simulations have been performed using the TCAD tool [9,10]. The 3-D simulator is well tuned to predict the leakage current such as GIDL by applying the Hurkx band-to-band tunneling model.

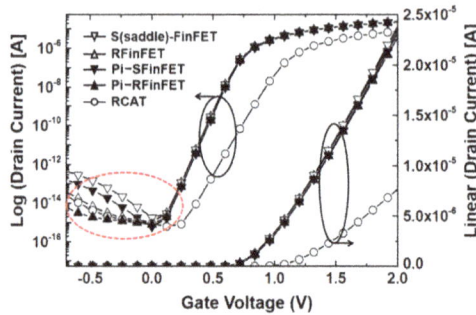

Figure 2. I_{DS}-V_{GS} characteristic for S-FinFET, RFinFET, Pi-FinFETs, and RCAT at V_{DS} = 1.5 V. $I_{off(S\text{-}FinFET)}$ = 1.62 × 10^{-13} A, $I_{off(RFinFET)}$ = 7.80 × 10^{-15} A, $I_{off.Pi\text{-}SFinFET}$ = 5.19 × 10^{-14} A, $I_{off.RFinFET}$ = 1.92 × 10^{-15} A at V_{GS} of -0.5 V.

Figure 3 shows the simulated potential contour near the drain region of conventional S-FinFET and Pi-FinFET at V_{DS} of 1.5 V and V_{GS} of -0.5 V. We defined the gap of equipotential lines as the potential drop width (PDW). In particular, dotted arrows in Figure 3a,b indicate PDW from 1.8 V to -0.4 V. As shown in Figure 3a, the conventional S-FinFET has intensive and narrow PDW from 1.8 V to -0.4 V near the drain/body (D/B) junction. On the other hand, PDW is mostly limited by the buried insulator of Pi-FinFET, as shown in Figure 3b. Namely, the rather constant electric field in the buried insulator of Pi-FinFET induces a wider PDW near the drain region [12]. The dotted red ellipse in Figure 3b shows that the PDW is wider in the silicon layer compared to the conventional S-FinFET. In other words, Pi-FinFET can reduce the electric field affecting GIDL in the silicon layer between the gate and drain regions. Moreover, the penetration of the electric field from the drain to the source affecting drain-induced barrier lowering (DIBL) is minimized in Pi-FinFET.

Figure 4a shows that I_{off} and I_{on} decrease as L_{in} decreases when L_g is 30 nm. If the ratio of L_{in} to L_g is as small as 50% or less, the lateral PDW of the silicon layer induced by the buried insulator will become narrow. This is because the deep penetration of the buried insulator limits the lateral PDW in the narrow silicon layer between the gate and drain regions. As a result, it induces abnormal I_{off} increase and I_{on} decrease, as shown in Figure 4a. Therefore, in order to maintain a relatively high I_{on} while maintaining a low I_{off}, it is desirable to define the ratio of L_{in} to L_g to 50 to 83.3%, respectively. As shown in Figure 4a, I_{off} is minimized when the ratio of L_{in} to L_g is 66.7%. As shown in Figure 4b, DIBL and SS characteristics are consistently improved because of wider PDW when L_{in} is decreased. Figure 5 shows I_{off} and DIBL characteristics according to L_g at V_{DS} of 1.5 V and V_{GS} of -0.5 V. L_{in} values in Figure 5 were set to maintain 60% of L_g. Pi-FinFETs consistently exhibited improved I_{off} and DIBL characteristics compared to conventional S-FinFETs regardless of L_g. Especially, Pi-FinFETs exhibited improved DIBL characteristics with opposite tendency compared to conventional S-FinFETs as L_g decreases, as shown in Figure 5b.

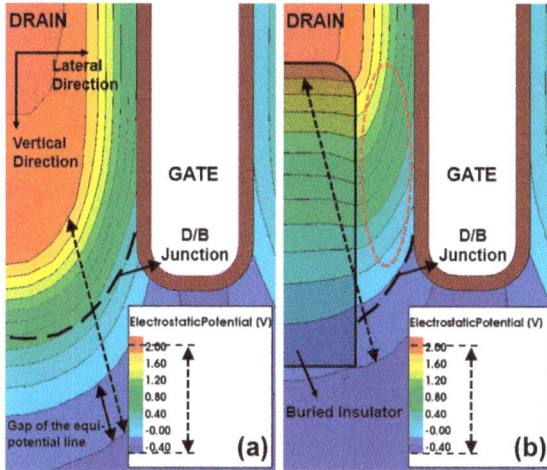

Figure 3. Potential contour indicating equipotential line distribution near the drain region at V_{DS} = 1.5 V, V_{GS} = −0.5 V. (**a**) Conventional S-FinFET, (**b**) Pi-FinFET. Dotted arrows indicate potential drop width (PDW) from 1.8 V to −0.4 V, while dotted red ellipse shows wider PDW compared to conventional S-FinFET.

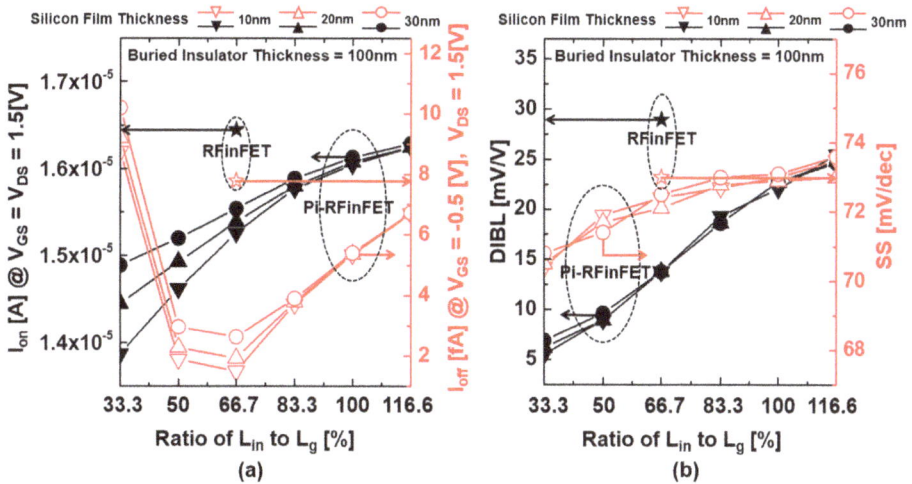

Figure 4. (**a**) I_{on}, I_{off} and (**b**) drain-induced barrier lowering (DIBL), subthreshold slope (SS) characteristics of Pi-RFinFET and Pi-FinFET according to the ratio of L_{in} to L_g when L_g is 30 nm.

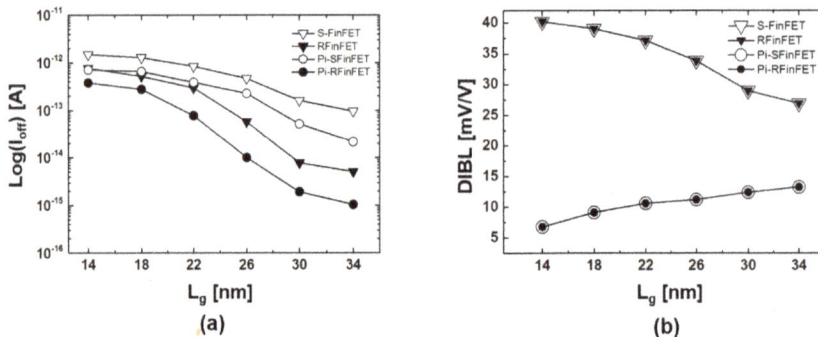

Figure 5. (a) I_{off} (V_{GS} = −0.5 V, V_{DS} = 1.5 V) and (b) DIBL ($V_{DS.low}$ = 0.05 V, $V_{DS.high}$ = 1.5 V) dependence on L_g of conventional S-FinFETs and Pi-FinFETs.

Figure 6 also shows the DIBL and $g_{m.max}$ characteristics of Pi-FinFETs and S-FinFETs according to the recessed depth. As shown in Figure 6, as the recessed depth decreases, DIBL characteristics of S-FinFETs increases sharply, whereas the DIBL of Pi-FinFETs is relatively constant because the PDW of Pi-FinFETs is wider. As the recessed depth decreases, the $g_{m.max}$, which represents the on-current characteristic, is improved. In other words, Pi-FinFETs have found that the recessed depth can be set more flexibly than conventional S-FinFETs.

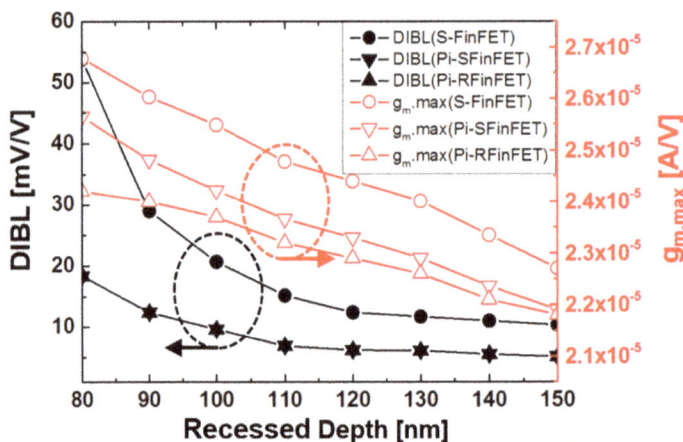

Figure 6. DIBL ($V_{DS.low}$ = 0.05 V, $V_{DS.high}$ = 1.5 V) and $g_{m.max}$ characteristics of Pi-FinFETs and S-FinFET according to the recessed depth when L_g is 30 nm.

4. Conclusions

We proposed a new device with a partial isolation region under the storage node of conventional saddle-FinFETs. This device can be classified as either Pi-SFinFET or Pi-RFinFET depending on the overlap area of the side gate and the S/D region. The proposed device not only maintains high DIBL characteristics regardless of the gate channel length, but also reduces the I_{off} by up to four times compared to the conventional saddle-FinFETs. It also minimizes I_{on} reductions. From the study, we concluded that Pi-FinFET is a promising candidate for sub 30 nm DRAM technology.

Author Contributions: Conceptualization, Y.K.K.; visualization, J.S.L. and G.K.; resources, Y.P.C., H.J.K and Y.J.P; supervision, T.P. and M.J.L.

Funding: This research was supported in part by the Korea Electric Power Corporation (Grant number: R17XA05-78), in part by the National Research Foundation of Korea (NRF) grant funded by the Korea government (MSIT) (Gant number: 2018R1A2B6008216), in part by the R&D Special Zone Development Project (Technology Transfer Business Project: 17GJI003) funded by the Ministry of Science, and the ICT & INNOPOLIS Foundation.

Conflicts of Interest: The authors declare no conflict of interest.

References

1. Kim, L.J.Y.; Oh, H.J.; Lee, D.S.; Kim, D.H.; Kim, S.E.; Ha, G.W.; Kim, H.J.; Kang, N.J.; Park, J.M.; Hwang, Y.S.; et al. S-RCAT (sphere-shaped-recesschannel-array transistor) technology for 70 nm DRAM feature size and beyond. In Proceedings of the VLSI symposium, Kyoto, Japan, 14–16 June 2005; pp. 34–35.

2. Chung, S.W.; Lee, S.D.; Jang, S.A.; Yoo, M.S.; Kim, K.O.; Chung, C.O.; Cho, S.Y.; Cho, H.J.; Lee, L.H.; Hwang, S.H.; et al. Highly scalable saddle-fin (S-Fin) transistor for sub 50 nm DRAM technology. In Proceedings of the VLSI symposium, Honolulu, HI, USA, 13–15 June 2006; pp. 147–148.

3. Crupi, G.; Schreurs, D.; Raskin, J.P.; Caddemi, A. A comprehensive review on microwave FinFET modeling for progressing beyond the state of art. *Solid State Electron.* **2013**, *80*, 81–95. [CrossRef]

4. Poljak, M.; Jovanovic, V.; Suligoj, T. Improving bulk FinFET DC performance in comparison to SOI FinFET. *Microelectron. Eng.* **2009**, *86*, 2078–2085. [CrossRef]

5. Lee, M.J.; Jin, S.H.; Baek, C.K.; Hong, S.M.; Park, S.Y.; Park, H.H.; Lee, S.D.; Chung, S.W.; Jeong, J.G.; Hong, S.G.; et al. A proposal on an optimized device structure with experimental studies on recent devices for the DRAM cell transistor. *IEEE Trans. Electron Devices* **2007**, *54*, 3325–3335. [CrossRef]

6. Ryu, S.W.; Min, K.; Shin, J.; Kwon, H.; Nam, D.; Oh, T.; Jang, T.S.; Yoo, M.; Kim, Y.; Hong, S. Overcoming the reliability limitation in the ultimately scaled DRAM using silicon migration technique by hydrogen anneling. In Proceedings of the 2017 IEEE International Electron Devices Meeting (IEDM), San Francisco, CA, USA, 2–6 December 2017; pp. 21.6.1–21.6.4.

7. Lee, M.J.; Cho, J.H.; Lee, S.D.; Ahn, J.H.; Kim, J.W.; Park, S.W.; Park, Y.J.; Min, H.S. Partial SOI type isolation for improvement of DRAM cell transistor characteristics. *IEEE Electron Device Lett.* **2005**, *26*, 332–334.

8. Kim, Y.K.; Lee, J.S.; Kim, G.; Park, T.; Kim, H.J.; Cho, Y.P.; Park, Y.J.; Lee, M.J. Simulation analysis in sub-0.1 μm for partial isolation Field-Effect Transistors. *Electronics* **2018**, *7*, 227. [CrossRef]

9. Sysnopsis Inc. *TCAD Sentaurus Manual, Sysnopsis®*, version D-2013.03; Sysnopsis Inc.: Mountain View, CA, USA, 2013.

10. Hurkx, G.A.M.; Klaassen, D.B.M.; Knuvers, M.P.G. A new recombination model for device simulation including tunneling. *IEEE Trans. Electron Devices* **1992**, *39*, 331–338. [CrossRef]

11. Jin, S.; Lee, M.J.; Yi, J.H.; Choi, J.H.; Kang, D.G.; Chung, I.Y.; Park, Y.J.; Min, H.S. A new direct evaluation method to obtain the data retention time distribution of DRAM. *IEEE Trans. Electron Devices* **2006**, *53*, 2344–2350. [CrossRef]

12. Jang, E.; Shin, S.; Jung, J.W.; Kim, K.R. Gate induced drain leakage reduction with analysis of gate fringing field effect on high-κ/metal gate CMOS technology. *Jpn. J. Appl. Phys.* **2015**, *54*, 06FG10-1-4. [CrossRef]

© 2018 by the authors. Licensee MDPI, Basel, Switzerland. This article is an open access article distributed under the terms and conditions of the Creative Commons Attribution (CC BY) license (http://creativecommons.org/licenses/by/4.0/).

electronics

MDPI

Article

A New Method of the Pattern Storage and Recognition in Oscillatory Neural Networks Based on Resistive Switches

Andrei Velichko *, Maksim Belyaev, Vadim Putrolaynen and Petr Boriskov

Institute of Physics and Technology, Petrozavodsk State University, 31 Lenina str., Petrozavodsk 185910, Russia; biomax89@yandex.ru (M.B.); vputr@petrsu.ru (V.P.); boriskov@psu.karelia.ru (P.B.)
* Correspondence: velichko@petrsu.ru; Tel.: +7-8142-63-5773

Received: 10 September 2018; Accepted: 18 October 2018; Published: 22 October 2018

Abstract: Development of neuromorphic systems based on new nanoelectronics materials and devices is of immediate interest for solving the problems of cognitive technology and cybernetics. Computational modeling of two- and three-oscillator schemes with thermally coupled VO_2-switches is used to demonstrate a novel method of pattern storage and recognition in an impulse oscillator neural network (ONN), based on the high-order synchronization effect. The method allows storage of many patterns, and their number depends on the number of synchronous states N_s. The modeling demonstrates attainment of N_s of several orders both for a three-oscillator scheme $N_s \sim 650$ and for a two-oscillator scheme $N_s \sim 260$. A number of regularities are obtained, in particular, an optimal strength of oscillator coupling is revealed when N_s has a maximum. Algorithms of vector storage, network training, and test vector recognition are suggested, where the parameter of synchronization effectiveness is used as a degree of match. It is shown that, to reduce the ambiguity of recognition, the number coordinated in each vector should be at least one unit less than the number of oscillators. The demonstrated results are of a general character, and they may be applied in ONNs with various mechanisms and oscillator coupling topology.

Keywords: oscillatory neural networks; pattern recognition; higher order synchronization; thermal coupling; vanadium dioxide

1. Introduction

Usage of artificial neural networks [1] for information processing allows mastering the problems that arise when traditional computation schemes are applied in such areas as pattern and speech recognition [2], and data computation and encoding [3]. Therefore, the important research trends include studying the modes of oscillator neural network (ONN) operation and training, implementation of associative memory modes based, for example, on weakly coupled phase oscillators (Kuramoto model) [4] or impulse oscillators [5,6]. The effect of synchronization plays a crucial role in ONN operation, and is often used as a marker of ONN action, for example, in pattern recognition event.

There is a class of ONNs based on relaxation oscillators that generate subsequent pulses (spikes). These oscillators, in turn, are composed of electronic components with resistive switching effect, for example, VO_2-switches [7,8], 1T-TaS_2 charge density wave devices [9–11], thyristors [12], tunneling diodes [13], resistive memory elements [14], spin-torque nano-oscillators [15]. Such ONNs appear to be interesting because of hardware solution simplicity, as well as compactness and energy efficiency of the developed micro- and nanoelectronic self-oscillators. VO_2-based oscillators, as the elements of ONNs, have been chosen because they ensure rapid electric switching (~10 ns) [16], manufacturability with high degree of nanoscaling [17] and, above all, because of the pronounced effect of thermal coupling that simplifies ONN assembly and circuit engineering of galvanically isolated oscillators.

Consequently, VO_2-oscillators started being used as the prototypes of neuro-oscillators for cognitive technology [8,16,18].

In ONNs, the system demonstrates frequency and phase synchronization [19–22] and, also, synchronization of high order [17,23] at certain control parameters, such as parameters of an oscillator scheme or coupling strengths between the oscillators. The method proposed, here, of pattern storage and recognition, is based on the effect of high-order synchronization, that has been experimentally demonstrated through thermally coupled VO_2-oscillators [17]. In many studies [22,24,25], patterns to be stored are expressed through a set of vectors. Vector coordinates contain information about the pattern and unambiguously associate it with one of possible variants. For instance, the vector of the object's color in RGB coordinates (white color—RGB (255,255,255)) may be used as a 3-dimension vector. There are some methods of vector storage based on oscillator elements' synchronization in ONNs, and one of them is presented in paper [20]. To store **E** vector, a phase-shift keying method of a test vector **T** is specified by weight matrix setting; at the second stage, the weights are sharply changed to the initial values (corresponding to the stored vectors), and the system arrives at one of the stable combinations of phase shift **E**. However, this phase method has the following drawbacks: N^2 couplings with tunable weights and a two-stage procedure of pattern recognition.

A second known method of vector storage is a frequency-shift keying method of encoding, based on synchronized frequency shift [22]. According to this method, vector **E** is stored through oscillator frequency shifts against the central frequency of oscillator array F^0 synchronization (on the first harmonic) per the values corresponding to the vector coordinates $\mathbf{E} = (\delta_{\omega 1}, \delta_{\omega 2}, \ldots, \delta_{\omega N})$. Recognition of test vector **T** occurs at the reverse shift of frequencies and, in the case when the vectors coincide $\mathbf{T} \approx \mathbf{E}$, the synchronization, indicating the fact of pattern recognition, takes place. This method allows usage of an oscillator star configuration and only N couplings, however, the disadvantage of this method is that just one vector is stored.

The present work suggests a conceptually new method of vector (pattern) storage and recognition when an array of coupled oscillators enables storage of a multitude of vectors. This result is achieved due to the effect of high-order synchronization in our ONN, that has many oscillator synchronous states and, also, because of specific algorithm of the network training and identification of the degree of match for the tested objects.

2. Materials and Methods

2.1. General Principle

An oscillator neural network is a system of N coupled oscillators, which may be connected via electric (by resistors and capacitors) [7,26], thermal [17], and optic [27] couplings, depending on the physical mechanism of oscillator interaction. In the general case, there is a matrix of coupling strengths $\Delta_{i,j}$ (weights), where i, j are the numbers of interacting oscillators, and $\Delta_{i,j}$ denotes the value of the i-th oscillator effect on the j-th one. Oscillator networks may form various topologies: fully connected—all-to-all; and not fully connected—bus, star, and ring. Figure 1a–c show examples of two and three oscillators connections using topologies "star", "all-to-all", and an example of N oscillator connections using a mixed topology (Figure 1d).

It is known [17,23] that oscillators in a network may undergo the effect of synchronization and, besides synchronization on the first harmonic, synchronous modes of high order may be observed if the signal spectra possess several harmonics. To evaluate synchronization, in this work, we use a family of metrics that consists of two parameters (SHR is the value of high order synchronization and η is the effectiveness of synchronization). A detailed method of its determination is given in Section 2.3 and in [17].

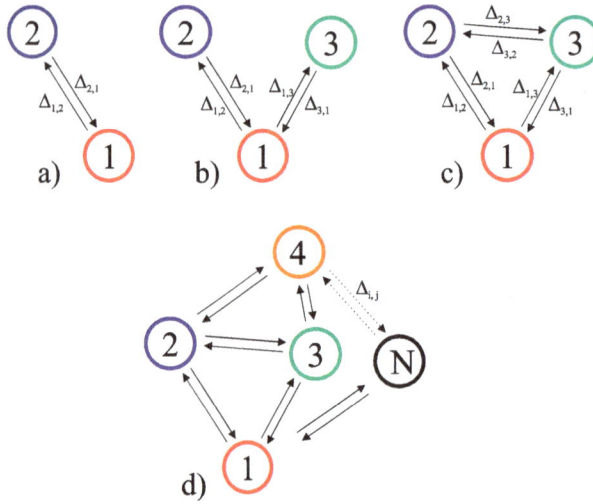

Figure 1. Examples of two (**a**) and three (**b**) oscillators connection into a neural network using topologies "star" (**b**), "all-to-all" (**c**), and *N*-oscillators using mixed topology (**d**), where $\Delta_{i,j}$ indicates the value of the *i*-th oscillator effect on the *j*-th one.

In the general case, high-order synchronization is determined by the ratio SHR = $k_1:k_2:k_3:...:k_N$, where k_N is a harmonic order of *N*-th oscillator at the common frequency of the network synchronization F_s, (SHR—subharmonic ratio). As an example, Figure 2 shows spectra of three electric oscillators that have synchronization of the order SHR = $k_1:k_2:k_3$ = 3:6:4. The following rule should be noted: if all paired oscillators have different synchronization frequencies, there is always a common synchronization frequency F_s for the whole system (all pairs), and the network synchronous state will also be determined by the ratio SHR = $k_1:k_2:k_3:...:k_N$ at frequency F_s (see Section 2.3).

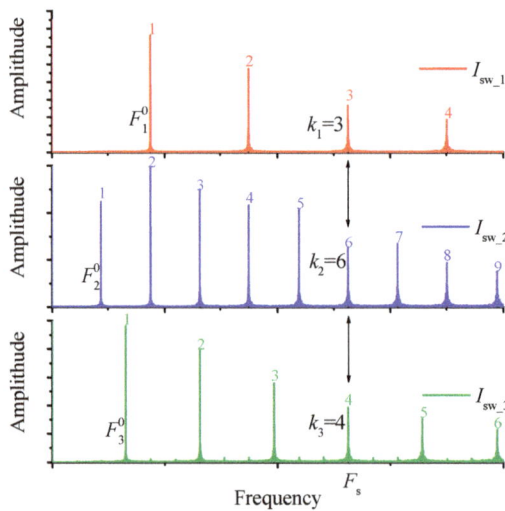

Figure 2. Example of oscillation spectra of three electric oscillators at synchronization order subharmonic ratio (SHR) = $k_1:k_2:k_3$ = 3:6:4, where I_{SW} is the current amplitude of a signal in an oscillator, F^0 is first harmonic, *k* is the harmonic number at the synchronization frequency F_s.

In addition to SHR, there is also a parameter of synchronization effectiveness η, that shows what share of oscillations of the whole time signal is synchronized. This parameter is expressed in percentages (see Section 2.3). If, at any point, η is less than the threshold value η_{th}, then SHR is absent, and the signal is considered conventionally non-synchronized.

Transition from one synchronous state into another is possible when the oscillator network control parameters are varied. For example, in electric oscillators, the main parameters may be oscillator feed currents I_p, their variation causes changes of the basic oscillation frequency F^0. Nevertheless, in some cases, transition between states may be achieved by variation of coupling forces or noise intensity.

The range of control parameters variation, where synchronization does not change its state, is called a synchronization area. There is a whole family of synchronization areas that are called Arnold's tongues (for the case of two oscillators). A schematic example of synchronization areas for a three-oscillator scheme is shown in Figure 3a. Here, the control parameters are oscillator feed currents. Each area has its own value of SHR. Besides, each area has its own distribution of the synchronization effectiveness value within $\eta_{th} < \eta < 100\%$, with a peaked curve.

The number of possible variants of synchronous states (synchronization areas), where the system may exist when the basic control parameters are varied, is denoted as N_s. The value of N_s depends on many parameters: the oscillator number N, the range of control parameters and their number, network topology, strength of coupling between oscillators, noise level in the system and on the threshold value of synchronization effectiveness η_{th}. We will cover the issue in detail later, nevertheless, we have shown in [28], that for a two-oscillator network, N_s has a maximum at certain values of coupling strengths between oscillators, and decreases when the system noise amplitude increases. When the coupling strength grows considerably, the value N_s decreases because of the nearby synchronization areas' integration.

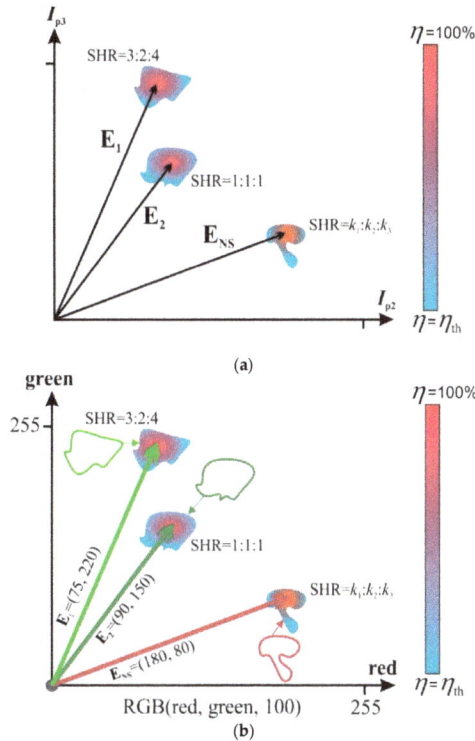

Figure 3. *Cont.*

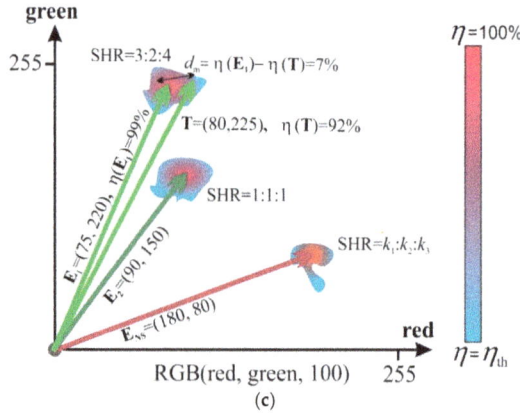

green

255 — SHR=3:2:4 $d_m = \eta\,(\mathbf{E}_1) - \eta\,(\mathbf{T}) = 7\%$

$\mathbf{T} = (80,225),\ \eta\,(\mathbf{T}) = 92\%$

SHR=1:1:1

$\mathbf{E}_1 = (75,220),\ \eta(\mathbf{E}_1) = 99\%$

$\mathbf{E}_2 = (90,150)$

$\mathbf{E}_{\mathrm{NS}} = (180,80)$

SHR=k_1:k_2:k_3

red

RGB(red, green, 100)   255

$\eta = 100\%$

$\eta = \eta_{\mathrm{th}}$

(c)

Figure 3. (**a**) Schematic representation of the synchronization areas for a three-oscillator scheme; (**b**) Examples of the vector and object's RGB color association, that illustrate the algorithm of the network training and recognition (**c**).

Vectors $\mathbf{E}_1$, $\mathbf{E}_2$, ..., $\mathbf{E}_{\mathrm{NS}}$, that connect the origin of coordinates with the points of synchronization effectiveness maximum η, can be associated with the synchronization areas. Thus, the system stores N_s of vectors, and the dimensionality of the stored vectors M is determined by the number of chosen control parameters. The coordinates determine the shift of oscillators' control parameters, for example, currents $\mathbf{E} = (\delta I_{p1}(1), \delta I_{p2}(2), \dots, \delta I_{pN}(M))$, against the origin of coordinates.

As we have mentioned in the introduction section, the patterns to be stored are usually expressed through a set of vectors. Vector coordinates contain information about the pattern and unambiguously associate it with one of possible variants. For example, Figure 3c shows storage of the object's colors in the RGB format (red, green, blue) through the coordinates of vectors $\mathbf{E}$, whose values give the information about the intensity of red and green colors $\mathbf{E}$ = (red, green), and parameter blue is fixed as blue = 100. This example, in Figure 3c, shows the intensities of RGB components on the axes that can be linearly transformed into the values of the oscillator currents and vice versa.

We suggest the following methods of pattern storage and recognition in a neural network, based on the high-order synchronization effect, and its general scheme is given in Figure 4.

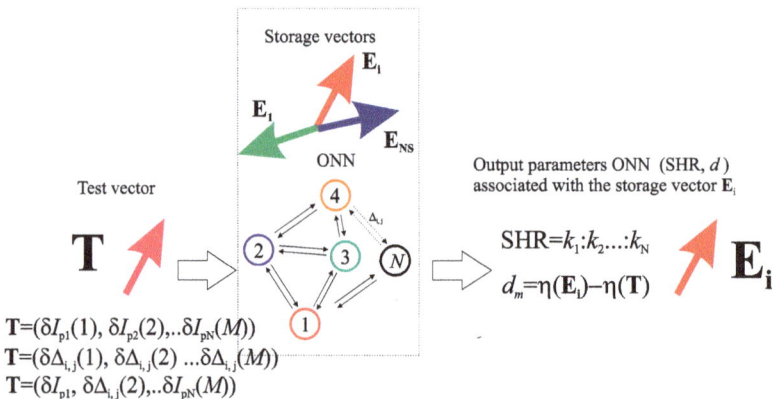

Storage vectors

$\mathbf{E}_1$

$\mathbf{E}_1$

$\mathbf{E}_{\mathrm{NS}}$

ONN

Test vector

Output parameters ONN (SHR, d) associated with the storage vector $\mathbf{E}_i$

$\mathbf{T}$

$\Delta_{i,j}$

SHR=k_1:k_2...:k_N

$d_m = \eta(\mathbf{E}_i) - \eta(\mathbf{T})$

$\mathbf{E}_i$

$\mathbf{T} = (\delta I_{p1}(1), \delta I_{p2}(2),..\delta I_{pN}(M))$
$\mathbf{T} = (\delta\Delta_{i,j}(1), \delta\Delta_{i,j}(2) ...\delta\Delta_{i,j}(M))$
$\mathbf{T} = (\delta I_{p1}, \delta\Delta_{i,j}(2),..\delta I_{pN}(M))$

Figure 4. Schematic representation of pattern recognition principle by using oscillator neural network (ONN), where M is the dimensionality of the test vector $\mathbf{T}$, N is the number of oscillators, N_s is the maximal number of the stored vectors $\mathbf{E}$.

2.1.1. Vector Storage and ONN Training

The general algorithm for vector storage and ONN training includes the following steps:

1. For storage, arbitrary vectors $\mathbf{E}_1$, $\mathbf{E}_2$, ... , $\mathbf{E}_i$, ..., $\mathbf{E}_{Ns}$ should be specified. If necessary, control parameters should be transformed into the corresponding coordinate system (for example, a color one, see Figure 3b,c). In general, vectors have dimensionality M and appear as a set of a network parameters that affects the system SHR. For example, they can be either currents, as shown in Figure 3 $\mathbf{E} = (\delta I_{p1}(1), \delta I_{p2}(2), \dots, \delta I_{pN}(M))$, or they can be coupling strengths between some definite oscillators $\mathbf{E} = (\delta \Delta_{i,j}(1), \delta \Delta_{i,j}(2), \dots, \delta \Delta_{i,j}(M))$, or mixed parameters $\mathbf{E} = (\delta I_{p1}(1), \delta \Delta_{i,j}(2), \dots, \delta \Delta_{i,j}(M))$ (see Figure 4).

2. Then, the network should be trained by the adjustment of the ONN parameters that are not used for the vectors' determination (coupling strengths, currents of other oscillators in the network, noise level, and synchronization effectiveness threshold η_{th}). The adjustment is performed until the synchronization areas coincide with the vectors' ends at the point of maximum value of synchronization effectiveness η (similar network training was used in the work [15]). The adjustment can be performed in two steps.

    a. First, by using random search until the vectors enter the synchronization area.

    b. Then, one of gradient methods [29] may be applied to search the maximum η. As a result, each stored vector corresponds to its unique value of SHR and maximum of $\eta(\mathbf{E})$.

3. If the training does not provide a positive result, one more oscillator should be included into the system and coupled with all oscillators already present, thus increasing the number of varied parameters and the number of possible synchronous states N_s. Then, the training should be repeated (see step 2).

2.1.2. Vectors Recognition

The algorithm of test vector $\mathbf{T}$ recognition includes the following steps:

1. Set the test vector $\mathbf{T}$ to the system input through applying shifts to the control parameters (see Figure 4). The vector's coordinates may be either shifts of currents, or coupling strengths or their combination, as it has been indicated above.

2. If one of the conditions is met ($\mathbf{T} \approx \mathbf{E}_1$ or $\mathbf{T} \approx \mathbf{E}_2$ or ... or $\mathbf{T} \approx \mathbf{E}_{NS}$), i.e., coordinates values of $\mathbf{T}$ are equal to one of the stored patterns, a transition to the synchronous state will occur and, actually, the act of the corresponding pattern recognition will take place. Which patterns have been exactly recognized can be determined by the value of SHR. The existence of the synchronization areas ensures the vector recognition even at its coordinates' insignificant displacement from the stored pattern.

3. The degree of match d_m between the objects may be such magnitude as the difference between the synchronization effectiveness of the stored and the test vectors $d_m = \eta(\mathbf{E}) - \eta(\mathbf{T})$. If the magnitude of $\eta(\mathbf{E})$ is not known, then to compare the degree of match, the formula $d_m = 100\% - \eta(\mathbf{T})$ can be used. The less d_m is, the closer vector $\mathbf{T}$ is to vector $\mathbf{E}$.

This method is a more complicated version of the method described in [22], where the analogy to the frequency-shift keying method of coding is used and, instead of setting the vector through frequencies $\mathbf{E} = (\delta\omega_1, \delta\omega_2, \dots, \delta\omega_N)$, in our method, the vector is set through the control parameters $\mathbf{E}_1 = (\delta I_{p1}, \delta I_{p2}, \dots, \delta I_{pN})$, that has the same meaning. The principle difference is that here, a high-order synchronization effect is used, thus allowing storage of a multitude of patterns in the ONN.

Besides, as described in the results section, it is more practical to use vector dimensions $\mathbf{E'} = (\delta I_1, \delta I_2, \dots, \delta I_{N-1})$ with one less than the number of oscillators ($N-1$).

2.2. Model Object

As a model object, we have chosen a neural network composed of three thermally coupled VO$_2$-oscillators, where each oscillator has the scheme of a relaxation oscillator. Our choice is conditioned by the fact that we have done some research in thermal coupling [17,30] and its modeling, however, the coupling may be an electric one (capacitive or resistive [7]). It is known that an electric switching effect is observed in VO$_2$ film-based structures, that is conditioned by a phase metal–insulator transition (MIT) at the moment when the temperature reaches T_t ~340 K, because of Joule heating by the passing current I_{sw} [16]. This gives high-impedance (OFF) and low-impedance (ON) branches on I–V characteristics with threshold voltages (*OFF→ON*) U_{th} ~5 V and holding voltages (*ON→OFF*) U_h ~1.5 V (see Figure 5a). Both branches of I–V characteristics are reasonably well approximated by f_{sw} curve, consisting of two linearized regions with dynamic resistance R_{off} ~9.1 kΩ and R_{on} ~615 Ω:

$$I_{sw} = f_{sw}(U) \approx \begin{cases} \frac{U}{R_{off}}, & State = OFF \\ \frac{(U - U_{bv})}{R_{on}}, & State = ON \end{cases} \tag{1}$$

where U_{bv} ~0.82 V is bias voltage of a low-impedance region, and *State* is a switch state.

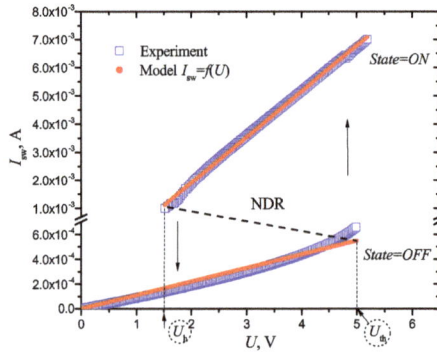

(a)

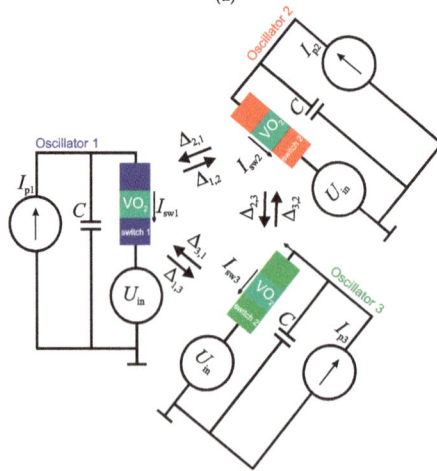

(b)

Figure 5. Experimental and model I–V characteristics of VO$_2$-switch (**a**); a model scheme of a neural network based on three oscillators circuits with VO$_2$-switches interacting via thermal coupling (**b**).

One of three topologies presented in Figure 1 may be realized, depending on coupling strength magnitudes Δ. At non-zero $\Delta \neq 0$, the topology is "all-to-all" (Figure 1c); at $\Delta_{2,3} = \Delta_{3,2} = 0$, the topology is "star" (Figure 1b); and at $\Delta_{2,3} = \Delta_{3,2} = \Delta_{1,3} = \Delta_{3,1} = 0$, the scheme turns into a two-oscillator one (Figure 1a). The control parameters here are source currents I_{p1}, I_{p2}, I_{p3}, and their variation leads to alteration of the fundamental oscillation frequency F^0 of oscillators.

Variations of each oscillator are described by the equation of Kirchhoff's law:

$$C\frac{dU_i(t)}{dt} = I_{p(i)} - I_{sw(i)}(t),\tag{2}$$

where $U_i(t)$ is the output voltage taken from the capacitor (C = 100 nF), $I_{sw(i)}(t) = f_{sw}(U_i(t) - U_{in})$ is the current passing through a switch, determined by I–V characteristics (1), $I_{p(i)}$ is the i-th oscillator supply current, respectively, U_{in} is the amplitude of switch internal noise, and i is the oscillator's number.

Thermal interaction between the i-th VO$_2$-oscillator and the neighbor ones (($i+$)—clockwise and ($i-$)—counterclockwise of the scheme in Figure 5b) is realized according to the rule

$$U_{th(i)} = \begin{cases} U_{th(i)} - \Delta_{(i+),i}, & \text{if } State_{(i+)} = ON \\ U_{th(i)} - \Delta_{(i-),i}, & \text{if } State_{(i-)} = ON \\ U_{th(i)} - \Delta_{(i-),i} - \Delta_{(i+),i}, & \text{if } State_{(i-)} = State_{(i+)} = ON \end{cases}.\tag{3}$$

If the states of oscillators $State_{(i+)}$ and $State_{(i-)}$ are on the OFF branch of I–V characteristics, then the threshold voltage of the i-th VO$_2$-oscillator does not change: $U_{th(i)} = U_{th}$. Rule (3) is the same for all oscillators (with regard to cyclic permutation).

Oscillograms of oscillations with ~250,000 points and time interval δt = 10 µs were simulated using Equations (1)–(3). After that, the oscillograms were automatically processed, the synchronization order was determined, and cross-sections of oscillator synchronization areas were built.

The switch parameters did not change in numerical simulation of the results, but current intensities I_p, coupling strength Δ, and noise amplitude U_{in} varied.

2.3. Method of Calculating a Family of Metrics

To define the synchronization order, we used the family of metrics described above, that consists of two parameters SHR and η.

The problem of finding the high-order synchronization value determined by the ratio of integers SHR = $k_1{:}k_2{:}k_3{:}...{:}k_N$ (see Section 2.1) may be solved in several ways. For example, by direct analysis of all oscillation spectra, or by searching the synchronization order of each pair of oscillators based on the method which we suggested in [17].

It should be noted that, at synchronous state, the frequency sets of fundamental (first) harmonics of oscillators (F_1^0, F_2^0, F_3^0, ... , F_N^0) must be commensurable. This is evident because at the synchronous state, there is a common synchronization frequency F_s, and the equality ($F_s = F_1^0 \cdot k_1 = F_2^0 \cdot k_2 = ... = F_N^0 \cdot k_N$) is fulfilled. If we divide F_1^0 into all frequencies in the set (F_1^0, F_2^0, F_3^0, ... , F_N^0), then we will get (1, F_1^0/F_2^0, F_1^0/F_3^0, ... , F_1^0/F_N^0) = (1, k_2/k_1, k_3/k_1, ... , k_N/k_1), that is, a new set of rational numbers determining pair synchronization of all oscillators in regard to the first oscillator (see [17]).

Thus, the method of specifying all values of k and the synchronization order of the system consisting of N-oscillators comes down to determining the set of pair synchronization fractional values (in regard to the first oscillator) for N-pairs (m_1/d_1, m_2/d_2, ... , m_{N-1}/d_{N-1}), and to its reduction to a common denominator:

$$\left(\frac{m_1}{d_1}, \frac{m_2}{d_2}, \frac{m_3}{d_3} \cdots \frac{m_{N-1}}{d_{N-1}}\right) \rightarrow \left(\frac{k_2}{k_1}, \frac{k_3}{k_1}, \frac{k_4}{k_1} \cdots \frac{k_N}{k_1}\right).\tag{4}$$

For example, a set of pair synchronization for oscillator pairs (№1–№2) and (№1–№3) in Figure 2 looks like (2/1, 4/3), after reduction to a common denominator (4), we get (2/1, 4/3) → (6/3, 4/3), and SHR = $k_1:k_2:k_3$ = 3:6:4.

It should also be noted that the algorithm of pair synchronization definition is based on the search of current oscillation peaks, I_{sw}, synchronous in time [17].

The effectiveness of pair synchronization η is determined as the percentage of the durability of all N_{SHR} synchronous periods T_s with the definite SHR, to the whole durability of the processed oscillogram T_{all}:

$$\eta = \frac{N_{SHR} \cdot Ts}{T_{all}} \cdot 100\% \tag{5}$$

If there are several synchronization types with different SHR, then the resulting η is associated with the maximum which, in turn, is compared with the threshold value η_{th} (in our case 90%). Oscillations are considered synchronized when η exceeds the threshold $\eta \geq \eta_{th}$. If the system consists of more than two oscillators, then the total effectiveness η is calculated as the mean value of all oscillator pairs. It should be noted that the proposed methods of SHR and η identification may be used in oscillator systems with noise. It has been noted that the noise increase leads mainly to the decrease of η, while SHR does not normally change.

3. Results

The results of synchronization areas modeling for a two-oscillator scheme (see Figure 1a) are given in Figure 6a. Control parameters are oscillator feed currents I_{p1}, I_{p2}, and noise and coupling strength values are U_{in} = 40 mV and Δ = 0.2 V. It can be seen that there is a whole family of synchronization areas that are called Arnold tongues [23]. The number of possible variants of synchronous states, N_s, in which the system may exist while the control parameters are varied, is N_s = 9. The dimension of the stored vectors in this case is 2, and the coordinates determine current shifts $\mathbf{E}_1 = (\delta I_{p1}, \delta I_{p2})$, with respect to the origin of coordinates.

The problem here is that the synchronization areas are long-ranged (of Arnold's tongues shape) therefore, there is a wide range of stored pattern coordinates which bring the system into a certain synchronous state. The solution lies in narrowing the dispersion of stored pattern coordinates by using vectors $\mathbf{E}'_1 = (\delta I_1, \delta I_2, \ldots, \delta I_{N-1})$ of a dimension one less than the number of oscillators $(N-1)$, in this case, $\mathbf{E}'_1 = (\delta I_1)$. In practice, this means that we fix the current for one oscillator, and vary the currents for the others (see Figure 6a, dashed line I_{p2} = const). Thus, we eliminate the ambiguity of synchronization definition by one of the vector coordinates, and the areas of possible synchronization are narrowed.

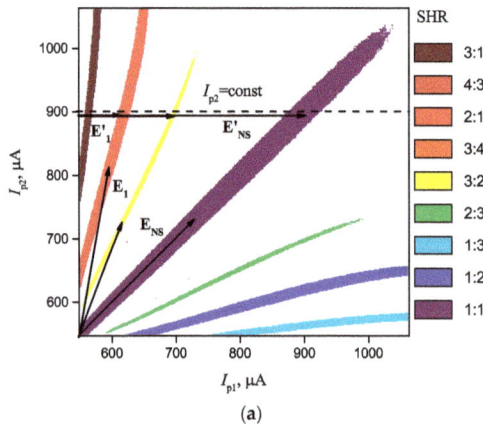

Figure 6. *Cont.*

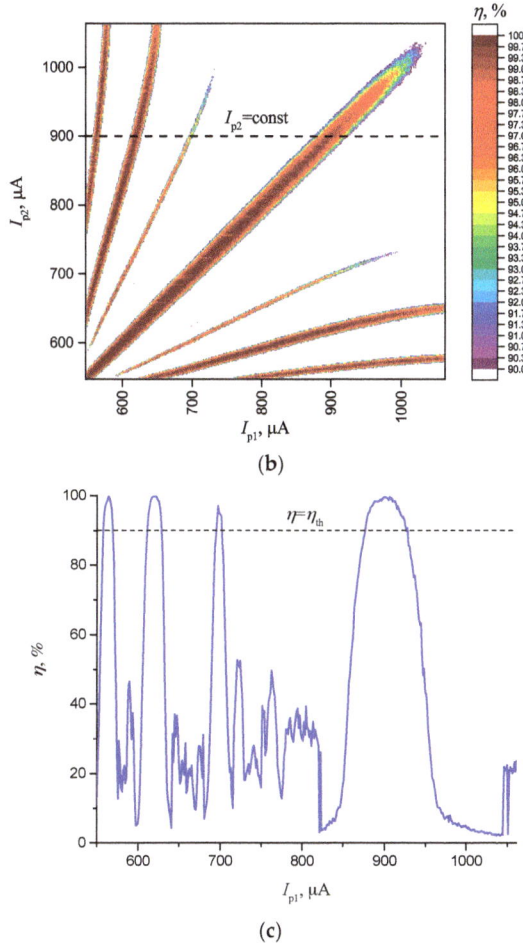

Figure 6. Example of synchronization areas for a two-oscillator scheme (**a**). The arrows show sampled vectors **E** and **E′**, in regard to the origin of coordinates. Distribution of η for a two-oscillator scheme (**b**). Cross-section η at $I_2 = 900$ µA (**c**).

Figure 6b shows the distribution of synchronization effectiveness inside the Arnold's tongues. It can be seen that η falls down to the edges of synchronization areas, whereas maximum of η has a progressive form in the area of control parameters, and is placed in the line in the center. When a cross-section is made (shown in Figure 6c), we can observe local maximums of η inside the synchronization areas and this is another argument to use the gradient search method in the algorithm of the network training. According to the above proposed method of recognition, the magnitude $d_{\mathrm{m}} = \eta(\mathbf{E'}) - \eta(\mathbf{T})$ may serve as the parameter of the degree of match between the test and stored vectors.

Figure 7 shows cross-sections of synchronization areas for a system consisting of three oscillators ("star", see Figure 1b) at fixed current on the first oscillator $I_{p1} = 950$ µA, and parameters $U_{\mathrm{in}} = 40$ mV and $\Delta = 0.2$ V, that are similar to a two-oscillator scheme.

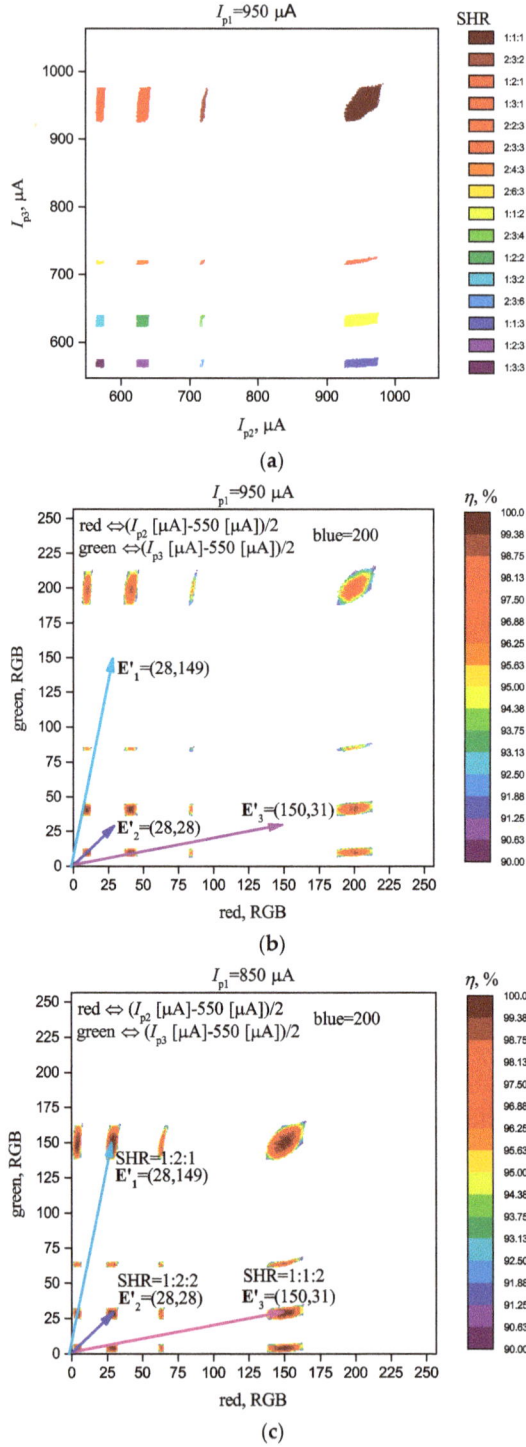

(a)

(b)

(c)

Figure 7. *Cont.*

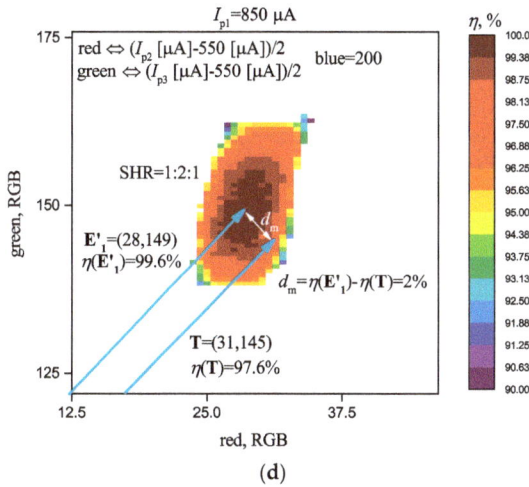

Figure 7. (**a**) Synchronization areas with SHR for a three-oscillator scheme "star" at $I_{p1} = 950\ \mu A$; (**b**) Distribution of synchronization effectiveness η with the set vectors E' at $I_{p1} = 950\ \mu A$; (**c**) Distribution of synchronization effectiveness η with the set vectors E' at $I_{p1} = 850\ \mu A$, on an enlarged scale with vector T (**d**); Levels of coupling $\Delta = 0.2$ V and noise $U_{in} = 40$ mV.

It can be seen that the synchronization areas are separate isolated regions that are suitable for setting vectors E' of dimension 2. In this case, with all other things being equal, N_s depends on the topology, and is $N_s = 16$ for a "star" connection and $N_s = 14$ for an "all-to-all" connection. The area shape also depends on the topology.

When comparing the values for two- and three-oscillator schemes with the same parameters, including the topology, we may propose a general rule stating that with the increase of the number of interacting oscillators, N_s, increases. Yet, this is evident as the number of freedom degree increases at determining the synchronization value SHR = $k_1:k_2:k_3:...:k_N$. Nevertheless, as we show below, at certain parameters, there are some exceptions from the general rule.

Distribution of synchronization effectiveness is shown in Figure 7b. Here, we can see that a local maximum η is present in each area, and it can be used for vector storage and recognition, according to the method described in Section 2.1.

Below, we will give an example of a vector storage and recognition (that determine colors RGB) by using the algorithm described in Section 2.1 for the scheme "star".

3.1. Vector Storage and ONN Training

Step 1: Suppose that we have to store three vectors that correspond to three colors RGB at the constant level of the component blue = 200, with coordinates $E'_1 = (28, 149)$, $E'_2 = (28, 28)$, and $E'_3 = (150, 31)$. It should be noted that the number of coordinates in each vector is one unit less than the number of oscillators, and is equal two. As it has been explained above, this is necessary to narrow the area of possible synchronization and to reduce the recognition ambiguity. Liner transformation of coordinates, from the current parameters into color parameters, should be thought over initially. In our case, we used the following formulas: red $\Leftrightarrow (I_{p2}\ [\mu A] - 550\ [\mu A])/2$ and green $\Leftrightarrow (I_{p3}\ [\mu A] - 550\ [\mu A])/2$. After the working area has been transformed, we set three vectors, as shown in Figure 7b.

Step 2: Then, we start the system training by adjusting parameters of the ONN (I_{p1} and coupling strength $\Delta_{i,j}$) in such a way that the synchronization areas are obtained on the vectors at the point of maximum η. A fine adjustment for the maximum η can be done by using the gradient search. As a result, we have found that at $\Delta_{i,j} = \Delta = 0.2$ V, and $I_{p1} = 850\ \mu A$, the system complies with the assigned task.

Step 3: As we have achieved a positive result after the network training, the vector recording may be considered completed at this step.

3.2. Vectors Recognition

When test vector $\mathbf{T} \approx \mathbf{E'}_1$ is supplied, the neural network is transformed into a synchronous state with the synchronization order 1:2:1. At $\mathbf{T} \approx \mathbf{E'}_2$, we get SHR = 1:2:2. At $\mathbf{T} \approx \mathbf{E'}_3$, we get SHR = 1:1:2. In each case, the degree of match $d_m = \eta(\mathbf{E'}) - \eta(\mathbf{T})$ is calculated, that determines the degree of match of vector $\mathbf{T}$ with the stored vectors. For example, at $\mathbf{T} = (31,145)$, the system transfers to SHR = 1:2:1 and the vector is recognized as the vector $\mathbf{E'}_1 = (28,149)$, with the degree of match $d_m = \eta(\mathbf{E'}_1) - \eta(\mathbf{T}) = 2\%$ (see Figure 7d).

Step 1: Set the test vector $\mathbf{T} = (31,145)$.

Step 2: Determine that the system transfers to SHR = 1:2:1, and the vector is recognized as the vector $\mathbf{E'}_1 = (28,149)$.

Step 3: Determine the degree of match $d_m = \eta(\mathbf{E'}_1) - \eta(\mathbf{T}) = 2\%$.

Thus, we have performed storage and recognition of three various patterns with RGB color, although the capacity of this system is considerably higher and enables storing up to 16 patterns simultaneously.

Figure 8 shows the dependence of the number of synchronous states N_s on Δ at three different configurations of a neural network at the constant noise level $U_{in} = 20$ mV. The existence of the main maximum N_s is evident at some optimal value Δ_{opt}, in this case, this value is roughly the same as $\Delta_{opt} \sim 0.1$ V for all configurations, and does not depend on the oscillator number. The existence of the curve maximum $N_s(\Delta)$ reduplicates our results obtained in [28] for a two-oscillator scheme. Inalterability of Δ_{opt} for a different number of oscillators N, with all other parameters being equal, might be explained by the fact that, with the increase of the number of freedom degrees for synchronization order, SHR = $k_1:k_2:k_3:...:k_N$, the value of N_s has the tendency to grow.

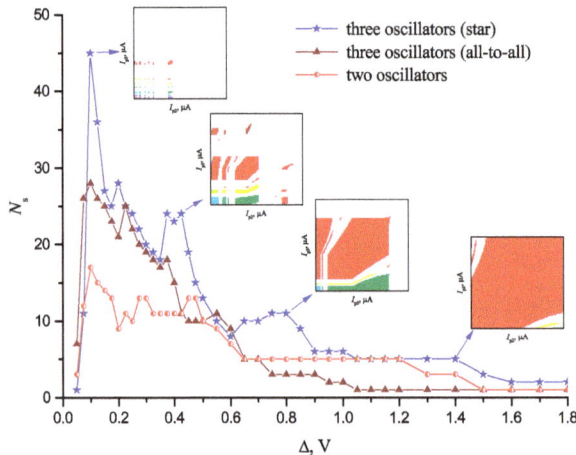

Figure 8. Dependence N_s on coupling strength value between oscillators Δ at various configurations of oscillator neural network and constant noise level $U_{in} = 20$ mV. The insertions show the evolution of synchronization areas and demonstrate the effect of their merging with Δ growth.

We should also note the general tendency for N_s to decrease when the coupling strength Δ grows above Δ_{opt} and, at its large values, the system tends to the lowest possible $N_s = 1$ with synchronization value 1:1:1. This is related to the fact that, with the increase of Δ, the surface of certain synchronization areas increase. Neighboring areas merge; in this case, the synchronization order of the resulting area predominantly consists of lower harmonic numbers. As the dimension of the control parameters is limited, such growth of synchronization area surfaces irrevocably results in a decrease of their number and value of N_s. The insertions in Figure 8 show the evolution of synchronization areas and demonstrate the effect of their merging with Δ growth.

Besides, we should note the existence of local maximums at $\Delta > \Delta_{opt}$. In turn, this is related to the fact that, in the presence of noise in the neural network, the increase of Δ may result in the development of new synchronization areas at the control parameters values that previously corresponded to the non-synchronous state of the system. Therefore, in a general case, the curve $N_s(\Delta)$ may have a complicated shape with several maximums, as we can see in Figure 8.

In addition, the initial sharp growth of all three curves at the plot should be noted when Δ increases from 0 to Δ_{opt}. The latter is due to synchronization effect degradation at $\Delta{\to}0$.

When comparing the curves, it may be noted that the increase of the oscillator number in the network leads to the increase of the maximum value of N_s in the system; this does not contradict the rule suggested above. For example, $N_{s\_max} = 17$ is for two oscillators, for three-oscillator schemes, and $N_{s\_max} = 28$ and $N_{s\_max} = 45$ are for the "all-to-all" and "star" schemes, respectively. At the same time, at certain values of coupling strength (for example, at $\Delta = 1.1$ V), the value of N_s for a two-oscillator scheme may be even higher. Also, the regularity that the "star" topology has higher N_s than the topology of "all-to-all" is observed. All of these things mean that the increase of the coupling number may contribute to the effect of system desynchronization and decrease of N_s, as it seems that oscillators prevent each other from synchronization.

Figure 9 shows the curve of N_s vs noise level in the system U_{in} at the same coupling strength $\Delta = 0.2$ V, for three configurations of an oscillator neural network. The general trend for the decrease of N_s at the noise amplitude increase is due to the decrease of the surface of synchronous areas which eventually disappear (see the insertions in Figure 9).

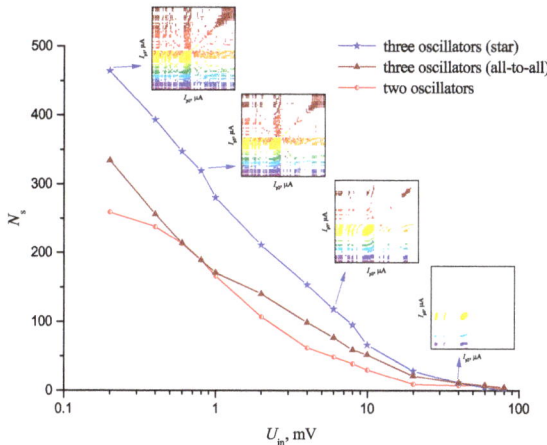

Figure 9. Dependence N_s on noise level U_{in} at various configurations of an oscillator neural network at the same coupling strength $\Delta = 0.2$ V.

It should also be noted that the general rule is observed: stating that the value of N_s for a three-oscillator configuration "star" is higher than that for a two-oscillator one. The shapes of curves $N_s(U_{in})$ are similar, which indicated that the physics of noise effect on the network is similar, and does not depend on the number of oscillators.

Comparing the above curves $N_s(\Delta)$ and $N_s(U_{in})$, it may be seen that the number of synchronization areas N_{s_max} in our models may reach $N_{s\_max} \sim 450$, at an optimal coupling strength value $\Delta = \Delta_{opt}$, and at lowered noise $U_{in} = 10$ µV, it increases to $N_{s\_max} \sim 650$.

4. Conclusions

A new method of pattern storage and recognition in an impulse oscillator neural network based on resistive switches and the high-order synchronization effect is presented, using computational modeling of two- and three-oscillator schemes with thermally coupled VO_2-switches.

Our method allows storage of a multitude of patterns N_s, where each state of the system is characterized by synchronization order SHR = $k_1:k_2:k_3:...:k_N$.

A general rule is suggested, stating that N_s increases with the increase of the number of interacting oscillators. The modeling demonstrates achievement of N_s of several orders: N_s ~650 for a three-oscillator scheme and N_s ~260 for a two-oscillator scheme.

Several regularities of functional characteristics of such ONNs have been obtained; in particular, the existence of an optimal coupling strength between oscillators has been revealed, when the number of synchronous states is maximal. A general tendency for N_s decrease with the increase of coupling strength and switches' inner noise amplitude, is also shown.

The algorithm of vector storage, network training, and test vector recognition has been proposed, where the parameter of synchronization effectiveness is used as the degree of match. It has been shown that it is more expedient to use the number of coordinates in each vector at least one unit less than the number of oscillators ($N - 1$), because it is necessary to narrow the area of possible synchronization and to lower the recognition ambiguity.

By contrast, for example, to the FSK method [22], such an approach to the problem of pattern storage and recognition allows one to significantly increase the information capacity, N_s, of a neural network using the minimum number of neural oscillators. In addition, the proposed concept of pulse synchronization definition, through calculation of a family of metrics, opens a natural way for gradient method application to an oscillator network training (optimization).

Although the research has been performed on a certain model object (VO_2 thermally coupled relaxation oscillators), the demonstrated method of pattern storage and recognition is sufficiently general, and the fundamental character of the obtained regularities may be the subject of further research of ONNs, of various mechanisms and oscillator-coupling topology.

Author Contributions: Conceptualization, A.V. and M.B.; methodology, V.P. and P.B.; software, A.V.; validation, P.B.; writing—original draft preparation, A.V., M.B. and V.P.; project administration, A.V.

Funding: This research was supported by Russian Science Foundation (grant no. 16-19-00135).

Acknowledgments: The authors express their gratitude to O. Dobrynina for some valuable comments in the course of the article translation.

Conflicts of Interest: The authors declare no conflict of interest.

References

1. Heaton, J. *Artificial Intelligence for Humans*; Createspace Independent Publishing: Scotts Valley, CA, USA, 2015; ISBN 9781505714340.
2. Bishop, C.M. *Neural Networks for Pattern Recognition*; Clarendon Press: New York, NY, USA, 1995; ISBN 0198538642.
3. Hopfield, J.J.; Tank, D.W. Computing with neural circuits: A model. *Science* **1986**, *233*, 625–633. [CrossRef] [PubMed]
4. Strogatz, S.H. From Kuramoto to Crawford: Exploring the onset of synchronization in populations of coupled oscillators. *Phys. D Nonlinear Phenom.* **2000**, *143*, 1–20. [CrossRef]
5. Vodenicarevic, D.; Locatelli, N.; Abreu Araujo, F.; Grollier, J.; Querlioz, D. A Nanotechnology-Ready Computing Scheme based on a Weakly Coupled Oscillator Network. *Sci. Rep.* **2017**, *7*, 44772. [CrossRef] [PubMed]
6. Nakano, H.; Saito, T. Grouping Synchronization in a Pulse-Coupled Network of Chaotic Spiking Oscillators. *IEEE Trans. Neural Netw.* **2004**, *15*, 1018–1026. [CrossRef] [PubMed]

7. Velichko, A.; Belyaev, M.; Putrolaynen, V.; Pergament, A.; Perminov, V. Switching dynamics of single and coupled VO2-based oscillators as elements of neural networks. *Int. J. Mod. Phys. B* **2017**, *31*, 1650261. [CrossRef]

8. Shukla, N.; Parihar, A.; Cotter, M.; Barth, M.; Li, X.; Chandramoorthy, N.; Paik, H.; Schlom, D.G.; Narayanan, V.; Raychowdhury, A.; Datta, S. Pairwise coupled hybrid vanadium dioxide-MOSFET (HVFET) oscillators for non-boolean associative computing. In Proceedings of the 2014 IEEE International Electron Devices Meeting, San Francisco, CA, USA, 15–17 December 2014; pp. 28.7.1–28.7.4.

9. Khitun, A.G.; Geremew, A.K.; Balandin, A.A. Transistor-Less Logic Circuits Implemented With 2-D Charge Density Wave Devices. *IEEE Electron. Device Lett.* **2018**, *39*, 1449–1452. [CrossRef]

10. Khitun, A.; Liu, G.; Balandin, A.A. Two-dimensional oscillatory neural network based on room-temperature charge-density-wave devices. *IEEE Trans. Nanotechnol.* **2017**, *16*, 860–867. [CrossRef]

11. Liu, G.; Debnath, B.; Pope, T.R.; Salguero, T.T.; Lake, R.K.; Balandin, A.A. A charge-density-wave oscillator based on an integrated tantalum disulfide-boron nitride-graphene device operating at room temperature. *Nat. Nanotechnol.* **2016**, *11*, 845–850. [CrossRef] [PubMed]

12. Ghosh, S. Generation of high-frequency power oscillation by astable mode arcing with SCR switched inductor. *IEEE J. Solid-State Circuits* **1984**, *19*, 269–271. [CrossRef]

13. Chen, C.; Mathews, R.; Mahoney, L.; Calawa, S.; Sage, J.; Molvar, K.; Parker, C.; Maki, P.; Sollner, T.C.L. Resonant-tunneling-diode relaxation oscillator. *Solid. State. Electron.* **2000**, *44*, 1853–1856. [CrossRef]

14. Sharma, A.A.; Bain, J.A.; Weldon, J.A. Phase Coupling and Control of Oxide-Based Oscillators for Neuromorphic Computing. *IEEE J. Explor. Solid-State Comput. Devices Circuits* **2015**, *1*, 58–66. [CrossRef]

15. Romera, M.; Talatchian, P.; Tsunegi, S.; Araujo, F.A.; Cros, V.; Bortolotti, P.; Yakushiji, K.; Fukushima, A.; Kubota, H.; Yuasa, S.; et al. Vowel recognition with four coupled spin-torque nano-oscillators. *arXiv*, 2018; arXiv:1711.02704.

16. Belyaev, M.A.; Boriskov, P.P.; Velichko, A.A.; Pergament, A.L.; Putrolainen, V.V.; Ryabokon', D.V.; Stefanovich, G.B.; Sysun, V.I.; Khanin, S.D. Switching Channel Development Dynamics in Planar Structures on the Basis of Vanadium Dioxide. *Phys. Solid State* **2018**, *60*, 447–456. [CrossRef]

17. Velichko, A.; Belyaev, M.; Putrolaynen, V.; Perminov, V.; Pergament, A. Thermal coupling and effect of subharmonic synchronization in a system of two VO2 based oscillators. *Solid State Electron.* **2018**, *141*, 40–49. [CrossRef]

18. Sakai, J. High-efficiency voltage oscillation in VO2 planer-type junctions with infinite negative differential resistance. *J. Appl. Phys.* **2008**, *103*, 103708. [CrossRef]

19. Hoppensteadt, F.C.; Izhikevich, E.M. Oscillatory Neurocomputers with Dynamic Connectivity. *Phys. Rev. Lett.* **1999**, *82*, 2983–2986. [CrossRef]

20. Hoppensteadt, F.C.; Izhikevich, E.M. Pattern recognition via synchronization in phase-locked loop neural networks. *IEEE Trans. Neural Netw.* **2000**, *11*, 734–738. [CrossRef] [PubMed]

21. Izhikevich, E.M. Weakly pulse-coupled oscillators, FM interactions, synchronization, and oscillatory associative memory. *IEEE Trans. Neural Netw.* **1999**, *10*, 508–526. [CrossRef] [PubMed]

22. Nikonov, D.E.; Csaba, G.; Porod, W.; Shibata, T.; Voils, D.; Hammerstrom, D.; Young, I.A.; Bourianoff, G.I. Coupled-Oscillator Associative Memory Array Operation for Pattern Recognition. *IEEE J. Explor. Solid-State Comput. Devices Circuits* **2015**, *1*, 85–93. [CrossRef]

23. Pikovsky, A.; Rosenblum, M.; Kurths, J. *(Jürgen) Synchronization: A Universal Concept in Nonlinear Sciences*; Cambridge University Press: Cambridge, UK, 2003; ISBN 9780521533522.

24. Theodoridis, S.; Koutroumbas, K. *Pattern Recognition*; Academic Press: Cambridge, MA, USA, 2009; ISBN 9781597492720.

25. Kumar, A.; Mohanty, P. Autoassociative Memory and Pattern Recognition in Micromechanical Oscillator Network. *Sci. Rep.* **2017**, *7*, 411. [CrossRef] [PubMed]

26. Shukla, N.; Parihar, A.; Freeman, E.; Paik, H.; Stone, G.; Narayanan, V.; Wen, H.; Cai, Z.; Gopalan, V.; Engel-Herbert, R.; et al. Synchronized charge oscillations in correlated electron systems. *Sci. Rep.* **2014**, *4*, 4964. [CrossRef]

27. Hoppensteadt, F.C.; Izhikevich, E.M. Synchronization of laser oscillators, associative memory, and optical neurocomputing. *Phys. Rev. E* **2000**, *62*, 4010–4013. [CrossRef]

28. Velichko, A.; Putrolaynen, V.; Belyaev, M. Effects of Higher Order and Long-Range Synchronizations for Classification and Computing in Oscillator-Based Spiking Neural Networks. *arXiv*, **2018**; arXiv:1804.03395.

29. Gill, P.E.; Murray, W.; Wright, M.H. *Practical Optimization*; Emerald Group Publishing: Bingley, UK, 1982; ISBN 0122839528.

30. Velichko, A.; Belyaev, M.; Putrolaynen, V.; Perminov, V.; Pergament, A. Modeling of thermal coupling in VO$_2$-based oscillatory neural networks. *Solid State Electron.* **2018**, *139*, 8–14. [CrossRef]

© 2018 by the authors. Licensee MDPI, Basel, Switzerland. This article is an open access article distributed under the terms and conditions of the Creative Commons Attribution (CC BY) license (http://creativecommons.org/licenses/by/4.0/).

![electronics logo] *electronics*

MDPI

Article

CMOS Compatible Bio-Realistic Implementation with Ag/HfO$_2$-Based Synaptic Nanoelectronics for Artificial Neuromorphic System

Lin Chen *, Zhen-Yu He, Tian-Yu Wang, Ya-Wei Dai, Hao Zhu, Qing-Qing Sun * and David Wei Zhang

State Key Laboratory of ASIC and System, School of Microelectronics, Fudan University, Shanghai 200433, China; 17212020012@fudan.edu.cn (Z.-Y.H.); 16210720087@fudan.edu.cn (T.-Y.W.); 14110720046@fudan.edu.cn (Y.-W.D.); hao_zhu@fudan.edu.cn (H.Z.); wzhang@fudan.edu.cn (D.W.Z.)

* Correspondence: linchen@fudan.edu.cn (L.C.); qqsun@fudan.edu.cn (Q.-Q.S.);
  Tel.: +86-21-6564-3150 (L.C.); +86-21-6564-2389 (Q.-Q.S.)

Received: 2 May 2018; Accepted: 22 May 2018; Published: 25 May 2018

Abstract: The emerging resistive switching devices have attracted broad interest as promising candidates for future memory and computing applications. Particularly, it is believed that memristor-based neuromorphic engineering promises to enable efficient artificial neuromorphic systems. In this work, the synaptic abilities are demonstrated in HfO$_2$-based resistive memories for their multi-level storage capability as well as being compatible with advanced CMOS technology. Both inert metal (TaN) and active metal (Ag) are selected as top electrodes (TE) to mimic the abilities of a biological synapse. HfO$_2$-based resistive memories with active TE exhibit great advantages in bio-realistic implementation such as suitable switching speed, low power and multilevel switching. Moreover, key features of a biological synapse such as short-term/long-term memory, "learning and forgetting", long-term potentiation/depression, and the spike-timing-dependent plasticity (STDP) rule are implemented in a single Ag/HfO$_2$/Pt synaptic device without the poorly scalable software and tedious process in transistors-based artificial neuromorphic systems.

Keywords: memristor; synaptic device; spike-timing-dependent plasticity; neuromorphic computation

1. Introduction

Biological solid-state devices have attracted more and more interest in recent years due to their promising advantages such as massive parallelism, power efficiency, adaptivity to complex non-linear computations, and high tolerance towards defects and variability [1,2]. For brain science, the synapse is the fundamental unit in neuromorphic systems, and therefore, to study the imitation of biological synapse is a critical first step toward neuromorphic computing. In previous research, conventional von Neumann digital computers software simulation and silicon complementary metal-oxide-semiconductor (CMOS) logic circuits were selected to implement artificial neural networks. However, these architectures lacked synapses' intrinsic advantages in processing well-defined biological behavior, and instead were incomparable with biological systems in terms of space, speed and energy efficiency [3,4]. Large chip area and high power dissipation became blocks to the application of transistor-based neuromorphic system architectures, since ten transistors are required to mimic the function of each biological synapse at least. Additionally, data storage and computation are separated in von Neumann digital architecture and result in intrinsic delay during data transfer, which influence the computation speed. Therefore, it is more feasible to emulate the biological synapse in single solid-state devices with their inherent memory and computing abilities.

Recently, memristor has been widely investigated for next-generation nonvolatile memory (NVM) in embedded and stand-alone applications due to its advantages such as simple device structure, excellent scalability, fast program/erase speed and low power consumption. It is worth noting that the typical two-terminal metal-insulator-metal (MIM) structure of memristor is similar to synapses in a biological brain, and its high-density three-dimensional (3D) cross-point integrated architecture is compatible with neuromorphic systems in the biological brain. In addition, memristors with bipolar switching behaviors could analog memory characteristic by offering multi-level conductance states. These memristors have been proposed as attractive candidates to emulate synaptic functions in biologically inspired neuromorphic systems due to their multi-level conductance states [5–7]. So far, a wide variety of materials including metal oxides, metal nitrides, metal-oxide nano-laminates and organics have shown synaptic behaviors [8–13]. Among these materials, few of them were compatible with CMOS process. A systematic study of the memristor operating as a biological synapse with CMOS compatible materials and fabrication processes are still lacking. HfO_2, as a high-k materials with resistive switching characteristics, has the potential to be used as the functional layer of synaptic devices [14–16], and could well be compatible with the traditional CMOS process and play an important role in the process of device scaling.

In this paper, we develop comprehensive abilities of biological synapse in HfO_2-based synaptic devices. Highly uniform HfO_2 were fabricated by atomic layer deposition (ALD) method, which enable device sizes capable of scaling to nanoscale. Besides, HfO_2 has superior electrical properties and excellent thermal and chemical stability, as well as being one of the preferred high-k gate dielectric materials in CMOS technology. High-density HfO_2-based memristive devices can be successfully integrated with conventional CMOS technology. According to this work, synaptic devices with Ag active electrodes show comprehensive biological synapse abilities with lower power consumption. Short-term plasticity (STP), long-term potentiation/depression, spike-timing-dependent plasticity (STDP) and other properties of biological synapses were achieved with a series of electrical diagnosis scheme. These precisely electronic synapse behaviors are derived from the oxidation and reduction of metallic Ag between electrodes. This CMOS compatible HfO_2-based synaptic devices show high potential to form an artificial neuromorphic computing system.

2. Materials and Methods

Synaptic Devices Fabrication. To fabricate the HfO_2-based synaptic device (Figure 1a), a 50 nm Pt bottom electrode was deposited by reactive magnetron sputtering onto the cleaned SiO_2/Si substrate. Then, a 10 nm HfO_2 layer was deposited using ALD derived from TEMAH and H_2O precursors at 250 °C. Finally, 50 nm Ag or TaN top electrode was fabricated by photolithography and followed by reactive magnetron sputtering, respectively.

Electrical Measurements. Direct-current measurements of the HfO_2-based synaptic devices were performed using Agilent B1500A semiconductor parameter analyzer with biased top and grounded bottom electrodes at room temperature. Current compliance was applied to protect the devices and to control the filament size. Pulse measurements were performed using Agilent B1500A semiconductor device parameter analyzer and two semiconductor pulse generator units (SPGU).

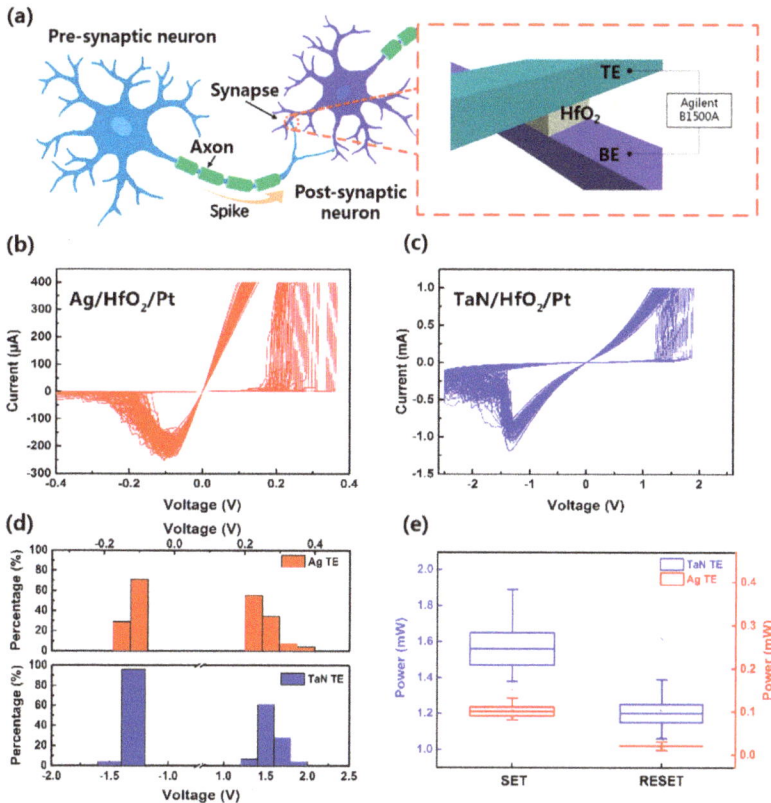

Figure 1. (**a**) Illustration of biological synapse connecting two neurons and schematic structure of the equivalent HfO_2-based synaptic device; (**b**) repetitive I-V characteristics in $Ag/HfO_2/Pt$ devices; and (**c**) $TaN/HfO_2/Pt$ devices; (**d**) histograms of the RESET/SET operating voltage in $Ag/HfO_2/Pt$ and $TaN/HfO_2/Pt$ devices during cycling test; (**e**) Statistical charts of the power consumption in $Ag/HfO_2/Pt$ and $TaN/HfO_2/Pt$ devices.

3. Result and Discussion

A synapse is the fundamental unit in the biological brain for connection between adjacent neurons. Figure 1a illustrates a biological synapse connecting two neurons and the schematic structure of equivalent an HfO_2-based synaptic device. The action potentials (spikes) are generated by one neuron, propagate through the axon and are transmitted to the next neuron through the synapses. The typical two-terminal HfO_2-based metal-insulator-metal (MIM) structure is similar to synapse in a biological neuromorphic system, and top and bottom electrodes of a synaptic device can be regarded as pre-synaptic and post-synaptic neuron, respectively.

The devices with Ag top electrode and TaN top electrode were evaluated through direct-current voltage sweeping. The initial synaptic devices show high resistance state (HRS). After an electroforming process, repetitive I-V characteristics were measured under proper compliance current. As shown in Figure 1b,c, both Ag and TaN top electrode exhibit superior bipolar switching characteristics. Stable cycling switching ability with an ON-OFF ratio and over 10-year data retention are obtained (Figure S1). The set and reset operation voltage distribution and power consumption are presented as histograms and statistical charts in Figure 1d,e, respectively. The $Ag/HfO_2/Pt$ devices exhibit much lower SET/RESET operation voltage and power consumption. The SET and RESET voltage ranges of the Ag/HfO$_2$/Pt devices were from 0.21 to 0.37 V and from −0.1 to −0.15 V, while those of the TaN/HfO$_2$/Pt devices were from 1.3 to 1.8 V and from −1.3 to −1.5 V. The average power consumption was 0.1/0.02 mW in Ag/HfO$_2$/Pt devices, while it was 1.6/1.2 mW in TaN/HfO$_2$/Pt devices.

Moreover, a pulse-driven switching test was also implemented in TaN/HfO$_2$/Pt devices; these devices exhibit excellent pulse endurance properties at the nanosecond level (Figure S2). As superior data storage reliability and ultra-fast programming speed, it is clear that HfO$_2$-based devices with TaN top electrode seems more suitable for NVM application, while Ag/HfO$_2$/Pt devices exhibit lower power consumption and multi-level resistance states (Figure S3), which offer more advantages in neuromorphic computing application.

To further clarify the switching mechanism of the two types, HfO$_2$ based devices, the I-V curves were analyzed in Figure 2a–d. The current transportation in both low resistance state (LRS) and high resistance state (HRS) exhibit ohmic current at positive and negative electrical field in Ag/HfO$_2$/Pt devices. For TaN/HfO$_2$/Pt devices, the LRS exhibits quasi-ohmic current (slope is approximately equal to 1) at positive electric field. However, the HRS exhibits ohmic, quasi-ohmic and space charge limited current along with an increasing electric field. Similar current transport mechanisms were observed with negative voltages. These results indicated that the stochastic and avalanching generation and recombination of oxygen ion and oxygen vacancy in the HfO$_2$ layer result in the forming and rupture of conducting filaments in TaN/HfO$_2$/Pt devices, which induce ultra-fast switching characteristics in TaN/HfO$_2$/Pt devices. In Ag/HfO$_2$/Pt devices, the forming and rupture of conducting filaments derive from the oxidation and reduction of metallic Ag, which can be driven by a much lower electric field.

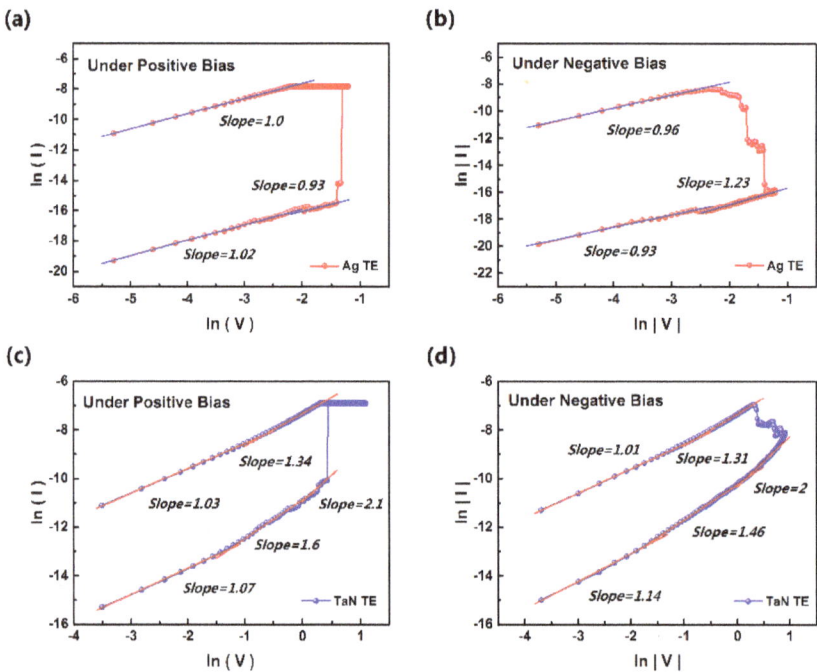

Figure 2. The current fitting of the Ag/HfO$_2$/Pt devices at (**a**) positive and (**b**) negative electric field; the current fitting of the TaN/HfO$_2$/Pt devices at (**c**) positive and (**d**) negative electric field.

Figure 3 shows the schematic illustration of the evolution of the metallic Ag filament formation and rupture in Ag/HfO$_2$/Pt synaptic devices. These Ag/HfO$_2$/Pt synaptic devices correspond to conventional electrochemical memory (ECM) cells [17]. Under a positive bias, the active Ag is oxidized into Ag$^+$ ions at the Ag top electrode, and then, the migration and accumulation of those Ag$^+$ ions happened toward the Pt cathode. Finally, the Ag$^+$ ions are reduced to Ag conductive filaments across the HfO$_2$ dielectric layer. Therefore, these Ag/HfO$_2$/Pt synaptic devices exhibit an ohmic transport mechanism in the LRS state (set process). The diameter of Ag filament can be modulated by regulating the value of compliance current, thus multi-level resistance states can be realized in Ag/HfO$_2$/Pt synaptic devices (Figure S3). When a negative bias is applied to the top Ag electrode, the oxidation of Ag filaments into Ag$^+$ ions and reduction at the Ag electrode occurred, which leading to the rupture of the filament conduction path and switching these devices to an OFF-state (reset process).

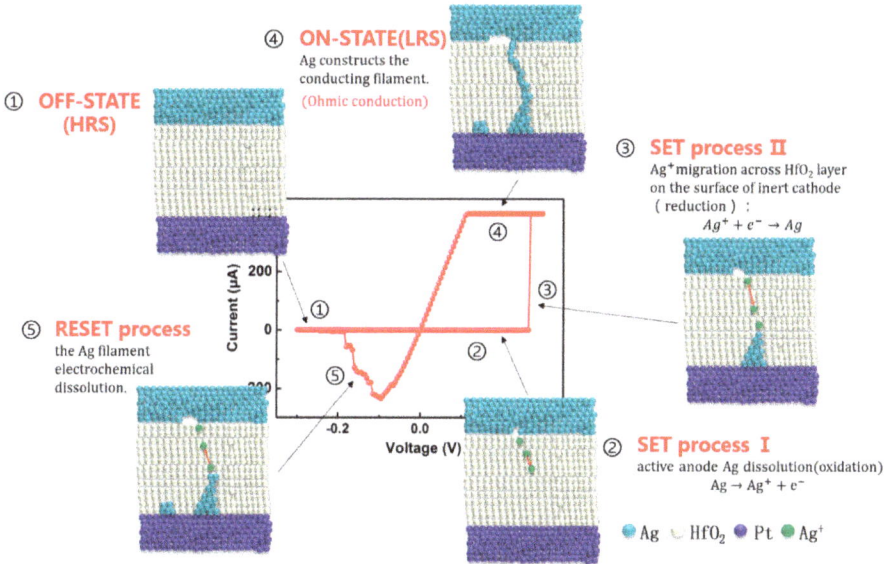

Figure 3. schematic illustrations of the evolution of the metallic Ag filament formation and rupture in Ag/HfO$_2$/Pt devices.

Synaptic plasticity is widely considered as the major cellular mechanism during learning and memorization in the biological brain. To mimic the abilities of synaptic plasticity comprehensively is the key feature of electron synaptic devices [18–21]. Synaptic plasticity illustrates the dynamic change of connection strength (synaptic weight) between biological neurons under electric spikes, and it can be classified into short-term and long-term plasticity (STP and LTP). STP means the duration of the potentiation of synaptic weight ranging from a few seconds to a few minutes. While for LTP, the potentiation of synaptic weight can last from hours to weeks or even years. However, STP can be converted to LTP through repeated rehearsals as well.

Figure 4a shows the STP properties of the Ag/HfO$_2$/Pt synaptic devices. To mimic the synaptic behaviors, 300 mV amplitude input pulse sequences with 100 ms pulse width and 1 s repetition interval were applied. An increase of conductance can be observed when a pulse arrived, and the conductance value reaches 5.7 μS (corresponding to a 1.7 μA output current). However, these devices cannot maintain its high conductance state and the conductance quickly decays to its initial conductance value during the relaxation duration of 1 s. This behavior corresponds to the STP behaviors in biological synapses. It is worth noting that the conductance value under the input pulse is lower than 77.5 μS, which is calculated approximately as the conductance value of one quantized channel. This phenomenon is attributed to the incomplete formation of Ag conductive filament. In order to further demonstrate this phenomenon, a DC I-V characteristic with a compliance current of 1 μA was measured. These devices show volatile memory behavior (shown in Figure 4b). Figure 4c shows the I-V characteristic in LRS and HRS. The current transportation in HRS exhibits an ohmic current, which is consists to the conclusion previous noted in this work. On the other hand, the current in LRS across the device is a space charge limited current. The threshold between different transport mechanisms in the LRS of the device is the conductance of one quantized channel. When the compliance current increase and the device conductance is over a constant value (77.5 μS), the devices exhibit an ohmic transport mechanism again in the LRS (shown in Figure S3). Thus, STP behaviors converts to LTP ability in Ag/HfO$_2$/Pt synaptic devices.

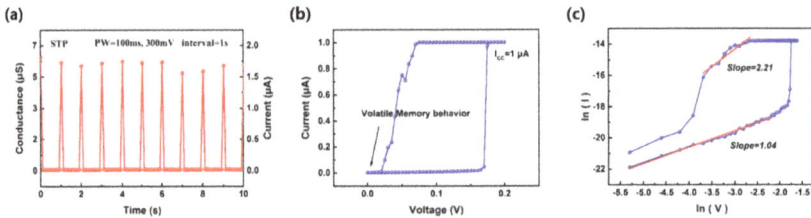

Figure 4. (**a**) STP ability in Ag/HfO$_2$/Pt synaptic devices. Input pulse sequences with a pulse amplitude of 300 mV, a pulse width of 100 ms and repetition intervals of 1 s were applied to mimic the STP ability; (**b**) Ag/HfO$_2$/Pt synaptic devices exhibit volatile memory behavior with a compliance current of 1 μA; (**c**) current fitting of HRS and LRS under positive electric field of Ag/HfO$_2$/Pt synaptic device. The current in HRS is an ohmic current while in LRS, it exhibits space charge limited current.

The "learning and forgetting" behaviors of a biological brain were also observed in Ag/HfO$_2$/Pt synaptic devices. As shown in Figure 5a, input pulse sequences with a pulse number of 60, amplitude of 300 mV, 100 ms pulse width, and 100 ms intervals were applied to the fresh synaptic devices. The conductance gradually increases with the number of pulses, and saturates at 140 μS after the first "learning process". When the applied pulse sequences are removed, a spontaneous decay of conductance occurs. The decay rate of conductance is fast at the initial state and then gradually slows down. Such a changing trend corresponds to the "human-memory forgetting curve" (Figure 5b). Insets in Figure 5a,b exhibit the "learning-experience" behavior on semi log coordinate. It is worth noting that the conductance of the synaptic devices does not relax to the initial state but rather stabilizes at an intermediate level. In the following "learning process", less input pulse sequences are needed to produce the same amount of memory as the first "learning process".

From a filamentary view, during the initial "learning process", the conductance gradually increases due to the formation of a metallic Ag filament in the HfO$_2$ layer. After learning, the Ag filament in Ag/HfO$_2$/Pt synaptic devices is unstable, which is similar to the incomplete forming process during NVM application. After the "learning process", the Ag filament is not dissolved completely. Thus, the conductance of Ag/HfO$_2$/Pt synaptic devices remains in an intermediate level, and it can re-learn the forgotten information much easier through rehearsals.

Figure 6a shows the measured long-term potentiation (LTP) and depression (LTD) curves using potentiating/depressing pulse sequences with an amplitude of 300 mV/−200 mV, a pulse width of 100 ms and a period of 250 ms. A read pulse with an amplitude of 50 mV/−50 mV, a pulse width of 100 ms was applied to measure the cell conductance followed every potentiating and depressing pulse, respectively. The conductance of Ag/HfO$_2$/Pt synaptic devices gradually increases with the stress of each potentiating pulse, and then gradually decreases with the following depressing pulses. The evolution of the conductance of Ag/HfO$_2$/Pt synaptic devices exhibits nonlinear behavior, and the change rate of conductance in the potentiating process is faster than in the depressing process. This could be due to the potentiating process being a positive feedback between the formation rate of metallic Ag filament, the temperature and local field strength [22,23], while the depressing process is a negative feedback and a slow decay of conductance can be observed.

Figure 6b shows the implementation of the STDP rule in Ag/HfO$_2$/Pt synaptic devices. The STDP rule is one of the important aspects in synaptic plasticity in a biological brain. The STDP rule can be understood as the change of synaptic weight dependence on the relative timing of pre-synaptic and post-synaptic spikes. If the pre-synaptic spike precedes a post-synaptic spike in arriving at the synapse ($\Delta t > 0$), the synaptic weight potentiation, corresponds to the increase of conductance. If the post-synaptic spike precedes pre-synaptic spike ($\Delta t < 0$), the synaptic weight depression, corresponds to the decrease of conductance. Furthermore, as the absolute value of Δt decreases, the amplitude of the change of synaptic weight increases. Both pre- synaptic and post-synaptic spikes contain a negative pulse with an amplitude of -200 mV in the beginning followed by a positive pulse with an amplitude of 300 mV. Each input pulse alone cannot modulate the conductance of Ag/HfO$_2$/Pt synaptic devices. When the pre-synaptic spike reaches the device earlier than the post-synaptic ($\Delta t > 0$), a conductance increase is observed. If $\Delta t < 0$, the conductance of Ag/HfO$_2$/Pt synaptic devices will decrease as well. The change of the conductance in the depression area is smaller than the potentiation one; it could be due to the inherent large resistive window in Ag/HfO$_2$/Pt devices which interfere with the visual presentation of the STDP rule on some level.

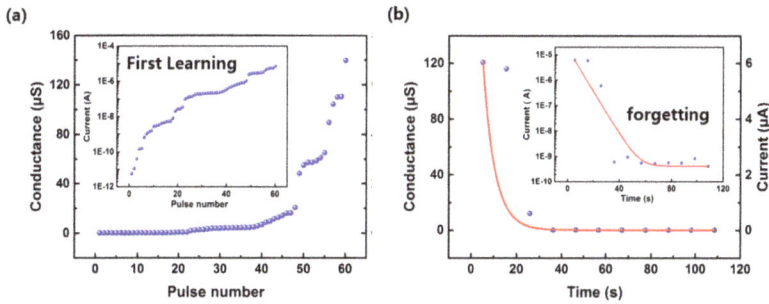

Figure 5. "learning and forgetting" ability in Ag/HfO$_2$/Pt synaptic devices. (**a**) Nonlinear increase of the conductance by consecutive input pulses during the first "learning process"; (**b**) the spontaneous decay of the device conductance, which is similar to the "human-memory forgetting curve". The red lines are the exponential fits to the experimental data.

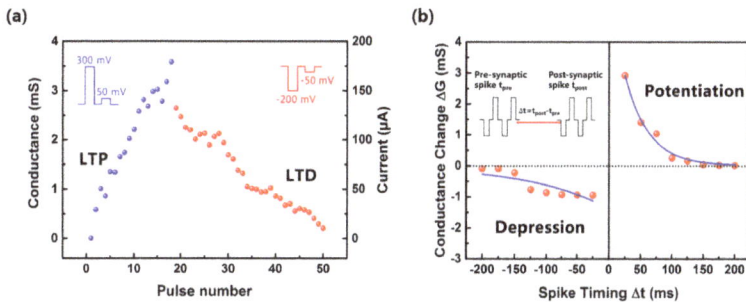

Figure 6. (**a**) The implementation of LTP and LTD in Ag/HfO$_2$/Pt synaptic devices. The change of the conductance during potentiating/depressing process was recorded using a read voltage pulse with an amplitude of 50/-50 mV and a pulse width of 100 ms, respectively. Potentiation pulses: 300 mV, 100 ms; depression pulses: -200 mV, 100 ms; (**b**) STDP rule scheme in Ag/HfO$_2$/Pt synaptic devices. The blue lines are the exponential fits to the experimental data.

4. Conclusions

In summary, the results of the present study demonstrated the synaptic abilities in CMOS compatible HfO_2-based resistive memories that highly resemble its biological counterpart. Both the cells with TaN top electrode with immovable ions and Ag top electrode with movable ions show superior bipolar switching characteristics. In general, HfO_2-based devices with Ag top electrodes exhibit great advantages in bio-realistic implementation due to its ECM switching mechanism. Short-term/long-term memory, "learning and forgetting" behavior, LTP/LTD ability and STDP rule are synthetically implementation in a single $Ag/HfO_2/Pt$ synaptic device without the need for the complicated software and multi-transistor-based framework. These results suggest that an HfO_2-based synaptic device has definite potential in high scalability and low power dissipation neuromorphic computing applications. Future studies should explore how HfO_2-based neuromorphic circuits level implementation into the efficiency of bio-inspired computing systems.

Supplementary Materials: The following are available online at http://www.mdpi.com/2079-9292/7/6/80/s1, Figure S1: Device reliability evaluations in HfO_2-based synaptic devices, Figure S2: AC pulse response in $TaN/HfO_2/Pt$ devices, Figure S3: Current transport analysis in $Ag/HfO_2/Pt$ devices.

Author Contributions: Data curation, L.C., Z.-Y.H. and T.-Y.W.; Formal analysis, Z.-Y.H. and T.-Y.W.; Investigation, L.C. and Y.-W.D.; Methodology, L.C. and H.Z.; Resources, D.W.Z.; Supervision, Q.-Q.S.

Acknowledgments: The work was supported by the NSFC (61704030, 61376092 and 61427901), 02 State Key Project (2017ZX02315005), Shanghai Rising-Star Program (14QA1400200), Shanghai Educational Development Foundation, Program of Shanghai Subject Chief Scientist (14XD1400900), the S&T Committee of Shanghai (14521103000, 15DZ1100702, 15DZ1100503), and "Chen Guang" project supported by Shanghai Municipal Education Commission and Shanghai Education Development Foundation.

Conflicts of Interest: The authors declare no competing financial interest.

References

1. Bessonov, A.A.; Kirikova, M.N.; Petukhov, D.I.; Allen, M.; Ryhänen, T.; Bailey, M. Layered memristive and memcapacitive switches for printable electronics. *Nat. Mater.* **2014**, *14*, 199–204. [CrossRef] [PubMed]

2. Jo, S.H.; Chang, T.; Ebong, I.; Bhadviya, B.B.; Mazumder, P.; Lu, W. Nanoscale memristor device as synapse in neuromorphic systems. *Nano Lett.* **2010**, *10*, 1297–1301. [CrossRef] [PubMed]

3. Du, C.; Ma, W.; Chang, T.; Sheridan, P.; Lu, W.-D. Biorealistic Implementation of Synaptic Functions with Oxide Memristors through Internal Ionic Dynamics. *Adv. Funct. Mater.* **2015**. [CrossRef]

4. Kuzum, D.; Yu, S.; Philip Wong, H.-S. Synaptic electronics: Materials, devices and applications. *Nanotechnology* **2013**, *24*, 382001. [CrossRef] [PubMed]

5. Matveyev, Y.; Egorov, K.; Markeev, A.; Zenkevich, A. Resistive switching and synaptic properties of fully atomic layer deposition grown $TiN/HfO_2/TiN$ devices. *J. Appl. Phys.* **2015**, *117*, 044901. [CrossRef]

6. Tian, Y.; Guo, C.; Guo, S.; Yu, T.; Liu, Q. Bivariate-continuous-tunable interface memristor based on Bi_2S_3 nested nano-networks. *Nano Res.* **2014**, *7*, 953–962. [CrossRef]

7. Wu, Y.; Yu, S.; Wong, H.-S.P.; Chen, Y.-S.; Lee, H.-Y.; Wang, S.-M.; Gu, P.-Y.; Chen, F.; Tsai, M.-J. AlO_x-based resistive switching device with gradual resistance modulation for neuromorphic device application. In Proceedings of the 2012 4th IEEE Internationl Memory Workshop IMW, Milan, Italy, 20–23 May 2012.

8. Mandal, S.; El-Amin, A.; Alexander, K.; Rajendran, B.; Jha, R. Novel synaptic memory device for neuromorphic computing. *Sci. Rep.* **2014**, *4*, 5333. [CrossRef] [PubMed]

9. Karunarathne, M.C.; Knisley, T.J.; Tunstull, G.S.; Heeg, M.J.; Winter, C.H. Exceptional thermal stability and high volatility in mid to late first row transition metal complexes containing carbohydrazide ligands. *Polyhedron* **2013**, *52*, 820–830. [CrossRef]

10. Wang, Y.-F.; Lin, Y.-C.; Wang, I.-T.; Lin, T.-P.; Hou, T.-H. Characterization and Modeling of Nonfilamentary $Ta/TaOx/TiO_2/Ti$ Analog Synaptic Device. *Sci. Rep.* **2015**, *5*, 10150. [CrossRef] [PubMed]

11. Prezioso, M.; Merrikh-Bayat, F.; Hoskins, B.D.; Adam, G.C.; Likharev, K.K.; Strukov, D.B. Training and operation of an integrated neuromorphic network based on metal-oxide memristors. *Nature* **2015**, *521*, 61–64. [CrossRef] [PubMed]

12. Wang, Z.Q.; Xu, H.Y.; Li, X.H.; Yu, H.; Liu, Y.C.; Zhu, X.J. Synaptic learning and memory functions achieved using oxygen ion migration/diffusion in an amorphous InGaZnO memristor. *Adv. Funct. Mater.* **2012**, *22*, 2759–2765. [CrossRef]

13. Gao, B.; Liu, L.; Kang, J.-F. Investigation of the synaptic device based on the resistive switching behavior in hafnium oxide. *Prog. Nat. Sci. Mater. Int.* **2015**, *25*, 47–50. [CrossRef]

14. Ambrogio, S.; Balatti, S.; Nardi, F.; Facchinetti, S.; Ielmini, D. Spike-timing dependent plasticity in a transistor-selected resistive switching memory. *Nanotechnology* **2013**, *24*, 384012. [CrossRef] [PubMed]

15. Covi, E.; Brivio, S.; Serb, A.; Prodromakis, T.; Fanciulli, M.; Spiga, S. Analog Memristive Synapse in Spiking Networks Implementing Unsupervised Learning. *Front. Neurosci.* **2016**, *10*, 482. [CrossRef] [PubMed]

16. Yang, X.; Fang, Y.; Yu, Z.; Wang, Z.; Zhang, T.; Yin, M.; Lin, M.; Yang, Y.; Cai, Y.; Huang, R. Nonassociative learning implementation by a single memristor-based multi-terminal synaptic device. *Nanoscale* **2016**, *8*, 18897. [CrossRef] [PubMed]

17. Valov, I.; Waser, R.; Jameson, J.R.; Kozicki, M.N. Electrochemical Metallization Memories Fundamentals, Applications, Prospects. *Nanotechnology* **2011**, *22*, 254003. [CrossRef] [PubMed]

18. Barbera, S.L.; Vuillaume, D.; Alibart, F. Filamentary Switching: Synaptic Plasticity through Device Volatility. *ACS Nano* **2015**, *9*, 941–949. [CrossRef] [PubMed]

19. Kim, S.; Du, C.; Sheridan, P.; Ma, W.; Choi, S.; Lu, W.D. Experimental Demonstration of a Second-Order Memristor and Its Ability to Biorealistically Implement Synaptic Plasticity. *Nano Lett.* **2015**, *15*, 2203–2211. [CrossRef] [PubMed]

20. Hu, S.G.; Liu, Y.; Liu, Z.; Chen, T.P.; Yu, Q.; Deng, L.J.; Yin, L.; Hosaka, S. Synaptic long-term potentiation realized in Pavlov's dog model based on a NiOx-based memristor. *J. Appl. Phys.* **2014**, *116*, 214502. [CrossRef]

21. Ohno, T.; Hasegawa, T.; Tsuruoka, T.; Terabe, K.; Gimzewski, J.K.; Aono, M. Short-term plasticity and long-term potentiation mimicked in single inorganic synapses. *Nat. Mater.* **2011**, *10*, 591–595. [CrossRef] [PubMed]

22. Guan, X.; Yu, S.; Philip Wong, H.-S. On the switching parameter variation of metal-oxide RRAM—Part I: Physical modeling and simulation methodology. *IEEE Trans. Electron Devices* **2012**, *59*, 1172–1182. [CrossRef]

23. Yu, S.; Guan, X.; Philip Wong, H.-S. On the Switching Parameter Variation of Metal Oxide RRAM—Part II: Model Corroboration and Device Design Strategy. *IEEE Trans. Electron Devices* **2012**, *59*, 1183–1188. [CrossRef]

© 2018 by the authors. Licensee MDPI, Basel, Switzerland. This article is an open access article distributed under the terms and conditions of the Creative Commons Attribution (CC BY) license (http://creativecommons.org/licenses/by/4.0/).

electronics

MDPI

Article

Analog Memristive Characteristics and Conditioned Reflex Study Based on Au/ZnO/ITO Devices

Tiedong Cheng [1,2], Jingjing Rao [2], Xingui Tang [1,*], Lirong Yang [2] and Nan Liu [2]

1   School of Physics & Optoelectric Engineering, GuangDong University of Technology, Guangzhou 510006, China; chengtiedong@126.com
2   School of Electrical Engineering & Automation, JiangXi University of Science and Technology, Ganzhou 341000, China; raojingjin@163.com (J.R.); candy_yang_jx@163.com (L.Y.); liunanzilu@163.com (N.L.)
*   Correspondence: xgtang@gdut.edu.cn; Tel.: +86-20-3932-2265

Received: 5 July 2018; Accepted: 1 August 2018; Published: 8 August 2018

Abstract: As the fourth basic electronic component, the application fields of the memristive devices are diverse. The digital resistive switching with sudden resistance change is suitable for the applications of information storage, while the analog memristive devices with gradual resistance change are required in the neural system simulation. In this paper, a transparent device of ZnO films deposited by the magnetron sputtering on indium tin oxides (ITO) glass was firstly prepared and found to show typical analog memristive switching behaviors, including an I–V curve that exhibits a 'pinched hysteresis loops' fingerprint. The conductive mechanism of the device was discussed, and the LTspice model was built to emulate the pinched hysteresis loops of the I–V curve. Based on the LTspice model and the Pavlov training circuit, a conditioned reflex experiment has been successfully completed both in the computer simulation and the physical analog circuits. The prepared device also displayed synapses-like characteristics, in which resistance decreased and gradually stabilized with time under the excitation of a series of voltage pulse signals.

Keywords: memristive device; ZnO films; conditioned reflex

1. Introduction

Since the laboratory of Hewlett-Packard (HP) Development Company (Palo Alto, CA, USA) announced that the 'missing' circuit elements were found in 2008, memristive devices have attracted a great deal of attention worldwide as a potential element for future device applications, such as nonvolatile information storage and artificial neural network. Nonvolatile storage, characterized by abrupt resistance change and high ON/OFF ratio resistance states, is the basic feature of the digital memristor or resistive switching devices. However, the artificial neural network invariably needs analog memristive devices to implement synapses [1]. Smooth current–voltage curves with gradual change of the device resistance are the fingerprints of analog memristive devices, which are always known as analog resistive switching or analog memristor. There are considerable studies of analog memristor-based neural networks, because the gradual resistance provides a way to mimic synaptic strength and synaptic plasticity [2], and the advantages of a simple structure can be compared with pure software implementations [3,4]. Some studies have experimentally demonstrated memristor devices and showed that the hybrid system composed of complementary metal oxide semiconductor (CMOS) neurons and analog memristor synapses can support important synaptic functions, such as spike-timing-dependent plasticity [4,5]. Moreover, with the deepening of the memristor on the application of the integrated circuit, various mathematic and physical models related to memristors also have been a research hotspot in electronic engineering application. The first simulation program with integrated circuit emphasis (SPICE) model of memristor was put forward and built by Szmanda [6], and later on, McCreery successfully [7] modeled and simulated the characteristic curves of the HP memristor according to the data and modeling equation provided by the HP laboratory in 2004. Batas et al. [8] first introduced a behavior model of a memristive solid-state device for simulation with a SPICE compatible circuit. An original SPICE macro-model of the physically implemented memristor was presented by Rak [9], whose model offered a tool for electrical engineers to design and conduct new circuits with a memristor. Hu [10] recently presented a novel mathematical model for the TiO_2 thin-film memristor device discovered by the HP laboratory, and the proposed model could be well applied to neural networks.

Although memristors have been widely studied in two aspects of material physics and modeling, the two researching fields were mostly in the state of independent development, and relatively little attention has been devoted to systematic research. Consequently, there are few reports that investigate the material characteristics and circuit model in a same work. In this paper, zinc oxide (ZnO) was chosen to fabricate the memeristive device because ZnO is an attractive memristive material [11,12] and semiconductor material with excellent magnetic, optical, and radiation-resisting advantages. Concretely, transparent substrate indium tin oxides (ITO) were used to construct an invisible ZnO/ITO/glass device, and the material properties were studied. Then a linear technology simulation program with integrated circuit emphasis (LTspice) model was built for the prepared device, and a Pavlov training circuit for neural network learning application was setup to verify the validity in the conditioned reflex.

2. Materials and Methods

ZnO thin films were prepared by magnetron sputtering on commercially available ITO glass, and the sputtering target was made of ZnO ceramic target with 99.99% purity. The films were deposited in a mixing atmosphere (argon:oxygen = 40:10, 0.5 Pa) and subsequently annealed in an oxygen atmosphere at 400 °C for 30 min. The gold (Au) top electrode layers were subsequently deposited on the ZnO thin film patterned by a shadow mask via Direct Current sputtering at room temperature. A source meter (Keithley 2400, Cleveland, OH, USA) was applied to the I–V measurement, and an oscilloscope (Tektronix, RSA306B, Beaverton, OR, USA) was used to record the physical waveform. The microstructure of the ZnO/ITO/glass structure was analyzed by a transmission electron microscope (TEM, FEI Tecnai G2F20, Hillsboro, OR, USA). The plan-view and cross-sectional TEM images are shown in Figure 1a and 1b, respectively. An interplanar distance of 0.263 nm was determined from Figure 1a, and this value matches the standard crystal face (002) of ZnO. An X-ray diffraction (XRD, Rigaku, Toyko, Japan) was also performed, and the resulting patterns shown in Figure S1 of the Supplementary Materials further prove the presence of wurtzite ZnO-related peaks corresponding to reflection (002) planes. About 120-nm-thick ZnO layers, which are clearly seen in Figure 1b, were believed to be well crystallized.

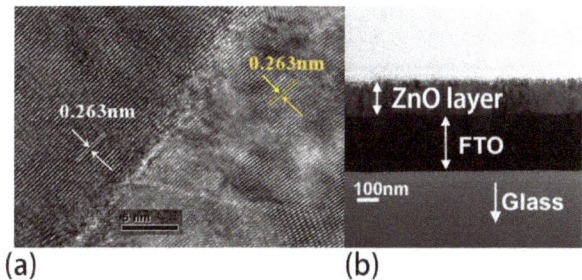

Figure 1. Transmission electron microscope (TEM) images of the zinc oxide (ZnO)/indium tin oxide (ITO)/glass structure (**a**) plan-view and (**b**) cross-sectional image.

3. Results and Discussion

3.1. Confirmation of Memristor and I–V Characteristics

The I–V characteristics were examined by applying the output voltage of the source meter on ZnO/ITO, and positive bias was defined as the current flow from ZnO to ITO (the Au top electrode is positively biased, while the ITO bottom electrode was negatively biased). As shown in Figure 2a, the ZnO/ITO structure exhibited rectifying characteristics, and the reason behind it will be discussed below. In the positive bias region, a reproducible pinched hysteresis loop was recorded in Figure 2a for a single cycle at room temperature, as voltage was swept in the '1→2' sequence with a rate of 0.1 V/step, and the time interval was 50 ms. In the reported literature about memristors, the Au/ZnO [11], $SrTiO_3$:Nb/$LaMnO_3$/Pt [13], and Ti/TiO_2/Pt structures [14] also had the similar hysteresis loop I–V characteristic curve as that appeared in this work.

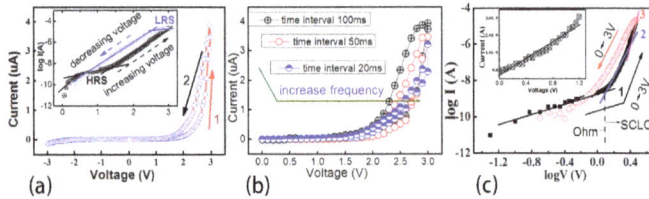

Figure 2. I–V characteristic curve (**a**) I–V curve at room temperature; (**b**) frequency-independent I–V curve; and (**c**) logI–logV characteristics curve.

For a device to be a memristor, it should exhibit three characteristic fingerprints as follows: (1) the signature of pinched hysteresis loop in the voltage–current plane, (2) the hysteresis lobe area should decrease monotonically as the excitation frequency increases, and (3) the pinched hysteresis loop should shrink to a single-valued function when the frequency tends to infinity [15]. The prepared ZnO/ITO/glass, whose I–V curves are shown in Figure 2a, presents a pinched hysteresis loop similar to a Lissajous Figure as discussed in the literature [15]. To prove the frequency characteristic as the second fingerprint mentioned, we measured the I–V curves with different excitation frequencies, and the results in Figure 2b clearly demonstrate a shrinking loop as the frequency increases. This effect is consistent with fingerprints (2) and (3). In addition, the hysteresis direction of the first quadrant (the first half) is counterclockwise, which also coincides with the features of a typical memristor as proposed by Chua [16]. Hence, the experimental evidence suggests that the as-prepared ZnO/ITO/glass is a representative memristor device. As also clearly shown by the semilogarithmic current scale in the inset of Figure 2a, a set process was observed at about 1.3 V, while a reset process occurred at a voltage of about 2.3 V. This I–V curve can still be repeated after 100 cycles, indicating that the device has good reproducibility. We can find that the ratio of the highest resistance state (HRS) to the lowest resistance stage (LRS) is about 10^3. The current is only a few microamperes, which is two orders of magnitude smaller than most recently reported results [17]. In addition, no significant degradation of the device was observed when 6-month-old devices were tested, indicating good stability. The best reported structures of ZnO-based memristors having a good combination of low power, endurance, and retention performance so far is Ag/ZnO/Pt for electrochemical metallization memory devices [18]. To make a comparison, the major ZnO-based device parameters as a function of different metal electrodes are summarized in Table 1 [17].

Table 1. ZnO-based resistive random access memory fabricated with various metal electrodes.

No	Structure	Current	Endurance	Roff/Ron	Retention
1	Pt/ZnO/Pt	30 mA	100	10^3–10^4	not available
2	Ru/ZnO/Pt	10 mA	200	61	10^4
3	Ag/ZnO/Pt	10 mA	40	100	not available
4	Cu/ZnO/ITO	not available	300	>20	not available

In general, conducting mechanisms of memristive devices can be categorized into filamentary and homogeneous switching [19]. I–V curve of continuous resistance decreasing/increasing with gradual voltage sweep shown in Figure 2a, suggesting a homogeneous resistive switching or typical analog property instead of filamentary resistive switching or so-called digital resistive memristor. The main modes of carrier transport in memristive devices are ohmic and space charge limited (SCL) conduction and the I–V characteristics of SCL were defined by modified Child's law [20] as shown in Equation (1):

$$I = (\frac{9s\varepsilon_0\varepsilon_r\mu_n}{8d^3})\theta V^2 \tag{1}$$

where s, ε, μ_n, and d are area, the dielectric constant, electron mobility, and the thickness of the device's thin films, respectively. θ is the ratio of free carriers to trapped carriers. Figure 2c shows the log I–V curve for the positive sweep region. In the '0 V→3 V' region, the curve is obviously divided into three stages (1, 2, and 3). It can be seen from the inset of Figure 2c that the I–V relation is linear under the application of a low electric field (0–1.2 V) as shown in stage 1. When the applied voltage reaches V_{SET} (i.e., 1.3 V), the I–V curve is quadratic. When the voltage is increased to about 2.6 V, then the current rises rapidly, and the slope is six times that of stage 1. The low electric field should be defined as the Ohm zone, where the thermal electric conduction is dominated by an intrinsic carrier. When applied voltage reaches V_{SET}, the I–V curve changes from Ohmic to SCL conduction, and the carrier transit time is equal to the Ohmic relaxation time [20]. In stage 3, because the injected excess carriers dominate the thermally generated carriers, it leads to a greater ratio of free carriers to trapped carriers, and the current behavior switches to the trap-free SCL conduction. In the '3 V→0 V' region, the reason for the hysteresis is because the trapped carriers are not released completely with the reducing voltage [20].

The typical rectifying characteristics shown in the I–V curve of Figure 2a are due to the contact formed at the Au/ZnO/ITO structure interface. As we know, the work function of ITO is about W_{ITO} = 4.4 eV [21], the electron affinity of ZnO is $E_{ZnO} \approx$ 4.45 eV (ZnO is usually an N-type semiconductor) [22], and the work function of Au is W_{Au} = 5.1 eV. Thus, ITO shows low barrier height when contacting ZnO, and ohmic contact of the ZnO/ITO interface can form easily. However, because $W_{Au} \gg E_{ZnO}$ and Au was considered to be the shared material to construct the Schottky contact with the N-type ZnO, a rectifying effect could be formed between the interface of Au and ZnO. Once a rectifying contact is formed, the electrons in the ZnO films enter Au through the interface where the space-charge zone was constituted, and a built-in potential barrier that points from the ZnO surface to the top-electrode Au was constructed to stop the electrons from diffusing into the Au. Owing to impurity and defects caused in the process, the actual values of built-in potential are different, and the difference is reflected in the value of V_{SET}. When the voltage applied on the top Au electrode is negatively biased, the built-in potential was enhanced with an external electric field, and the current value was almost zero.

3.2. Memorizing Characteristics

To explore memorizing properties of the device in a physical circuit system, the same measurement configuration as described in Section 3.1 was used, and continuous voltage pulses (duty cycle is 1) with different amplitudes were applied to study the memristive characteristics of the device. Figure 3a shows the current–time (I–t) curve of the device undergoing 50 continuous voltage pulses with an amplitude of 1 V. Because the applied voltage did not exceed V_{SET} voltage (1.3 V), the corresponding current was at 10^{-10} A order, and this result exhibited an excellent endurance of HRS. Figure 3b shows the I–t curve under the application of 50 voltage pulses with an amplitude of 2 V. Because the voltage amplitude has exceeded the V_{SET} value, the current value was larger than that under the application of 1 V pulse, and the resistance shows a decreasing trend with the continuous pulse application. Figure 3c shows the I–t curve under the application of 50 voltage pulses with an amplitude of 3 V. It is obvious that, in the applying process of the first several pulses, the device presents high impedance. However, with the further increasing of the number of pulses, the current rises to 10^{-6} A order rapidly, and the resistance dropped sharply. Figure 3d shows the I–t curve of the device under the continuous application of 100 voltage pulses with an amplitude of 4 V. The experimental results showed that, after the sample changed to LRS from HRS, the resistance did not keep in a stable LRS. Instead of dropping with the continuous application of pulses, this indicated that the device probably broke down.

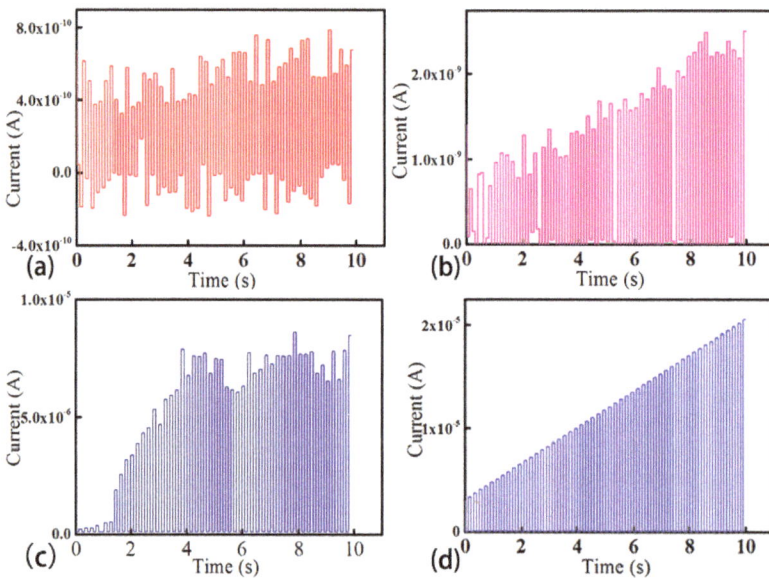

Figure 3. I–t curve of the device undergoing 50 voltage pulses with (**a**) an amplitude of 1 V; (**b**) an amplitude of 2 V; (**c**) an amplitude of 3 V; and (**d**) an amplitude of 4 V (100 pulses).

As mentioned in the introduction, the analog memristive device was always considered to be the electronic component that is most similar to cerebral synapses [23]. In Figure 3c, when V_{max} of the circular voltage exceeds the set voltage, the conductance of the device shows progressive increasing with the numbers of pulse and exhibits a saturation feature. This characteristic is similar to the function of synapses in neuromorphic engineering which means that the reinforcement and inhibition functions of synapses can be simulated through the device [24]. If voltage pulses imposed on the top electrode and bottom electrode of the device were regarded as pre-stimulus and post-stimulus of synapses, spike-timing-dependent-plasticity (STDP), as a crucial form of Hebbian learning, could be simulated through the device by programmable voltage pulse. A resistive switching of Ti/ZnO/Pt memristive devices was prepared and based on which STDP learning rule was implemented [25].

3.3. LTspice Models and Simulation

Common mathematical models of memristors include the HP model [26], the threshold model [27], and the matching model [28]. In particular, the matching model, proposed by Chris Yakopcic, has the characteristics of other two models and could accurately simulate multiple memristors (such as TiO_2, Ge_2Se_3, etc.) by adjusting parameters. Moreover, the model has outstanding advantages in the memristor-based neural network simulation. Lai [29] further proposed the general thin-film memristor model based on the matching model. The above two matching models have a similar understructure, whose basic formula [30] is as follows in Equation (2):

$$I(t) = \begin{cases} a_1\omega(t)\sinh(b_1V(t)) & \text{If } V(t) \geq 0 \\ a_2\omega(t)\sinh(b_2V(t)) & \text{If } V(t) \leq 0 \end{cases} \tag{2}$$

where $I(t)$ is the current through the memristor. $V(t)$ is the voltage crossing the memristor. a_i and b_i are constants that are larger than zero, are respectively used to control the direction of current $I(t)$, and indicate the effect of voltage $V(t)$. ω is a variable that represents the internal states of the memristor as shown in Equation (3):

$$\frac{d\omega}{dt} = \begin{cases} c_1\sinh(d_1V(t)) & \text{If } V(t) \geq 0 \\ c_2\sinh(d_2V(t)) & \text{If } V(t) \leq 0 \end{cases} \tag{3}$$

where $d\omega/dt$ are directly proportional to the drifting velocity of charged particles (such as oxygen vacancies) in the memristor. c_i and d_i are constants that are larger than zero and are set according to the conductance of samples. In the actual physical process of the memristor, the drifting velocity of the particle and the applied electric field show a nonlinear relationship. In order to describe such a nonlinear relationship in mathematics, usually, a nonlinear window function [31,32] was appended on Equation (3). Among those proposed window functions, Prodromakis et al. presented a fine window function, as follows in Equation (4), which kept the continuity of the function and showed commendable ability in regulation:

$$f(x) = j(1 - \left[x - 0.5^2 + 0.75\right]^p) \tag{4}$$

where j and x are the scalar parameter and current, respectively. p acts as any positive-real number and is employed to characterize the nonlinearities of the function. The nonlinearity is inversely proportional to the value of p: the smaller the p, higher the nonlinearity of the window function. Hence, Equation (3) can be transformed as Equation (5):

$$\frac{d\omega}{dt} = \begin{cases} c_1\sinh(d_1V(t))f(x(t)) & \text{If } V(t) \geq 0 \\ c_2\sinh(d_2V(t))f(x(t)) & \text{If } V(t) \leq 0 \end{cases} \tag{5}$$

According to the above mathematical model and the I–V experimental data, the optimized parameters were obtained by carefully adjusting, and the LTspice code for the memristor is shown in Section S1 of the Supplementary Materials. With the LTspice code, a simulation I–V curve, which in line with the experimental results, is shown in Figure 4a, and we can find that V_{SET}, V_{RESET}, and the current order show good agreement with the experimental results. The simulation schematic is shown in the inset of Figure 4a, where M1 is the memristor and V1 is a step-wave voltage source with 0.05 V voltage-step and 50 ms time interval.

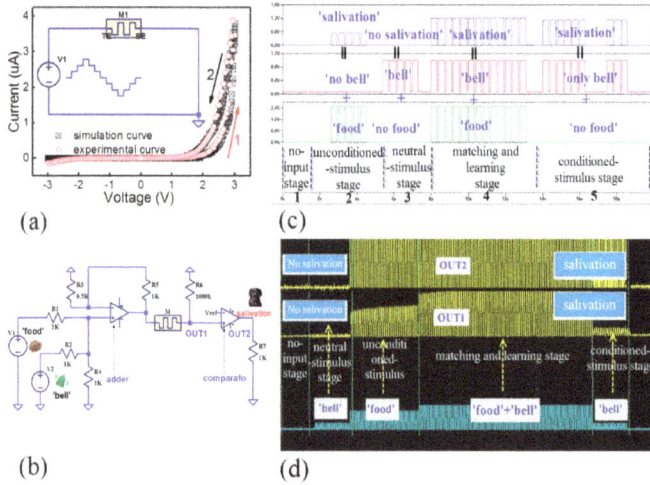

Figure 4. Simulation and experimental results (**a**) Simulation results of the device using the LTspice model; (**b**) Pavlov training circuit; (**c**) Resulting simulation of Pavlov training circuit; (**d**) Physical experiment results. M1 is the memristor and V1 is a step-wave voltage source with 0.05 V voltage-step and 50 ms time interval.

3.4. Conditioned Reflex Simulation and Circuit Experiments

In order to verify the potential significance of the memristor on the conditioned reflex, we employed the prepared ZnO/ITO/glass device to design a Pavlov training circuit [33,34] as shown in Figure 4b and then conducted a simulation and real physical experiments. In Pavlov's experiments, food for the dog was used to activate an unconditioned stimulus (UCS) and the ring of a bell represents the neutral stimulus (NS). Once the food signal was received, Pavlov's dog started to salivate. However, at first, the ring alone did not lead to any salivation by Pavlov's dog. By feeding and ringing at the same time, Pavlov's dog learned to associate the NS with the UCS. As a result, Pavlov's dog salivated by even hearing the bell ring alone [34]. In the circuit of Figure 4b, V1 and V2 were used to generate voltage pulses for simulating food and bell signals, respectively. A comparator was used for signal shaping. The 'OUT1' and 'OUT2' in the circuit of Figure 4b represent 'salivation' before and after signal shaping.

Based on the LTspice mode and the designed Pavlov training circuit, a simulation results with five distinct stages, were shown in Figure 4c. In stage 1 (i.e., the no-input stage), there is no outputting signal, because the adder outputs a low level that induces the memristor M to stay in HRS, which limits OUT1 and OUT2, holding them in the low level. As shown in stage 2 (the UCS stage), when the voltage pulse, which represents the 'food' signal, holds for several periods (the amplitude is higher than V_{SET}), the adder will output high-level pulses, which leads the M to change from HRS to LRS, as demonstrated in Figure 3c. As a result, a pulse signal that represents the 'salivation' appears on both OUT1 and OUT2 of the Pavlov training circuit. If there is only one input of the 'bell' signal (the amplitude is lower than V_{SET}), as shown in stage 3, OUT2 also could not output a pulse signal, because the 'bell' voltage amplitude is too low to cause the M to go into LRS, resulting in OUT1 and OUT2 staying in the low-level state. After the 3-stage experiments stated above, in the event that both the 'food' and 'bell' signals are imputed at the same time, the adder performs a superposition operation on the two signals and outputs the resulting pulse, which leads the M to change from HRS to a retaining LRS. OUT1 and OUT2 hold in the high level as a consequence of the LRS of M. This can be called the matching and learning stage, as shown in stage 4 of Figure 4c. Afterwards, even if there is 'no food' input, the 'dog' can also secrete 'salivation' as the CS stage (stage 5) described, because the M is kept in LRS.

A set of input–output characteristics in Figure 4d were recorded for the Pavlov training circuit at room temperature for a ZnO/ITO/glass memristor as a synapse. Both the 'bell' and 'food' input signals were represented by voltage pulse signals with a frequency of 100 Hz and amplitude of 1.5 V. The commercial operational amplifiers LM358 (Motorola, Chicago, IL, USA) and TL072 (Texas Instruments Incorporated, Dallas, TX, USA) were used as adder and comparator, respectively. Observe that the output is exactly the same as the result of the simulation, which confirms that the application of the memristive device in a conditioned reflex has been achieved.

4. Conclusions

In this paper, we employed the magnetron sputtering deposition method to successfully fabricate a memristor with ITO/ZnO/glass structure. Our results show that such devices possess the typical analog memristive device characteristics. Based on a mathematical model, a LTspice circuit model was simulated successfully for the device, and a Pavlov training circuit using a memristor as synapses was designed to accomplish the physical process of matching and learning. In addition, because the current in the devices under 3 V turns out to be in the order of μA, the proposed ZnO-based memristor has great potential for commercial applications of artificial intelligence.

Supplementary Materials: The following are available at the following link: http://www.mdpi.com/2079-9292/7/8/141/s1, Figure S1: XRD pattern of ZnO/ITO/glass structure; Section S1: LTspice code of as prepared memristor.

Author Contributions: Conceptualization, T.C. and X.T.; Mathematical Model, J.R. and N.L.; and Circuit Design, L.Y.

Funding: This research was funded by "the National Natural Science Foundation of China 11574057", "the Science Foundation of Department of Education of JiangXi Province [GJJ15331]", "the Guangdong Provincial Natural Science Foundation of China [2016A030313718]" and "the Foundation of JXUST [NSFJ2015-K11]".

Conflicts of Interest: The authors declare no conflict of interest.

References

1. Lv, F.; Yang, R.; Guo, X. Analog and digital reset processes observed in Pt/CuO/Pt memristive devices. *Solid State Ion.* **2017**, *303*, 161–166. [CrossRef]
2. Ravichandran, V.; Li, C.; Banagozar, A.; Yang, J.J.; Xia, Q.F. Artificial neural networks based on memristive devices. *Sci. China Inform. Sci.* **2018**, *61*, 1–14. [CrossRef]
3. Pershin, Y.V.; Ventra, D.M. Practical approach to programmable analog circuits with memristors. *IEEE Trans. Circuits Syst. I Reg. Pap.* **2010**, *57*, 1857–1864. [CrossRef]
4. Jo, S.H.; Chang, T.; Ebong, I.; Bhadviya, B.B.; Mazumder, P.; Lu, W. Nanoscale memristor device as synapse in neuromorphic systems. *Nano Lett.* **2010**, *10*, 1297–1301. [CrossRef] [PubMed]
5. Kim, H.; Sah, M.P.; Yang, C.; Roska, T.; Chua, L.O. Neural synaptic weighting with a pulse-based memristor circuit. *Trans. Circuits Syst. I Reg. Pap.* **2012**, *59*, 148–158. [CrossRef]
6. Ouyang, J.; Chu, C.W.; Szmanda, C.R.; Ma, L.; Yang, Y. Programmable polymer thin film and non-volatile memory device. *Nat. Mater.* **2004**, *3*, 918–922. [CrossRef] [PubMed]
7. Wu, J.; Mccreery, R.L. Solid-state electrochemistry in molecule/TiO_2 molecular heterojunctions as the basis of the TiO_2 "memristor". *J. Electrochem. Soc.* **2009**, *156*, 29–37. [CrossRef]
8. Batas, D.; Fiedler, H. A memristor spice implementation and a new approach for magnetic flux controlled memristor modeling. *IEEE Trans. Nanotechnol.* **2011**, *10*, 250–255. [CrossRef]
9. Rak, A.; Cserey, G. Macromodeling of the memristor in SPICE. *IEEE Trans. Comput.-Aided Des. Integr. Circuits Syst.* **2010**, *29*, 632–636. [CrossRef]
10. Hu, X.; Feng, G.; Li, H.; Chen, Y.; Duan, S. An adjustable memristor model and its application in small-world neural networks. In Proceedings of the 2014 International Joint Conference on Neural Networks (IJCNN), Beijing, China, 6–11 July 2014.

11. Sun, Y.; Yan, X.; Zheng, X.; Liu, Y.; Shen, Y.; Ding, Y.; Zhang, Y. Effect of carrier screening on ZnO-based resistive switching memory devices. *Nano Res.* **2017**, *10*, 77–86. [CrossRef]

12. Murali, S.; Rajachidambaram, J.S.; Han, S.Y.; Chang, C.H.; Herman, G.S.; Conley, J.F.J. Resistive switching in zinc–tin-oxide. *Solid State Electron.* **2013**, *79*, 248–252. [CrossRef]

13. Xu, Z.; Jin, K.; Gu, L.; Jin, Y.; Ge, C.; Wang, C.; Guo, H.; Lu, R.; Zhao, H.; Yang, G. Evidence for a crucial role played by oxygen vacancies in LaMnO$_3$ resistive switching memories. *Small.* **2012**, *8*, 1279–1284. [CrossRef] [PubMed]

14. Fullam, S.; Ray, N.J.; Karpov, E.G. Cyclic resistive switching effect in plasma electrolytically oxidized mesoporous Pt/TiO$_2$ structures. *Superlatt. Microstruct.* **2015**, *82*, 378–383. [CrossRef]

15. Adhikari, S.P.; Sah, M.P.; Kim, H.; Chua, L. Three fingerprints of memristor. *Trans. Circuits Syst. I Reg. Pap.* **2013**, *60*, 3008–3021. [CrossRef]

16. Chua, L. Resistance switching memories are memristors. *Appl. Phys. A* **2011**, *102*, 765–781. [CrossRef]

17. Simanjuntak, F.M.; Panda, D.; Wei, K.H.; Tseng, T.Y. Status and prospects of ZnO-based resistive switching memory devices. *Nanoscale Res. Lett.* **2016**, *11*, 368–385. [CrossRef] [PubMed]

18. Huang, Y.; Shen, Z.; Wu, Y.; Wang, X.; Zhang, S.; Shi, X.; Zeng, H. Amorphous ZnO based resistive random access memory. *RSC Adv.* **2016**, *6*, 7867–7872. [CrossRef]

19. Huang, C.H.; Huang, J.S.; Lai, C.C.; Huang, H.W.; Lin, S.J.; Chueh, Y.L. Manipulated transformation of filamentary and homogeneous resistive switching on ZnO thin film memristor with controllable multistate. *ACS Appl. Mater. Interfaces* **2013**, *5*, 6017–6023. [CrossRef] [PubMed]

20. Shang, D.S.; Wang, Q.; Chen, L.D.; Dong, R.; Li, X.M.; Zhang, W.Q. Effect of carrier trapping on the hysteretic current-voltage characteristics in Ag/La$_{0.7}$Ca$_{0.3}$MnO$_3$/Pt heterostructures. *Phys. Rev. B* **2006**, *73*, 245427. [CrossRef]

21. Nüesch, F.; Forsythe, E.W.; Le, Q.T.; Gao, Y.; Rothber, L.J. Importance of indium tin oxide surface acido basicity for charge injection into organic materials based light emitting diodes. *J. Appl. Phys.* **2000**, *87*, 7973–7980.

22. Fang, Y.J.; Sha, J.; Wang, Z.L.; Wan, Y.T.; Xia, W.W.; Wang, Y.W. Behind the change of the photoluminescence property of metal-coated ZnO nanowire arrays. *Appl. Phys. Lett.* **2011**, *98*, 033103. [CrossRef]

23. Ielmini, D. Brain-inspired computing with resistive switching memory (RRAM): Devices, synapses and neural networks. *Microelectron. Eng.* **2018**, *190*, 44–53. [CrossRef]

24. Chang, T.; Jo, S.H.; Kim, K.H.; Sheridan, P.; Gaba, S.; Lu, W. Synaptic behaviors and modeling of a metal oxide memristive device. *Appl. Phys. A* **2011**, *102*, 857–863. [CrossRef]

25. Pan, R.; Li, J.; Zhuge, F.; Zhu, L.; Liang, L.; Zhang, H.; Gao, J.; Cao, H.; Fu, B.; Li, K. Synaptic devices based on purely electronic memristors. *Appl. Phys. Lett.* **2016**, *108*, 382001. [CrossRef]

26. Biolek, Z.; Biolek, D.; Biolkova, V. SPICE model of memristor with nonlinear dopant drift. *Radioengeering* **2009**, *18*, 210–214.

27. Pershin, Y.V.; Ventra, M.D. Experimental demonstration of associative memory with memristive neural networks. *Neural Netw.* **2010**, *23*, 881–886. [CrossRef] [PubMed]

28. Yakopcic, C.; Taha, T.M. Subramanyam, G.; Pino, R.E.; Rogers, S. A memristor device model. *IEEE Electron Device Lett.* **2011**, *32*, 1436–1438. [CrossRef]

29. Laiho, M.; Lehtonen, E.; Russel, A.; Dudek, P. Memristive synapses are becoming reality. *Neuromorphic Eng.* **2010**. [CrossRef]

30. Lehtonen, E.; Poikonen, J.; Laiho, M.; Wei, L. Time-dependency of the threshold voltage in memristive devices. *IEEE Int. Symp. Circuits Syst.* **2011**, *19*, 2245–2248.

31. Yu, J.; Mu, X.; Xi, X.; Wang, S. A memristor model with piecewise window function. *Radioengineering* **2013**, *22*, 969–974.

32. Prodromakis, T.; Peh, B.; Papavassiliou, C.; Toumazou, C. A versatile memristor model with nonlinear dopant kinetics. *IEEE Trans. Electron Devices* **2011**, *58*, 3099–3105. [CrossRef]

33. Wang, F.Z.; Helian, N.; Wu, S.; Yang, X.; Guo, Y.K.; Lim, G.; Rashid, M.M. Delayed switching applied to memristor neural networks. *J. Appl. Phys.* **2012**, *111*, 07E317. [CrossRef]

34. Ziegler, M.; Soni, R.; Patelczyk, T.; Ignatov, M.; Bartsch, T.; Meuffels, P.; Kohlstedt, H. An electronic version of pavlov's dog. *Adv. Funct. Mater.* **2012**, *22*, 2744–2749. [CrossRef]

© 2018 by the authors. Licensee MDPI, Basel, Switzerland. This article is an open access article distributed under the terms and conditions of the Creative Commons Attribution (CC BY) license (http://creativecommons.org/licenses/by/4.0/).

electronics

MDPI

Article

Electron Affinity and Bandgap Optimization of Zinc Oxide for Improved Performance of ZnO/Si Heterojunction Solar Cell Using PC1D Simulations

Babar Hussain [1,*], Aasma Aslam [2], Taj M Khan [3,4], Michael Creighton [1] and Bahman Zohuri [2]

[1]   Intel Corporation, Rio Rancho, NM 87124, USA; michaelcreighton38@gmail.com
[2]   Department of Electrical and Computer Engineering, The University of New Mexico, Albuquerque, NM 87131, USA; aasmaaslam@yahoo.com (A.A.); zohurib@unm.edu (B.Z.)
[3]   School of Physics, Trinity College Dublin, Dublin 2, Ireland; tajakashne@gmail.com
[4]   National Institute of Lasers and Optronics, Islamabad 45650, Pakistan
[*]   Correspondence: babarhussain2002@hotmail.com

Received: 26 December 2018; Accepted: 14 February 2019; Published: 20 February 2019

Abstract: For further uptake in the solar cell industry, n-ZnO/p-Si single heterojunction solar cell has attracted much attention of the research community in recent years. This paper reports the influence of bandgap and/or electron affinity tuning of zinc oxide on the performance of n-ZnO/p-Si single heterojunction photovoltaic cell using PC1D simulations. The simulation results reveal that the open circuit voltage and fill factor can be improved significantly by optimizing valence-band and conduction-band off-sets by engineering the bandgap and electron affinity of zinc oxide. An overall conversion efficiency of more than 20.3% can be achieved without additional cost or any change in device structure. It has been found that the improvement in efficiency is mainly due to reduction in conduction band offset that has a significant influence on minority carrier current.

Keywords: zinc oxide; silicon; ZnO/Si; electron affinity; bandgap tuning; conduction band offset; heterojunction; solar cells; PC1D

1. Introduction

Zinc oxide (ZnO) is an emerging material in the semiconductor industry due to its abundance and being environmentally friendly. The only major drawback of ZnO that hinders its use in the fabrication of a homojunction device is that it cannot be p-doped reliably and reproducibly. However, n-ZnO has been found to have applications in several optoelectronic devices, such as photovoltaic cells [1]. Since the proposed use of n-ZnO as an emitter layer and antireflection (AR) coating, several researchers have employed n-ZnO thin films to fabricate potentially high efficiency and low-cost solar cells [2–4]. Apart from several other properties which make ZnO a unique wide bandgap material, its bandgap and electron affinity can be tuned over a large range by doping or alloying. Recently, nickel (Ni) doped ZnO thin films were prepared using spray pyrolysis and an optical bandgap decrease from 3.47 eV for the undoped ZnO film to 2.87 eV for 15% Ni doping was achieved [5]. In 2010, Mayer et al. demonstrated that the bandgap of ZnO prepared using pulsed laser deposition can be narrowed down to 2 eV by Se incorporation [6]. Later, the same research group reported effects of growth parameters on the electron affinity of ZnO [7]. Furthermore, there are various reports available demonstrating significant reduction in conduction band offset (or electron affinity) by incorporating magnesium (Mg) in ZnO. We have previously reported synthesis of ZnO thin films using metal organic chemical vapor deposition (MOCVD) with optimized parameters for the fabrication of a n-ZnO/p-Si solar cell [1]. It was anticipated that an overall conversion efficiency of 19% and fill factor of 81% can be achieved using the proposed structure. Following this, several groups reported different experimental results [8–12] for the n-ZnO/p-Si solar cell with the best efficiency of 14% achieved by Pietruszka et al. recently [13]. Also, few groups have recently reported simulation results optimizing different parameters of ZnO/Si solar cell. Chen et al. investigated the influence of ZnO thickness, buffer layer and work function of electrodes on the performance of ZnO/Si solar cell using AFORS-HET tool and proved that the conversion efficiency of 17.16% can be achieved [10]. Ziani and Belkaid reported similar performance of ZnO/Si solar cell using SCAPS-1D software [14]. Very recently, Vallisree et al. used Silvaco ATLAS simulator and reported that efficiency up to 14.46% can be achieved by incorporation of Mg in ZnO [15]. Askari et al. provided an interesting study on the interface properties of ZnO/Si heterojunction and computed ~14% efficiency of ZnO/Si solar cell using TCAD simulations [16]. The ZnO/Si heterojunction solar cell is a relatively new idea and there is no study available that reports optimized values of electron affinity and bandgap of ZnO, which dictate conduction- and valence-band offsets, to achieve best performance of the solar cell.

In this paper, we report simulation based optimization of bandgap and electron affinity of ZnO to enhance the conversion efficiency of a ZnO/Si single heterojunction solar cell. The schematic of the solar cell structure is depicted in Figure 1. The effects of valence-band and conduction-band off-set engineering on the open circuit voltage (V_{OC}), short circuit current density (J_{SC}), fill factor (FF), and overall conversion efficiency (η) have been investigated using PC1D software. The simulations prove that the conversion efficiency as high as 20.3% can be obtained from ZnO/Si solar cell.

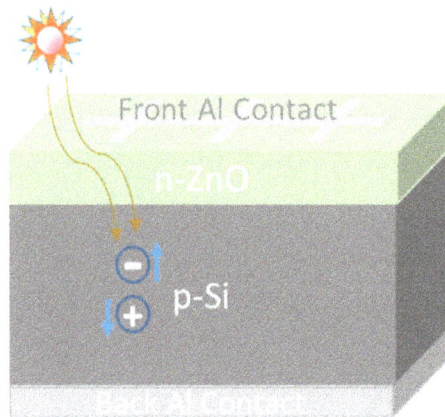

Figure 1. Schematic of the n-ZnO/p-Si single heterojunction solar cell structure.

2. Background

The energy band diagram of an ideal n-ZnO/p-Si heterojunction is depicted in Figure 2, which appears similar to a type-II heterojunction. The features of the band alignment are determined based on the Anderson energy-band rule also known as the electron affinity model. The conduction band offset (ΔE_C) and valence band offset (ΔE_V) according to Anderson's rule are given by

$$\Delta E_C = \chi_2 - \chi_1 \tag{1}$$

$$\Delta E_V = (\chi_2 + E_{g2}) - (\chi_1 + E_{g1}), \tag{2}$$

where χ is electron affinity, E_g is bandgap, and subscripts 1 and 2 correspond to Si and ZnO, respectively in our case. The electron affinity and bandgap of Si are well known to be 4.05 and 1.12 eV, respectively. These values vary significantly in literature for ZnO. Actually, the bandgap of ZnO can be tuned over a large range from 3 to 5 eV by alloying, which is considered as one of the unique advantages of ZnO. Sundaram et al. have reported the electron affinity of ZnO to be around 4.5 eV, which was calculated using I-V measurements and the Shottky-Mott model [17]. Since the ZnO films are highly doped, their fermi level almost overlaps with the conduction band edge. Therefore, the work function of ZnO is considered the same as electron affinity. These values result in ΔE_C and ΔE_V of ~0.4 eV and 2.55 eV, respectively. Sundaram et al. have also considered a very thin (1–2 nm) oxide layer at the interface that is likely to develop due to high energy process like magnetron sputtering [18,19]. The carrier flow across the junction is largely affected by oxide layer thickness, which dictates the tunneling coefficient [20].

There are two limitations of Anderson's rule [21]. The first is that Anderson's rule neglects the electron correlation effect. The correlation effect occurs when the electron moves to the vacuum level (according to the definition of electron affinity) and surrounding electrons rearrange themselves to reduce the total energy of the system. The magnitude of the correlation effect is generally very small. The second drawback is the lack of consideration of lattice mismatch and interface defects. Dangling chemical bonds at the interface of two semiconductors form interface states. Conduction and valence band discontinuities are affected by the dipole effect induced by electron transfer. Interestingly, the dipole effect is significantly reduced and becomes negligible if there is large lattice mismatch, as is the case with a ZnO/Si heterojunction [22]. Therefore, Anderson's rule is valid to determine the band edge offsets of a highly mismatched ZnO/Si heterojunction.

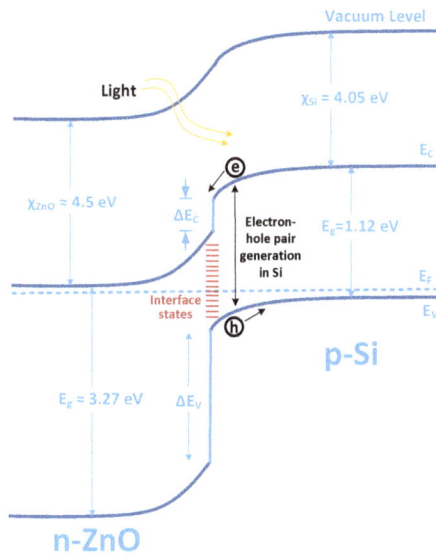

Figure 2. Schematic diagram of n-ZnO/p-Si heterojunction band-bending. χ denotes electron affinity of the material mentioned in the subscript.

When a photon gets absorbed in *p*-Si after being transmitted through a wide bandgap in *n*-ZnO, it generates an electron, a minority carrier, in the conduction band, as shown in Figure 2. The minority carrier current plays a vital role in solar cell performance. For an applied bias voltage V_b, the electron minority carrier current is given by [23]

$$J_n = \frac{J_{n0}}{(1+\upsilon)}\left(exp\left(\frac{qV_b}{kT}\right)\right),\qquad(3)$$

where

$$J_{n0} = kT\frac{\mu_{n1}}{L_{n1}}n_1,\qquad(4)$$

and μ_n, L_n, and n_1 denote minority-carrier mobility, diffusion length, and concentration, respectively. The subscript 1 represents the depletion region extended in p-type material (Si in our case). The factor 'v' appearing in Equation (3) defines the influence of the conduction band offset (ΔE_C) on the minority carrier electron current and is given by

$$v = \frac{1}{L_{n1}} \int_{x1}^{x2} \left(\frac{\mu_{n1} N_{c1}}{\mu_n N_c} \right) exp\left(-\frac{E_{c1} - E_c + q\Psi}{kT} \right) dx,$$ (5)

where $x_1 < 0$ to $x_2 > 0$ is the depletion region extended from Si to ZnO, and $\Psi(x)$ is the potential function. The subscripts 1 and 2 here represent Si and ZnO material respectively. The abrupt *pn* heterojunction with applied bias near-zero, Equation (5) can be solved to get

$$v = \frac{I_{v1}}{L_{n1}} + \frac{1}{L_{n1}} \left(\frac{\mu_{n1} N_{c1}}{\mu_{n2} N_{c2}} \right) I_{v2} exp\left(-\frac{E_{c1} - E_{c2}}{kT} \right),$$ (6)

where

$$I_{v1} = \int_{x1}^{0} exp\left(\frac{q\Psi(x)}{kT} \right) dx,$$ (7)

and

$$I_{v2} = \int_{0}^{x2} exp\left(\frac{q\Psi(x)}{kT} \right) dx.$$ (8)

The potential function can be supposed to be decoupled from the carrier transport equations and it satisfies Poisson's equation as

$$\frac{d^2\Psi}{dx^2} = -\frac{\rho(\Psi)}{\epsilon},$$ (9)

where ρ is charge distribution and ϵ is the permittivity of the medium. The electrostatic potential $\Psi(x)$ obtained from Equation (9) can be used to solve Equations (3)–(8) for a wide bandgap *n*-layer (ZnO) on a narrow bandgap *p*-layer (Si) to get [24]

$$J_n = kT \frac{\mu_{n1} N_{c2} n_{i1}^2}{I_{v2} N_{c1} N_A} exp\left(-\frac{|\Delta E_c|}{kT} \right) \left(exp\left(\frac{qV}{kT} \right) - 1 \right),$$ (10)

which proves that for a wide bandgap n-layer on a narrow bandgap p-layer heterojunction, the minority carrier current J_n prominently decreases due to the conduction band offset ΔE_C. The simulation results demonstrated in Section 3.1 (Figure 3) confirm this phenomena.

3. Results and Analysis

We prepared ZnO thin films using RF sputtering and performed detailed characterization. The experimental details have been reported elsewhere [25]. The photoluminescence and absorption measurements performed in our labs showed a bandgap value of 3.27 eV, which was used in the simulation. The most common value of electron affinity (4.5 eV) provided in literature was used initially. The absorption spectrum of ZnO of thickness ~0.5 μm measured in our lab using the Filmetrics tool was used in the simulation to investigate the effect of electron affinity. The details of the n-ZnO/p-Si solar cell modeling, structure schematic, and other optimized parameters for PC1D can be found in earlier reports [1,26].

3.1. Personal Computer One Dimensional (PC1D) Simulations

In the photovoltaic community, PC1D is a popular software package to simulate optical and electrical behavior of solar cell devices. This software was originally developed in the 1980s by Basore et al. [27] and has been continuously improved alongside progress in experimental work and theoretical models [28].

Figure 3 illustrates PC1D based simulation results of improvement in efficiency with reduction in electron affinity of ZnO. It is obvious that the conversion efficiency exceeds 20% by lowering electron affinity to 4.3 eV. The reason behind this phenomenon is reduction in conduction band offset ΔE_C that leads to an increase in minority-carrier current J_n as proved by Equation (10). Another factor contributing to improved efficiency is the decrease in the dark current. In other words, when a barrier for majority carrier electrons is formed by conduction band offset, it increases the probability of recombination via interface defects by the Shottky-Read-Hall (SRH) mechanism. The maximum efficiency of 20.34% can be achieved with ZnO having a bandgap of 3.27 eV and electron affinity of ~4.1 eV. Further reduction in electron affinity deteriorates cell efficiency. This can be theoretically confirmed by the band-bending diagram of the ZnO/Si junction as shown in Figure 4. Since the electron affinity of Si is ~4.05 eV, the electron affinity of ZnO below this value results in formation of a spike in the conduction band of the n-ZnO region. This spike acts as a potential barrier and blocks electron flow from the p-Si to the n-ZnO region. Therefore, it is difficult for the p-Si region to contribute to the photocurrent. This reasoning is supported by Figure 5 which depicts significant increase in V_{OC} with reduction in electron affinity. A negligible increase in J_{SC} can be attributed to the same reason.

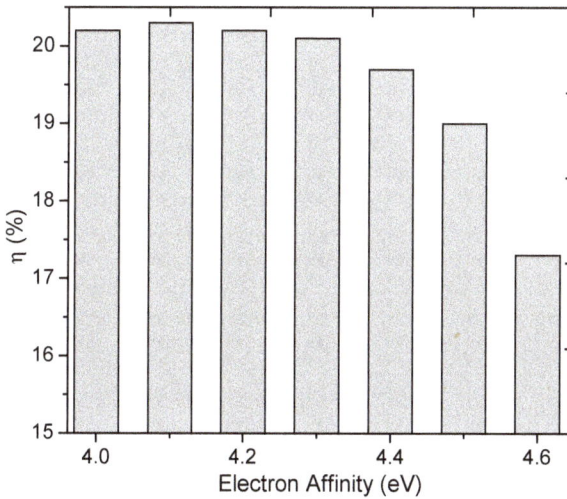

Figure 3. Influence of electron affinity of ZnO (bandgap: 3.27 eV) on the efficiency of n-ZnO/p-Si heterojunction solar cell using PC1D simulations.

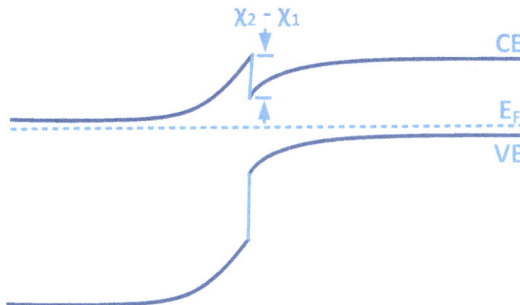

Figure 4. Schematic of the band-bending when electron affinity of ZnO (χ_1), at left, is lower than that of Si (χ_2), at right. CB: conduction band, VB: valence band, E_F: fermi level.

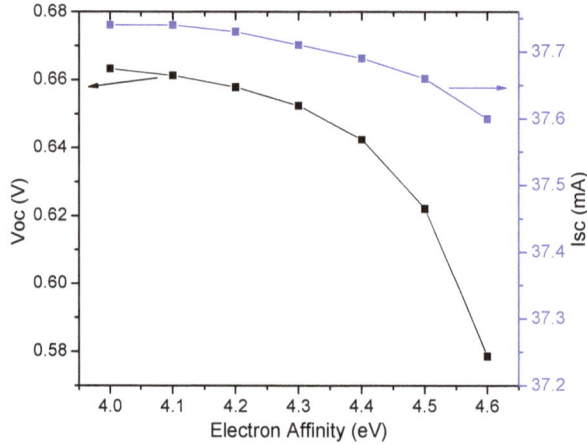

Figure 5. Effect of electron affinity of ZnO (bandgap: 3.27 eV) on the V_{OC} and I_{SC} of n-ZnO/p-Si solar cell using PC1D simulations.

The efficiency of the solar cell alters by modifying the bandgap as well. The change in conversion efficiency with a bandgap value of ZnO is shown in Figure 6 for three different values of electron affinity. The absorption spectrum was altered for values of bandgap other than 3.27 eV to get realistic results. The efficiency increases by decreasing the bandgap (or valence-band off-set). This improvement in efficiency cannot be explained using a band-bending diagram based on the famous Anderson's rule which ignores the effects of chemical bonding. The chemical bonding or electrical polarization due to interface states can alter the band bending significantly. Figure 6 also illustrates that the efficiency reduces significantly below a certain bandgap value. It is predictable because a considerable part of the solar spectrum gets absorbed in ZnO for such a small bandgap value. The ZnO layer is much thinner (0.5 μm) than Si (160 μm), but it has a higher absorption coefficient due to the direct bandgap of ZnO.

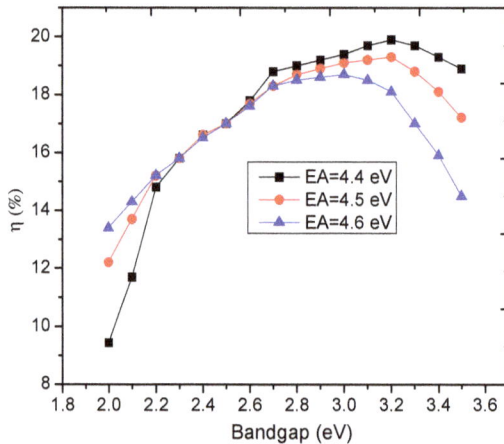

Figure 6. Change in overall conversion efficiency of n-ZnO/p-Si solar cell with modification of bandgap value of ZnO for three different values of electron affinity (EA). Few data points have been interpolated because numerical solution was not converging for those points in PC1D.

We have grown Ga rich ZnO:Ga films using MOCVD to examine the bandgap tuning. Trimethylgallium was used as the Ga source. The experimental details are provided elsewhere [1].

The photoluminescence (PL) spectra of both ZnO and ZnO:Ga were dominated by the near band edge (NBE) emission. The bandgap was blue-shifted by ~105 meV (12 nm). The molar ratio of Ga was ~50% during our growth process which lead to a bandgap of around 3.35 eV (370 nm). This is in accordance with the model reported by Zhao et al. [29]. We attribute this increase in the bandgap of ZnO to the well-known Burstein-Moss effect in which effective bandgap of a heavily doped semiconductor is increased as the absorption edge in the conduction band moves to higher energies because all states close to the conduction band edge are filled.

4. Conclusions

It is ascertained using PC1D simulations that the open circuit voltage of a n-ZnO/p-Si single heterojunction solar cell can be significantly improved by tuning the bandgap and/or electron affinity of ZnO by doping or alloying. The experimentally measured bandgap and absorption spectrum of ZnO was used in simulations using modified PC1D software. The major reasons for improvement in the solar cell efficiency are reduced conduction band offset that results in an increase in the minority carrier current and reduction in the dark current. The best values calculated for open circuit voltage, short circuit current density, fill factor, and conversion efficiency are 0.662 V, 37.7 mA/cm^2, 0.815, and 20.34%, respectively, for ZnO having a bandgap of 3.27 eV and electron affinity of 4.1 eV.

Author Contributions: Conceptualization, B.H.; methodology, B.H. and A.A.; software, B.H. and A.A.; validation, M.C. and T.M.K.; writing—original draft preparation, B.H. and A.A.; writing—review and editing, M.C., B.Z., and T.M.K.; visualization, B.Z. and A.A.

Funding: This research received no external funding.

Acknowledgments: The first author is thankful to Dr. Michael Fiddy from UNC Charlotte for his kind help and guidance during this research.

Conflicts of Interest: The authors declare no conflicts of interest.

References

1. Hussain, B.; Ebong, A.; Ferguson, I. Zinc oxide as an active n-layer and antireflection coating for silicon based heterojunction solar cell. *Sol. Energy Mater. Sol. Cells* **2015**, *139*, 95–100. [CrossRef]
2. Baturay, S.; Ocak, Y.S.; Kaya, D. The effect of Gd doping on the electrical and photoelectrical properties of Gd:ZnO/p-Si heterojunctions. *J. Alloys Compd.* **2015**, *645*, 29–33. [CrossRef]
3. Ren, X.; Zi, W.; Ma, Q.; Xiao, F.; Gao, F.; Hu, S.; Zhou, Y.; Liu, S.F. Topology and texture controlled ZnO thin film electrodeposition for superior solar cell efficiency. *Sol. Energy Mater. Sol. Cells* **2015**, *134*, 54–59. [CrossRef]
4. Zeng, X.; Wen, X.; Sun, X.; Liao, W.; Wen, Y. Boron-doped zinc oxide thin films grown by metal organic chemical vapor deposition for bifacial a-Si:H/c-Si heterojunction solar cells. *Thin Solid Films* **2016**, *605*, 257–262. [CrossRef]
5. Das, S.C.; Green, R.J.; Podder, J.; Regier, T.Z.; Chang, G.S.; Moewes, A. Band Gap Tuning in ZnO Through Ni Doping via Spray Pyrolysis. *J. Phys. Chem. C* **2013**, *117*, 12745–12753. [CrossRef]
6. Mayer, M.A.; Speaks, D.T.; Yu, K.M.; Mao, S.S.; Haller, E.E.; Walukiewicz, W. Band structure engineering of ZnO1-xSex alloys. In *Proceedings of SPIE Volume 7770, Solar Hydrogen and Nanotechnology V, San Diego, CA, USA, 24 August 2010*; SPIE: Bellingham, WA, USA. [CrossRef]
7. Mayer, M.A.; Yu, K.M.; Haller, E.E.; Walukiewicz, W. Tuning structural, electrical, and optical properties of oxide alloys: ZnO1−xSex. *J. Appl. Phys.* **2012**, *111*, 113505. [CrossRef]
8. Untila, G.; Kost, T.; Chebotareva, A. Bifacial 8.3%/5.4% front/rear efficiency ZnO:Al/n-Si heterojunction solar cell produced by spray pyrolysis. *Sol. Energy* **2016**, *127*, 184–197. [CrossRef]
9. Ahmmed, S.; Aktar, A.; Kuddus, A.; Ismail, A.B.M. Fabrication of Thin-Film Solar Cell using Spin Coated Zinc Oxide and Silicon Nanoparticles Doped Cupric Oxide Heterojunction. In Proceedings of the 2018 International Conference on Computer, Communication, Chemical, Material and Electronic Engineering (IC4ME2), Rajshahi, Bangladesh, 8–9 February 2018; pp. 1–4.

10. Chen, L.; Chen, X.; Liu, Y.; Zhao, Y.; Zhang, X. Research on ZnO/Si heterojunction solar cells. *J. Semicond.* **2017**, *38*, 054005. [CrossRef]
11. Shokeen, P.; Jain, A.; Kapoor, A. Embedded vertical dual of silver nanoparticles for improved ZnO/Si heterojunction solar cells. *J. Nanophoton* **2017**, *11*, 1. [CrossRef]
12. Hussain, B. Improvement in open circuit voltage of n-ZnO/p-Si solar cell by using amorphous-ZnO at the interface. *Prog. Photovolt: Res. Appl.* **2017**, *25*, 919–927. [CrossRef]
13. Pietruszka, R.; Witkowski, B.; Zielony, E.; Placzek-Popko, E.; Godlewski, M.; Gwozdz, K. ZnO/Si heterojunction solar cell fabricated by atomic layer deposition and hydrothermal methods. *Sol. Energy* **2017**, *155*, 1282–1288. [CrossRef]
14. Ziani, N.; Belkaid, M.S. Computer modeling zinc oxide/silicon heterojunction solar cells. *J. Nano Electron. Phys.* **2018**, *10*, 06002. [CrossRef]
15. Vallisree, S.; Thangavel, R.; Lenka, T.R. Modelling, simulation, optimization of Si/ZnO and Si/ZnMgO heterojunction solar cells. *Adv. Mater. Express* **2019**, *6*, 1–20. [CrossRef]
16. Askari, S.S.A.; Kumar, M.; Das, M.K. Numerical study on the interface properties of a ZnO/c-Si heterojunction solar cell. *Semicond. Sci. Technol.* **2018**, *33*, 1–8. [CrossRef]
17. Sundaram, K.B. Work function determination of zinc oxide films. *J. Vac. Sci. Technol. A: Vac. Surf. Films* **1997**, *15*, 428–430. [CrossRef]
18. Jiang, X.; Jia, C.L.; Szyszka, B. Manufacture of specific structure of aluminum-doped zinc oxide films by patterning the substrate surface. *Appl. Phys. Lett.* **2002**, *80*, 3090–3092. [CrossRef]
19. Lee, J.; Choi, Y.; Kim, J.; Park, M.; Im, S. Optimizing n-ZnO/p-Si heterojunctions for photodiode applications. *Thin Solid Films* **2002**, *403*, 553–557. [CrossRef]
20. Lee, J.; Choi, Y.; Choi, W.; Yeom, H.; Yoon, Y.; Kim, J.; Im, S. Characterization of films and interfaces in n-ZnO/p-Si photodiodes. *Thin Solid Films* **2002**, *420*, 112–116. [CrossRef]
21. Shih, J.-L. Zinc oxide-silicon heterojunction solar cells by sputtering. Master's Thesis, McGill University, Montreal, QC, Canada, November 2007; p. 118.
22. Ruan, Y.-C.; Ruan, Y.; Ching, W.Y. An effective dipole theory for band lineups in semiconductor heterojunctions. *J. Appl. Phys.* **1987**, *62*, 2885–2897. [CrossRef]
23. Gopal, V.; Singh, S.; Mehra, R. Analysis of dark current contributions in mercury cadmium telluride junction diodes. *Infrared Phys. Technol.* **2002**, *43*, 317–326. [CrossRef]
24. Kraut, E.A. The effect of a valence band offset on potential and current distributions in HgCdTe heterostructures. *J. Vac. Sci. Technol. A: Vac. Surf. Films* **1989**, *7*, 420–423. [CrossRef]
25. Hussain, B.; Ali, A.; Unsur, V.; Ebong, A. On structural and electrical characterization of n-ZnO/p-Si single heterojunction solar cell. In Proceedings of the 2016 IEEE 43rd Photovoltaic Specialists Conference (PVSC), Portland, OR, USA, 5–10 June 2016; pp. 1898–1901.
26. Hussain, B. Development of n-ZnO/p-Si single heterojunction solar cell with and without interfacial layer. Ph.D. Thesis, The University of North Carolina at Charlotte, Charlotte, NC, USA, 2017; p. 154.
27. Basore, P. Numerical modeling of textured silicon solar cells using PC-1D. *IEEE Trans. Electron Devices* **1990**, *37*, 337–343. [CrossRef]
28. Clugston, D.; Basore, P. PC1D version 5: 32-bit solar cell modeling on personal computers. In Proceedings of the Conference Record of the Twenty Sixth IEEE Photovoltaic Specialists Conference – 1997, Anaheim, CA, USA, 29 September–3 October 1997. [CrossRef]
29. Zhao, J.; Sun, X.W.; Tan, S.T. Bandgap-Engineered Ga-Rich GaZnO Thin Films for UV Transparent Electronics. *IEEE Trans. Electron Devices* **2009**, *56*, 2995–2999. [CrossRef]

© 2019 by the authors. Licensee MDPI, Basel, Switzerland. This article is an open access article distributed under the terms and conditions of the Creative Commons Attribution (CC BY) license (http://creativecommons.org/licenses/by/4.0/).

electronics

MDPI

Article

Ultraviolet Irradiation Effects on luminescent Centres in Bismuth-Doped and Bismuth-Erbium Co-Doped Optical Fibers via Atomic Layer Deposition

Rahim Uddin, Jianxiang Wen *, Tao He, Fufei Pang, Zhenyi Chen and Tingyun Wang *

Key laboratory of Specialty Fiber Optics and Optical Access Networks, Joint International Research Laboratory of Specialty Fiber Optics and Advanced Communication, Shanghai Institute for Advanced Communication and Data Science, Shanghai University, Shanghai 200444, China; decent12u@hotmail.com (R.U.); hetao@shu.edu.cn (T.H.); ffpang@shu.edu.cn (F.P.); zychen@mail.shu.edu.cn (Z.C.)

* Correspondence: wenjx@shu.edu.cn (J.W.); tywang@shu.edu.cn (T.W.)

Received: 27 September 2018; Accepted: 11 October 2018; Published: 18 October 2018

Abstract: The effects of ultraviolet irradiation on luminescent centres in bismuth-doped (BDF) and bismuth/erbium co-doped (BEDF) optical fibers were examined in this study. The fibers were fabricated by modified chemical vapor deposition combining with atomic layer deposition method. The fibers were exposed to irradiation from a 193 nm pulsed wave argon fluoride laser, and an 830 nm wavelength laser diode pump source was employed for excitation. The experimental results showed that, for the BDF, the transmission loss was slightly reduced and the luminescence intensity was increased at the bismuth-related active aluminum centre (BAC-Al). Then, for the BEDF, the transmission loss was increased a little and the luminescence intensity was also increased at the BAC-Al centre. However, the luminescence intensity was decreased at approximately 1420 nm of the bismuth-related active silica centre (BAC-Si) for all fiber samples. One possible formation mechanism for luminescence intensity changes was probably associated with the valence state transfer of bismuth ions. The other possible mechanism was that the ArF-driven two-photon process caused luminescence changes in BAC-Al and BAC-Si. It was very important to reveal nature of luminescence properties of Bi-doped and Bi/Er co-doped optical fiber.

Keywords: UV irradiation; luminescent centres; bismuth ions; two-photon process

---

1. Introduction

Bismuth-doped optical fibers are active laser media that have comprehensive broadband and near-infrared range (NIR) emission, covering approximately 150 nm and having an extended lifetime of approximately 1 ms [1]. By employing the surface-plasma chemical vapor deposition method, Bi-doped infused SiO_2 synthesized on silica substrate exhibited photoluminescence (PL) under laser irradiation of 193-nm argon-fluoride (ArF), 337-nm nitrogen and 248-nm krypton-fluoride (KrF). Three PL band peaks (650, 800 and 1400 nm) were observed [2]. By doping small concentration of Bi (0.1 mol %) pulsed lasing and continuous-wave were induced. It was noticed that, increasing the concentration of dopants above this quantity caused significant optical absorption (unsaturated), in the preparation of effective Bi-doped laser, which was not appropriate [3,4]. In high purity fused silica, with radiation-induced modification under a 193-nm pulsed laser irradiation, the stress birefringence and refractive index were observed to increase. The mechanisms were considered related to radiolytic atomic readjustment of the SiO_2 originated by the two-photon absorption [5].

In the boron co-doped germino-silicate fibers, the negative index gratings were noticed capable of functioning approximately 25 years at a temperature 300 °C without significant degradation [6]. The photo-induced decline in the absorption and emission bands of BACs under 244-nm radiation was recently revealed. It was confirmed that the degradation of BACs was a result of the photoionization of oxygen-deficient centres (ODC) [7]. Evidently, some of the experiments allowed the creation of the bismuth-related active centres (BACs), whereas unsaturated absorption-related centres are yet to be revealed. Hence, the characteristics of the active centres are significant problems that have to be understood. Their solutions would create means for enhancing the laser features of Bi-doped fibers [8]. Photo darkening in highly ytterbium-doped alumino-silicate and phosphor-silicate fibers, under a 488-nm irradiation, was presented. Both irradiation-induced excess loss and post-irradiation temporal loss evaluations revealed that the Yb-doped phosphor-silicate fiber is highly resistant to photodarkening [9]. It was observed that, Bi-doped high-germania-silica fibers had the most noticeable bleaching, whereas BACs in alumino-silicate fibers exhibited high level of stability against laser radiation [10]. The red luminescence photo activation in Bi-doped glass is because of its irradiation by femtosecond laser pulses [11]. Significant efforts have been made to understand the structure of the color centres and identify electronic transitions responsible for the infrared luminescence in the glass [12].

In this study, UV irradiation effects on luminescent centres in BDF and BEDF were investigated and new data pertaining to physical mechanism of these phenomena were collected. Further by acquiring the luminescence intensity and transmission spectra of the irradiated fiber samples, the rate of irradiation cycle and recovery phases in different time intervals were investigated. The aforementioned are of potential interest in understanding the nature of luminescence properties in Bi-doped and Bi/Er co-doped silica optical fibers. In Section 2, we describe the experimental setup, which consists of the fiber samples and equipment, as well as experimental parameters. In the experiment's methodology, we have specified the configuration for the irradiation effect measurement and experimental conditions. In Section 3, experimental results on transmission spectra of BDF and BEDF before and after irradiation and secondary luminescence spectra of BDF and BEDF before and after irradiation are presented. In Section 4, the paper is concluded with the discussion of the results of our study on UV irradiation effects on luminescent centres in Bi-doped and Bi/Er co-doped optical fibers.

2. Experiment Setup

2.1. Description of the Fibers and Equipment

The BDF and BEDF fibers were fabricated by modified chemical vapor deposition (MCVD) method combined with atomic layer deposition (ALD) method. The fiber samples were 8 cm in length, the preform material composition was measured by an electron probe micro-analyzer (JEOL JXA-8100, University of Lille 1, France). For the Bi-doped fiber, the elements Si, Ge, O, Bi and Al had mass ratio (wt.%) of 34.4, 19.0, 46.2, 0.15, and 0.05, respectively. In the core, the concentration of bismuth and erbium in Bi/Er co-doped fiber was approximately 200 ppm and 1200 ppm in the cladding layer region. A small quantity of Al was co-doped with the Bi/Er co-doped fiber to increase the activities of bismuth and the erbium ion. Its concentration was practically 100 ppm [13–15]. Originally, the laser beam power was measured to be 60 mW. However, after reaching the laser beam expander at the irradiation region, the power was 35 mW. The laser beam was 2.5 cm in length and 1.4 cm in width. It was enlarged using a laser beam expander. The output power of the pump source 830-nm laser diode was 25 mW. A 1310-nm coupler was employed as connector. An, ArF excimer laser with a 193-nm operating wavelength was used for irradiating the fiber samples, and an optical spectrum analyzer (Yokogawa, AQ-6315A, Tokyo, Iapan) was used to measure luminescence and transmission spectra.

2.2. Experiments Methodology

While observing the effects of pulsed wave UV irradiation on BDF and BEDF optical fibers, the luminescence spectra were tested by a backward pump system excited by pump source operating at 830-nm. The transmission spectra were tested by the cut-back technique employing a white light (Yokogawa, AQ4306, Tokyo, Japan) source and measurements were performed at room temperature.

The luminescence and transmission intensities of the fiber samples were measured before exposure to UV irradiation. After irradiation of the fiber samples at different time intervals, the measured intensities were analyzed in comparison with those measured before the irradiation process. A series of experiments were performed to understand UV irradiation effects on luminescent centres in Bi-doped and Bi/Er co-doped optical fibers. The laser emitted photon energy of 4.43–12.4 eV and 0.710–1.987 aJ with a relative power of 60 mW. The power of laser diodes was chosen to achieve a uniform population inversion of BACs along the active fiber length, thus, luminescence was collected from the whole length of the tested sample. As for the luminescence spectra measurement, the system was designed in a manner that the 830-nm pump was connected to 1310-nm coupler and to OSA. The fiber under test (FUT) is connected to 1310-coupler on the other end and set horizontally to the laser beam by removing the polymer coating over the fiber samples, as shown in Figure 1a. For the measurement of transmission spectra, the fiber was connected to a white light source on one end and OSA to the other end, via pigtail connector, with FUT aligned horizontally with the laser beam for irradiation, as shown in Figure 1b.

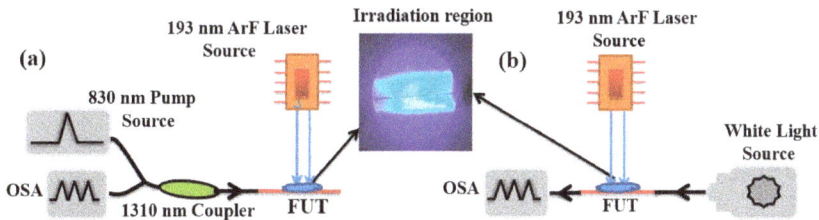

Figure 1. Irradiation effect measurement setup: (**a**) Luminescence spectra measurement system and (**b**) Transmission spectra measurement system.

3. Results and the Discussions

3.1. Transmission Spectra of BDF and BEDF Before and After UV Irradiation

The measured irradiation-induced transmission changes with the 35-mW laser power, the slight broadening of transmission intensity in BDF, and the enhanced characteristic bands in BAC-Al, as shown in Figure 2a, were all related to typical electronic transitions of bismuth ions.

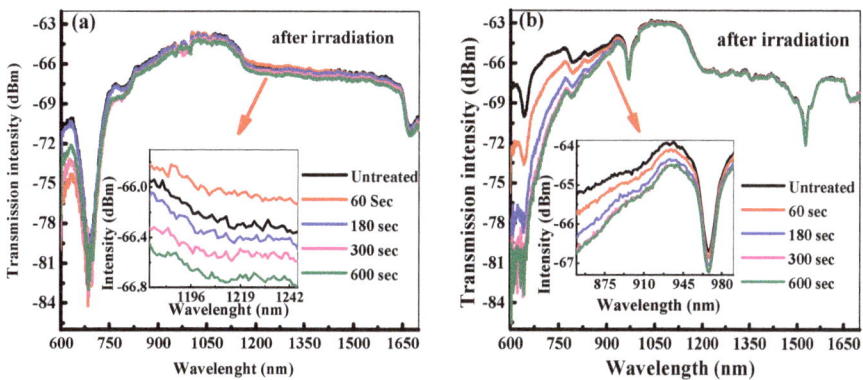

Figure 2. (**a**) Transmission spectra of Bi-doped fiber before and after irradiation as a function of time and the enlarged luminescence spectra, from 1170 nm to 1245 nm, is inserted. (**b**) Transmission spectra of Bi/Er co-doped fiber before and after irradiation, as a function of time and the enlarged luminescence spectra, from 850 nm to 1000 nm, is inserted.

In near-infrared and visible regions, a substantial variation in transmission was observed. Variations in transmission after irradiation of the fibers were because of structural modifications, which resulted from the damage or defects in glass. It can be assumed that photobleaching occurred in the region. In addition, it can also be assumed that photodarkening occurred in BAC-Si, where the intensity decreased after exposure to irradiation.

The mechanism for varying intensities indicated that photobleaching process influenced saturation at a slower rate than by the photodarkening rate, although its basis was not evident. The transmission loss reduced by about 0.24 dB with only 2.5 cm in fiber length (equivalent to 9.7 dB/m) at near-infrared region with the UV laser of 35 mW for 60 s. However, Figure 2b describes the variation of the optical transmittance intensity of BEDF in a region within the proximity 850 nm band as a function of the irradiation time. A considerable variation in transmission can be observed in that region, which indicated that there was no significant change in first excitation and ground states. However, the excitation of the state of bismuth ion population occurred after UV irradiation. The experimental evidence obtained after irradiation indicated that photobleaching was not involved to the first excitation, and the ground state transition of bismuth ions contributed to the creation of bismuth active centres. It was possible that the ground state of bismuth ion population changes, which contributed to the creation of BAC, was influenced by the irradiation [16].

3.2. Luminescence Spectra of BDF and BEDF Before and After UV Irradiation

The measurement of luminescence intensities were performed at different time intervals. Before irradiation was started, the luminescence intensity of the ideal fiber was measured through the excitation of the pump source. The irradiation effect of UV on BDF showed that at ~1100 nm the luminescence intensity (BAC-Al) was increasing firstly and then decreasing, whereas at ~1420 nm (BAC-Si) the intensity kept reducing, as shown in Figure 3a–c. Thus, that the variations observed under UV laser irradiation suggested the results were because of the two-photon process, where a number of blue photons had enhanced energies to match the energy of one UV photon. If it is supposed that the variation was because of the two-photon process, then the quadratic dependency of the amount of irradiation on the laser power intensity was probably the characteristic time of the irradiation and number of photons [17]. However, because the UV irradiation process is lower than the band gap of silica glass, exciting an electron from the valence band to the conduction band through a one-photon process is insufficient by the pulsed UV irradiation. When high-photon density and coherency absorptions because of multiple photons were considered, with a 6.4 eV photon energy, the laser operated at 50 Hz repetition rate and the laser energy was 4.1–4.2 mJ. Recent studies revealed that the irradiation of silica glasses by an ArF laser created E′ centres having intrinsic defects, which exhibited a characteristic behavior. The creation of E′ centres were saturated with a small dose of light. At room temperature the induced centres reduced quickly [18,19].

The change in the intensity can be ascribed to centres in the presence of Bi and Al ions, the luminescence at ~1100 nm corresponded to the BAC-Al was increased by 0.6 dB with only 2.5 cm in the fiber length (equivalent to 24 dB/m) at ~1100 nm by the UV irradiation for 60 s and the fluorescence intensity of BAC-Si at ~1420 nm was decreased obviously. In addition, it can be emphasized that the luminescence at the ~1100 nm band was allocated to the $^3P_1$, $^3P_2 \rightarrow ^3P_0$ transition of Bi^+ and $^2D_{3/2} \rightarrow {}^4S_{3/2}$ transition of Bi^0, and the luminescence at ~1420 nm band was assigned to the mixed valence states of Bi^{3+}/Bi^{5+}. Moreover, it was also observed that the intensity at ~1100 nm could be enhanced by changing the length of fiber and pump power, compared to that when the band was ~1420 nm [14,20].

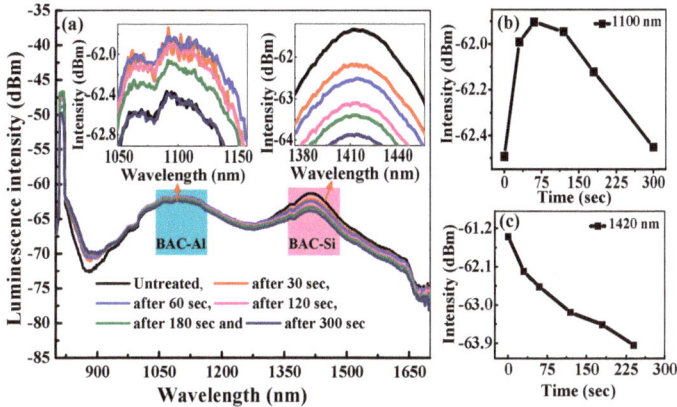

Figure 3. (a) Change of the luminescence spectra of Bi-doped fibers with an 830 nm pump excitation only after UV irradiation treatment. And the inserts are enlarged luminescence spectra from 1050 nm to 1150 nm (left) and 1380 nm to 1460 nm (right). (b) The change of luminescence intensity at ~1100 nm. (c) The change of luminescence intensity at ~1420 nm.

The irradiation changes can be expressed in relation to the decay curve of the luminescence intensity by considering a stretched exponential function. This function is physically interpreted as a continuous sum of a number of single exponential relaxation systems and extensively used to fit an integrated relaxation process in the disordered electronic and molecular system. It can be expressed by the following equation:

$$IA(t) = I(A), \infty(P) + I\beta, \infty(P)e^{-(\frac{t}{\tau(P)})^{\beta(P)}} \tag{1}$$

where $I\beta, \infty$ represents for the bleachable part of the luminescence, $IA(t)$ and $I(A)$; β is the stretched parameter; τ is the time constant and ∞ is the luminescence intensity at the time when the irradiation effect is saturated under the radiation power, P.

All measured luminescence spectra have the same exponential decay trend. However, the speed and degree of decay vary with the improvement of the induced power; the irradiation ratio increases and the time constant drops, exhibiting a faster and stronger irradiation effect.

After irradiation, BAC-Al and BAC-Si gained the same changes in BEDF fibers as shown in Figure 4. It can be observed from the figure that both irradiation ratio and decay rate were inclined to be saturated at a higher irradiation rate. Therefore, we related these two parameters with the ~1100 nm luminescence intensity in relation with the pump power in FUT. Furthermore, it was also observed the luminescence intensity was increased by 1.53 dB with only 2.5 cm in fiber length (equivalent to 61.32 dB/m) when the irradiation time was 60 s and tended to saturate, subsequently. And the peak at 1420 nm, which belonged to BAC-Si, dropped markedly with the increasing of the irradiation time. The trend can also be compared with the irradiation ratio and decay rate of photobleaching. The possible energy levels of BAC-Al and BAC-Si are shown in Figure 5a,b, respectively. These evidences allow the assumption that the principal mechanism responsible for effect on the BAC-Si was its loss of an excited electron. It can be stated that, firstly, BAC-Si absorbed the 193-nm photon and was agitated towards the second excitation level. Secondly, some of the BAC-Si fell to the lower excited state through the non-radiative transition and another release the electron to the acceptor site.

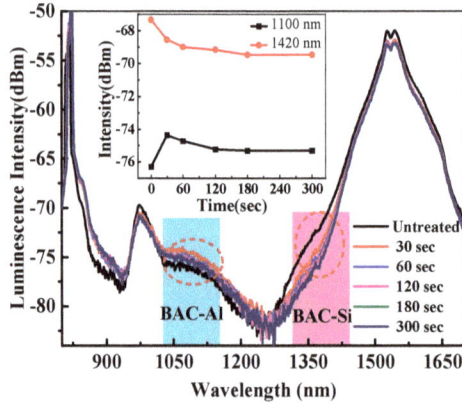

Figure 4. Luminescence spectra of Bi/Er co-doped fibers with an 830 nm pump excitation after UV irradiation treatment. And the changes of luminescence intensity at ~1100 nm (black line) and 1420 nm (red line) were inserted.

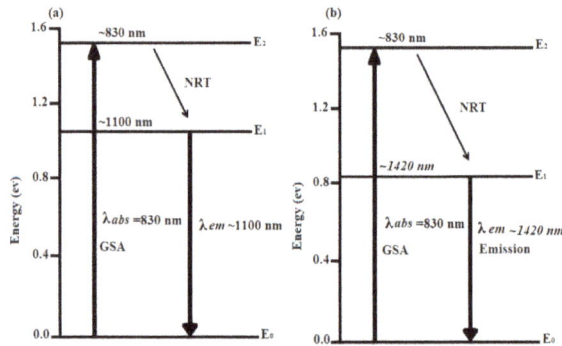

Figure 5. Possible energy levels with 830 nm pumping: (**a**) BAC-Al centre and (**b**) BAC-Si centre. (NRT: non-radiative transition; GSA: ground state absorption).

Accordingly, thermal vibrational energy generated by the laser beam aided in implementing the decay of luminescence and ground state absorption [20]. Finally, supported by thermal energy, the trapped electron still had possibility to returning to the bismuth ion, which can cause the recovery of photobleaching. All the acquired results suggest that the existence of the luminescence of Bi-doped fibers could not be entirely ascribed to some optical transitions of bismuth ion. The most substantial hypotheses for the source of the BAC with laser-active transitions were bismuth ions related to a structural defect, as previously proposed. The defect had greater probability of becoming an ODC with an absorption which approximated 5 eV band, creating an environment for the $Bi^{(n+)}$ ion. The schematic structure of BAC can be denoted as $(Bi^{(n+)} + ODC)$. The two types of ODC defects that are approximately twofold are as follows: corresponding silicon (Si + ODC) and corresponding germanium (Ge + ODC) atoms, which are similar to Si/Ge. The aforementioned possibly caused the formation of two types of BACs, namely, BAC-Si and BAC-Ge. The photoelectron progression because of the photoionization process can be confined by bismuth ions. However, no manifestation of new absorption or luminescence band was observed that could be related with bismuth ions. In the thermally-stimulated luminescence (TSL) of Bi-doped fibers, it was observed that bismuth related trap condition in germino-silicate glass fibers was not created. Although ODC was transformed in to an E' centre, the structural model was not a paired spin on a sp^3-hybridized molecular orbital of a threefold corresponding silicon/germanium atom. The transformation was most possibly shaped by the local structural readjustment of the glass caused the variation of the bismuth ion atmosphere [21].

4. Conclusions

The Bi-doped and Bi/Er co-doped optical fibers were fabricated by MCVD combined with ALD method. This study revealed a significant property of UV irradiation effects on luminescent centres in the optical fiber samples. At first, the transmission loss was slightly reduced by 0.24 dB in Bi-doped fibers at 1200 nm and the transmission loss of the Bi/Er co-doped fibers was increased a little. Then, the luminescence intensity was increased by 0.6 dB of the Bi-doped fibers and by 1.53 dB of Bi/Er co-doped fibers at ~1100 nm (BAC-Al) with UV laser irradiation treatment. In the meantime, the luminescence intensity was decreased at ~1420 nm (BAC-Si) of all fiber samples. In addition, an energy level diagram was constructed for BAC-Al and BAC-Si centres. It is very important to further understand the different processes of luminescence. The change mechanism for luminescence intensity was probably ascribed to valence state transfer of bismuth ions at the ~1100 nm band. And at ~1420 nm band, the mechanism was probably associated with the loss of an excited electron and ArF-driven two-photon process. These aided in understanding the nature of luminescence properties in Bi-doped and Bi/Er co-doped silica optical fibers. In the future, we will further investigate the relationship between active centres and valence states.

Author Contributions: R.U. conceived the presented idea, carried out the experiments and conducted the analysis. J.W. was in charge of direction and planning. T.H. assisted in the sample preparation and performing the experiments, F.P. and Z.C. provided critical feedbacks during the research and T.W. supervised the overall project.

Funding: This work is supported by National Natural Science Foundation of China (61520106014, 61475096, 61422507, 61635006, 61705126); Science and Technology Commission of Shanghai Municipality, China (15220721500).

Conflicts of Interest: The authors declare no conflict of interest.

References

1. Dianov, E.M. Amplification in extended transmission bands using bismuth-doped optical fibers. *J. Lightw. Technol.* **2013**, *31*, 681–688. [CrossRef]
2. Trukhin, A.; Teteris, J.; Bazakutsa, A.; Golant, K. Impact of fluorine admixture, hydrogen loading, and exposure to arf excimer laser on photoluminescence of bismuth defects in amorphous silica. *J. Non-Cryst. Solids* **2013**, *362*, 180–184. [CrossRef]
3. Bufetov, I.A.; Melkumov, M.A.; Firstov, S.V.; Riumkin, K.E.; Shubin, A.V.; Khopin, V.F.; Guryanov, A.N.; Dianov, E.M. Bi-doped optical fibers and fiber lasers. *IEEE J. Sel. Top. Quantum Electron.* **2014**, *20*, 111–125. [CrossRef]
4. Zlenko, A.S.; Mashinsky, V.M.; Iskhakova, L.D.; Semjonov, S.L.; Koltashev, V.V.; Karatun, N.M.; Dianov, E.M. Mechanisms of optical losses in Bi:SiO$_2$ glass fibers. *Opt. Express* **2012**, *20*, 23186–23200. [CrossRef] [PubMed]
5. Rothschild, M.; Ehrlich, D.J.; Shaver, D.C. Effects of excimer laser irradiation on the transmission, index of refraction, and density of ultraviolet grade fused silica. *Appl. Phys. Lett.* **1989**, *55*, 1276–1278. [CrossRef]
6. Dong, L.; Liu, W.F. Thermal decay of fiber bragg gratings of positive and negative index changes formed at 193 nm in a boron-codoped germanosilicate fiber. *Appl. Opt.* **1997**, *36*, 8222–8226. [CrossRef] [PubMed]
7. Firstov, S.; Alyshev, S.; Melkumov, M.; Riumkin, K.; Shubin, A.; Dianov, E. Bismuth-doped optical fibers and fiber lasers for a spectral region of 1600–1800 nm. *Opt. Lett.* **2014**, *39*, 6927–6930. [CrossRef] [PubMed]
8. Fujimoto, Y. Local structure of the infrared bismuth luminescent center in bismuth-doped silica glass. *J. Am. Ceram. Soc.* **2010**, *93*, 581–589. [CrossRef]
9. Sahu, J.K.; Yoo, S.; Boyland, A.; Basu, C.; Kalita, M.; Webb, A.; Sones, C.L.; Nilsson, J.; Payne, D.N. 488 nm irradiation induced photodarkening study of Yb-doped aluminosilicate and phosphosilicate fibers. In Proceedings of the Lasers and Electro-Optics/Quantum Electronics and Laser Science Conference and Photonic Applications Systems Technologies, San Jose, CA, USA, 4–9 May 2008.
10. Friebele, E.J.; Gingerich, M.E. Photobleaching effects in optical fiber waveguides. *Appl. Opt.* **1981**, *20*, 3448–3452. [CrossRef] [PubMed]

11. Kononenko, V.; Pashinin, V.; Galagan, B.; Sverchkov, S.; Denker, B.; Konov, V.; Dianov, E. Laser induced rise of luminescence efficiency in bi-doped glass. *Phys. Procedia* **2011**, *12*, 156–163. [CrossRef]
12. Meng, X.; Qiu, J.; Peng, M.; Chen, D.; Zhao, Q.; Jiang, X.; Zhu, C. Near infrared broadband emission of bismuth-doped aluminophosphate glass. *Opt. Express* **2005**, *13*, 1628–1634. [CrossRef]
13. Wang, P.; Wen, J.; Dong, Y.; Pang, F.; Wang, T.; Chen, Z. Bismuth-doped silica fiber fabricated by atomic layer deposition doping technique. In Proceedings of the Asia Communications and Photonics Conference, Beijing, China, 12–15 November 2013.
14. Wen, J.; Wang, J.; Dong, Y.; Chen, N.; Luo, Y.; Peng, G.; Pang, F.; Chen, Z.; Wang, T. Photoluminescence properties of Bi/Al-codoped silica optical fiber based on atomic layer deposition method. *Appl. Surf. Sci.* **2015**, *349*, 287–291. [CrossRef]
15. Liu, W.; Wen, J.; Dong, Y.; Pang, F.; Luo, Y.; Peng, G.-D.; Chen, Z.; Wang, T. Spectral characteristics of Bi/Er co-doped silica fiber fabricated by atomic layer deposition (ALD). In Proceedings of the Asia Communications and Photonics Conference, Hong Kong, China, 19–23 November 2015.
16. Firstov, S.; Alyshev, S.; Khopin, V.; Melkumov, M.; Guryanov, A.; Dianov, E. Photobleaching effect in bismuth-doped germanosilicate fibers. *Opt. Express* **2015**, *23*, 19226–19233. [CrossRef] [PubMed]
17. Arai, K.; Imai, H.; Hosono, H.; Abe, Y.; Imagawa, H. Two photon processes in defect formation by excimer lasers in synthetic silica glass. *Appl. Phys. Lett.* **1988**, *53*, 1891–1893. [CrossRef]
18. LaRochelle, S.; Ouellette, F.; Lauzon, J. Two-photon excitation and bleaching of the 400 nm luminescence band in germanium-doped-silica optical fibres. *Can. J. Phys.* **1993**, *71*, 79–84. [CrossRef]
19. Nuccio, L.; Agnello, S.; Boscaino, R. Annealing of radiation induced oxygen deficient point defects in amorphous silicon dioxide: evidence for a distribution of the reaction activation energies. *J. Phys. Matter* **2008**, *20*, 385215. [CrossRef] [PubMed]
20. Ding, M.; Wei, S.; Luo, Y.; Peng, G.-D. Reversible photo-bleaching effect in a bismuth/erbium co-doped optical fiber under 830-nm irradiation. *Opt. Lett.* **2016**, *41*, 4688–4691. [CrossRef] [PubMed]
21. Jain, S.; Duchez, J.B.; Mebrouk, Y.; Velazquez, M.M.A.N.; Mady, F.; Dussardier, B.; Benabdesselam, M.; Sahu, J.K. Thermally-stimulated emission analysis of bismuth-doped silica fibers. *Opt. Mater. Express* **2014**, *4*, 1361–1366. [CrossRef]

© 2018 by the authors. Licensee MDPI, Basel, Switzerland. This article is an open access article distributed under the terms and conditions of the Creative Commons Attribution (CC BY) license (http://creativecommons.org/licenses/by/4.0/).

![electronics logo] *electronics*

MDPI

Article

High Performance Graphene-Based Electrochemical Double Layer Capacitors Using 1-Butyl-1-methylpyrrolidinium tris (pentafluoroethyl) trifluorophosphate Ionic Liquid as an Electrolyte

Jacob D. Huffstutler [1,†], Milinda Wasala [1,†], Julianna Richie [1], John Barron [1], Andrew Winchester [1], Sujoy Ghosh [1], Chao Yang [2], Weiyu Xu [2], Li Song [2], Swastik Kar [3] and Saikat Talapatra [1,*]

[1] Department of Physics, Southern Illinois University, Carbondale, IL 62901, USA; tahfarce@siu.edu (J.D.H.); milinda.wasala@siu.edu (M.W.); jrichie@siu.edu (J.R.); john.barron95@siu.edu (J.B.); ajw1818@siu.edu (A.W.); sujoy.kittu@siu.edu (S.G.)
[2] National Synchrotron Radiation Laboratory, University of Science and Technology of China, Hefei 230029, China; ych1991@mail.ustc.edu.cn (C.Y.); xuweiyu@mail.ustc.edu.cn (W.X.); song2012@ustc.edu.cn (L.S.)
[3] Department of Physics, Northeastern University, Boston, MA 02115, USA; s.kar@northeastern.edu
* Correspondence: saikat@siu.edu; Tel.: +1-618-453-2270
† J.D.H. and M.W. contributed equally to this work.

Received: 6 August 2018; Accepted: 29 September 2018; Published: 2 October 2018

Abstract: There are several advantages to developing electrochemical double-layer capacitors (EDLC) or supercapacitors with high specific energy densities, for example, these can be used in applications related to quality power generation, voltage stabilization, and frequency regulation. In this regard, ionic liquids capable of providing a higher voltage window of operations compared to an aqueous and/or polymer electrolyte can significantly enhance the specific energy densities of EDLCs. Here we demonstrate that EDLCs fabricated using ionic liquid 1-butyl-1-methylpyrrolidinium tris (pentafluoroethyl) trifluorophosphate (BMP-FAP) as an electrolyte and few layer liquid-phase exfoliated graphene as electrodes show remarkable performance compared to EDLC devices fabricated with aqueous potassium hydroxide (6M) as well as widely used ionic liquid 1-butyl-3-methylimidazolium hexafluorophosphate (BMIM-PF6). We found that graphene EDLC's with BMP-FAP as an electrolyte possess a high specific energy density of ≈25 Wh/kg along with specific capacitance values as high as 200 F/g and having an operating voltage windows of >5 volts with a rapid charge transfer response. These findings strongly indicate the suitability of BMP-FAP as a good choice of electrolyte for high energy density EDLC devices.

Keywords: graphene; supercapacitor; energy storage; ionic liquid

---

1. Introduction

The ever-increasing energy storage needs of the world have presented a complex puzzle for energy scientists of all types. The large amount of energy desired to be stored in order to keep up with "dead zones" within the varied energy production methods (e.g., at night for solar energy) would suggest batteries as an ideal local component for the storage. They, however, cannot always meet the power demands of the energy sources. The power variance of the energy sources suggests a device with a large capability for handling high power output would be necessary, such as a capacitor. However, capacitors do not meet the stringent energy requirements of alternative energy as they would too quickly reach capacity. Clearly, it is imperative that a device capably handles both high energy and high-power output [1–4]; however, that is still not enough. In order to keep lifetime maintenance costs down and remain competitive, the device in question must be able to survive unmaintained for thousands if not tens of thousands or more full cycles. Electrochemical double-layer capacitors (aka EDLC/supercapacitor/ultracapacitor), composed of two electrodes separated by a porous membrane and soaked in an electrolyte sandwiched between current collectors, exhibit strong performance in all of these aspects [2]. Their range of energy densities typically outperforms traditional capacitors by 1–2 orders of magnitude. At the same time, their range of power densities outperforms batteries by 1–2 orders of magnitude [5]. Additionally, a vast majority of EDLC devices tested exhibit extraordinary lifecycles, where typically 80% or greater maximum capacitive behavior is retained even after tens of thousands of cycles or more. These behaviors perfectly place them in an ideal position to manage a vast variety of devices powered through alternative energy means, and to continue doing so without outside intervention or maintenance for extended periods of time [1–4].

In the recent past carbon-based materials [6], 2-D nanomaterials [7,8], conductive metal organic frameworks (MOFs) [9], as well as redox-active polymers [10], have been investigated as electrical energy storage systems/EDLCs. Among them, carbon-based materials (carbon nanotubes [2,11–14], graphene [15–21], activated carbon [2,5,22], hybrid structures [23–25], etc.) are preferred electrode materials for EDLC devices due to their high conductivity, low reactivity and extremely high specific surface area [2]. Past studies have shown that several electrolytes, for example, aqueous, polymer and ionic liquid electrolytes are suitable for carbon-based EDLC devices. However, an inherent technological bottleneck in carbon-based ELDC devices fabricated with aqueous, polymer and widely used ionic liquids, for example, 1-butyl-3-methylimidazolium hexafluorophosphate (BMIM)(PF6) electrolyte is in their limited operating voltage windows, which leads to low energy densities. The choice of electrolyte to be used in EDLCs, therefore, is crucial since the nature of the electrolyte can substantially enhance the performance of these devices.

In the choice of electrolyte lies a myriad of questions, each with different optimal answers. A balance must be struck between operating temperature, operating potential, electrochemical stability, conductivity, viscosity and ion size. Ionic liquids have become the preferred electrolyte due in no small part to their much wider operating potential windows and chemical stability [20,21]. This wider operating potential allows for the devices prepared from these samples to be used in a much more varied range of applications. Due to the issue of electrolysis with aqueous solutions, most are limited in range to 1 V. 1-Butyl-1-methylpyrrolidinium tris (pentafluoroethyl) trifluorophosphate (BMP-FAP) is a particularly good choice with its far wider 6.6–6.8 V window [26,27]. It will be discussed later that this full window is not entirely available for completely reversible charging and discharging, but it will be shown that even if a portion of this window is used, the devices will outperform standard aqueous electrolyte devices by a significant margin. BMP-FAP is also ideal in its chemical stability with smaller Faradaic peaks potentially being associated with impurities, rather than the breakdown of the electrolyte itself [26]. In contrast to aqueous electrolytes, the temperature stability of BMP-FAP allows for use within a wider range of temperatures, approximately $-50\,°C$ to ≈250–$277\,°C$ [26,27]. This widening of the allowable circumstances presents a strong argument for their use in extreme applications, such as military, space, and explorative research.

The liquid-phase exfoliation method in particular is ideal for exploratory testing of varying parameters such as this in devices as the production methods are quick and simple and produce consistent, few-layer dispersions of similarly-sized flakes. From this stage, the devices could be further improved through the use of functionalized materials, heterostructures or composite materials as electrodes. Here we demonstrate, that EDLCs fabricated using ionic liquid BMP-FAP as an electrolyte and few-layer liquid-phase exfoliated graphene (LPEG) as electrodes show remarkable performance compared to EDLC devices fabricated with aqueous 6 M potassium hydroxide (KOH), as well as widely used ionic liquid BMIM-PF6. Our results indicate that simple LPEG EDLCs utilizing BMP-FAP show an energy density of ≈25 Wh/kg which is on par with aqueous results for more complex electrode materials [28,29], while demonstrating power densities as much as 2 to nearly 5 times higher.

2. Materials and Methods

2.1. Structural Characterizations

The graphene samples for preparing the EDLC electrodes were obtained using a liquid-phase exfoliation technique, described in detail in the Supplementary Materials. In Figure 1, typical structural and spectroscopic characterization of the LPEG is shown. In Figure 1a, a digital image of a well-dispersed vial of LPEG in isopropyl alcohol (IPA) is shown. Figure 1b shows a typical transmission electron microscopy (TEM) image of a collection of few-layer graphene flakes. In Figure 1c, an atomic force microscopy (AFM) image of a thin film of the LPEG flakes (prepared on SiO_2 wafers via spin-coating) is shown. Detailed scans of various portions of this film showed that the flake step heights ranged from ≈2.7 nm to ≈5.2 nm (Figure 1d), which correspond approximately to ≈8–15 layers of graphene flakes assuming the van der Waals separation to be ≈0.335 nm for graphite [30]. This correlates well with the high resolution-transmission electron microscopy (HR-TEM) data in Figure 1f,g where 7 and 15 layers can be clearly identified. In Figure 1e, Raman measurement performed on films of LPEG is shown. Raman spectra serves as a useful tool for analyzing different carbon-based materials and can give significant insights about the quality of the samples. Samples for Raman measurements were also prepared using a spin-coating method onto SiO_2 wafers. Raman measurements were performed using 12 mW laser power at 532 nm, with five scans of 60 s being averaged for the LPEG samples. The D peak at 1356 cm^{-1} corresponding to the defect band of graphite, as well as a strong graphitic (E_{2g}) G peak at 1578 cm^{-1}, was observed in the LPEG samples. Additionally, a D′ peak at 1620 cm^{-1} was also observed in the LPEG samples (Note: this peak was absent in the un-exfoliated graphite powder, data not shown) perhaps due to edge defects generated by the sonication process used for exfoliation [31].

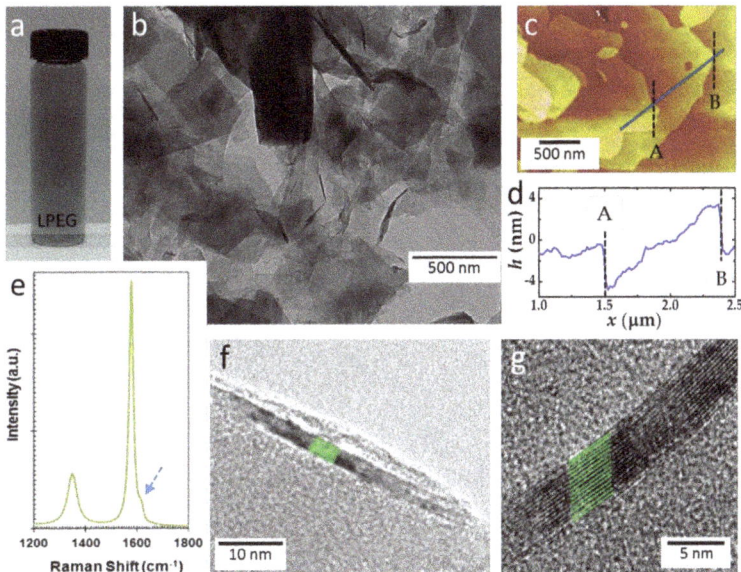

Figure 1. Optical characterization of the prepared liquid-phase exfoliated graphene (LPEG) sample. (**a**) Resulting LPEG solution following decanting of the supernatant liquid after centrifugation. (**b**) Transmission electron microscopy (TEM) image of dispersed few-layer flakes of prepared solution. (**c**) AFM image of an LPEG flake. (**d**) AFM height profile along the blue line shown in Figure 1c. (**e**) Raman spectroscopy results of LPEG. (**f**) HR-TEM image showing ≈8 layers at the edge of LPEG. (**g**) HR-TEM image showing ≈14 layers at the edge of the LPEG.

2.2. Electrochemical Characterizations

In order to characterize the electrochemical behavior of the sample produced, the samples were diluted to one-fourth of the concentration and vacuum-filtrated through a porous Teflon membrane filter (0.1 μm). The coated filter was then heated at 60 °C for fifteen minutes in order to evaporate any remaining IPA. The filter was weighed before and after the coating process in order to determine the weight of LPEG deposited. These filters were used for fabricating the EDLC electrodes. To prepare an EDLC using the filter, two square pieces of nearly equal size were cut from the filter containing LPEG samples to act as the electrode material. These two electrodes were separated with a larger square of filter paper, which acted as the porous separator. The stacked assembly was then transferred to a sealed stainless steel cell containing the electrolyte for electrochemical characterization.

We have tested three different electrolytes: 6 M KOH, BMIM-PF6 and BMP-FAP. Two-electrode cyclic voltammetry (CV) was performed at room temperature for all electrolytes over a broad range of scan rates. For example, for all samples tested with 6M KOH as the electrolyte, a baseline study was conducted in order to establish the behavior of the electrode material. This study consisted of a series of scans from –0.5 V to 0.5 V at scan rates of 1000 mV/s, 500 mV/s, 200 mV/s, 100 mV/s, 50 mV/s, 10 mV/s, and 1 mV/s at room temperature. Initial cell purge and equilibration times were kept at 10 s and 5 s for all CV testing. Similar protocol was also followed for measurements performed using BMIM-PF6. The data obtained from both these measurements are presented in Supplementary Material (Figure S3).

For all the samples tested with BMP-FAP as the electrolyte, different scan rates were again used with a variety of potential voltage windows in order to demonstrate the increased performance when the full range of this particular ionic liquid is utilized. Scans were taken about the origin with voltage windows taken in increasing 1 V steps. Measurements at 1 V, 2 V, 3 V, 4 V, 5 V and 6 V were taken in all. For each voltage window, scan rates of 1000 mV/s, 500 mV/s, 200 mV/s, 100 mV/s, 50 mV/s and 10 mV/s were taken at room temperature. Galvanostatic charge–discharge (CD) was also performed with ± 1 mA of current alternated every 1 s for 200 cycles to 2000 cycles in order to evaluate the longevity and stability of the devices prepared. The electrochemical measurements performed with BMP-FAP are shown in Figure 2.

3. Results and Discussion

The CV plots shown in Figure 2a demonstrated good reversibility and charge return, with little Faradaic (non-reversible) response aside from the 6 V scan voltage windows. The 5 V BMP-FAP scan window in particular demonstrated symmetric charge return and little to no Faradaic response when compared with its relative performance. The 6 V scan voltage window did demonstrate some Faradaic peaks, suggesting that while the 6 V window could be used to expand the device performance even further, where some device degradation would be seen over continued use. In Figure 2b, the specific capacitance (F/g) of the devices is plotted against the scan rate. The 10 mV/s scan rate test of the BMP-FAP device demonstrated a significant improvement upon the corresponding result from the 6 M KOH device (Figure 2d). For the 6 V window average, an over two-fold performance increase was observed, while the average of the 5 V window results still presented roughly a 64% increase in capacitive response. The stability and longevity of the device is presented over the course of 2000 cycles in Figure 2c. Once the initial cycling had occurred, the devices swiftly reach a stable response, demonstrating little variability across hundreds to thousands of cycles, with the resulting retention maintaining 82% of the initial performance. We believe that the capacitance drop with the increasing number of cycles is perhaps due to the adsorption of the electrolyte ions on defect sites and/or edges of the exfoliated graphene flakes. The process of exfoliation can produce structural defects in the form of dangling bonds on the edges, carbon vacancies, etc., on the graphene material. These structural defects are highly reactive and oftentimes acts as sites where electrolyte ions can get adsorbed strongly during the charging cycle and are not readily desorbed during the discharging. The overall effect is loss of active surface area of the electrode materials for some of the initial charge–discharge cycles, which manifests as loss of specific capacitance. Finally, Figure 2d presents the corresponding 10 mV/s CVs from each prepared device in terms of specific capacitance and the voltage window. From this plot, it is observed that EDLC performance of LPEG with BMP-FAP as an electrolyte was significantly better compared to other electrolytes used in this study. One of the key reasons for such an observation is perhaps due to the fact that the adsorption potentials for ionic liquid ions are extremely strong in the presence of flat graphene surfaces as suggested through detailed theoretical calculations [32]. Therefore, these ions are attracted closer to the flat electrode surface. This in turn makes the charge separation distance in the formed double layer smaller. A smaller charge separation distance (in parallel plate geometry) is perhaps one of the main causes of improved capacitance performance as seen in this investigation. A simple schematic showing an artist's rendition of formation of a double layer on a flat graphene surface is shown in the inset of Figure 2d.

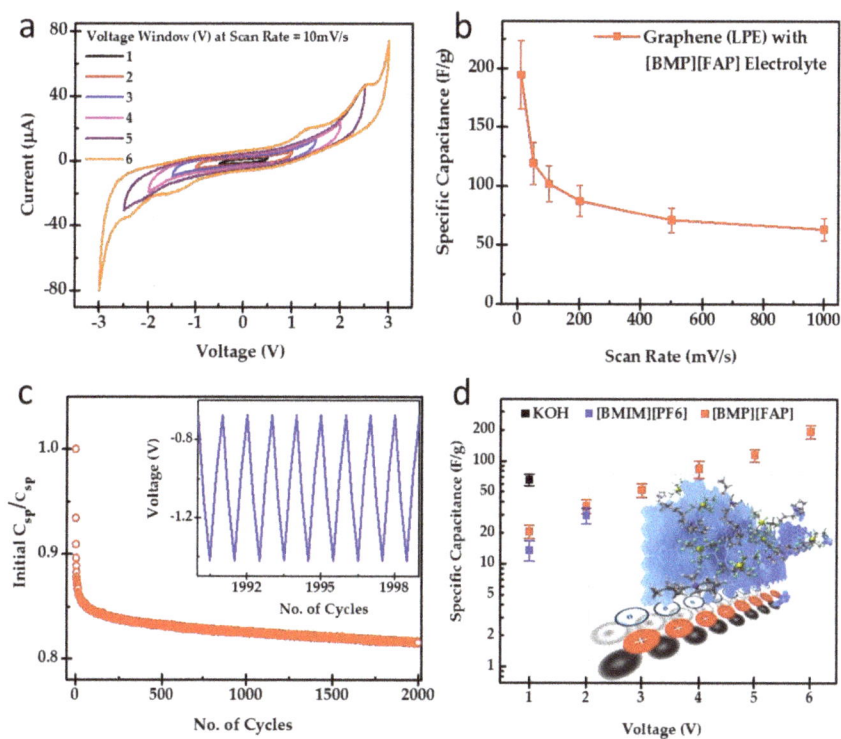

Figure 2. Electrochemical data for LPEG in BMP-FAP. (**a**) Cyclic voltammetry results for all voltage windows taken. (**b**) Specific capacitance vs scan rate for the 6 V window. (**c**) Capacitance retention taken over 2000 cycles of charging and discharging. (**c inset**) Sample capacitor charging and discharging curves. (**d**) Comparative results of specific capacitance at 10 mV/s scan rate for each electrolyte and voltage window. (**d inset**) Schematic diagram of formation of double layer.

To further characterize the EDLC system, electrochemical impedance spectroscopy (EIS) was used. The impedance spectrum was obtained within the frequency range of 126 mHz to 50 kHz with 10 mV root-mean-squared (RMS) voltage without a DC bias voltage. The resulting Nyquist plot of the real and the imaginary parts of the impedance is shown in Figure 3. By taking the intercept with the real axis of the impedance in the high frequency curve, the equivalent series resistance (ESR) of the system can be estimated. In order to understand the physical processes, such as the electrochemical kinetic reaction mechanism, double layer capacitance, pseudo-capacitance, etc., the impedance spectrum can be fit in to an equivalent circuit model [2,33–35]. Then the model can be understood in an analytical way by considering the different circuit elements and the parameters that are associated with those components.

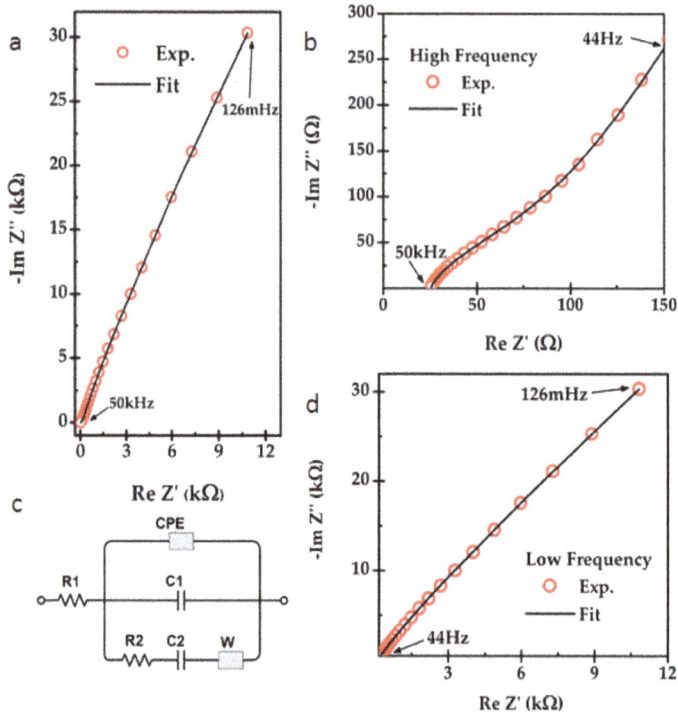

Figure 3. Electrochemical impedance spectroscopy (EIS) analysis results for LPEG in the BMP-FAP electrolyte. (**a**) Full-frequency fitted Nyquist data. (**b**) Prepared equivalent circuit. (**c**) High-frequency fitted Nyquist data. (**d**) Low-frequency fitted Nyquist data.

Figure 3a shows experimental and fitted Nyquist plots for the BMP-FAP ionic electrolyte and the Figure 3c shows the equivalent circuit model for the fitted data. From the high-frequency range (50 kHz–44 Hz) experimental data (Figure 3b), the system's ESR was estimated to be 25.57 Ω. In the circuit model, the R1 resistance corresponded to the ESR of the system, which mainly arose due to the series resistance of the electrolyte, filter papers and charge barriers between the stainless steel current collectors and the graphene electrode material. Presence of a constant phase element (CPE) corresponds to the non-ideal nature of the capacitor, which also represents the system behavior due to effects such as surface disorder, electrode porosity, and adsorption processes [33,34]. The impedance corresponding to a CPE (Z_{CPE}) can be written as:

$$Z_{CPE} = \frac{1}{(j\omega)^n Y_0} \tag{1}$$

where Y_0 is the numerical value of the admittance, n is the CPE parameter, and ω ($\omega = 2\pi f$) is the angular frequency at the lowest frequency (f) measured in the impedance spectrum. Capacitance associated with CPE (C_{CPE}) can be evaluated using the following equation:

$$C_{CPE} = \frac{Y_0}{\omega^{1-n} \sin(n\pi/2)} \tag{2}$$

Capacitance corresponding to the CPE has a major contribution to the total device capacitance of the system (Table 1). We believe that the presence of C1 and C2 pseudo-capacitances parallel to the CPE also indicate the presence of processes such as the adsorption of ions onto the electrodes. However, these pseudo-capacitances attribute only a small contribution to the total device capacitance [35]. The R2 resistor element is most likely due to a charge transfer barrier in the system. The Warburg element (W) in the circuit model represents the Warburg impedance, which arose due to the diffusion of ions.

Table 1. The individual circuit component characteristic values prepared from the electrochemical impedance spectroscopy (EIS) analysis.

Parameters	Value
Pseudo-capacitance (C1)/F	1.5959×10^{-6}
Pseudo-capacitance (C2)/F	9.5812×10^{-6}
Equivalent series resistance (ESR) (R1)/Ω	25.605
Faradaic resistance (R2)/Ω	155.17
Warburg component (W)	3029
Constant phase element (CPE) factor (Y0)	2.7059×10^{-5}
CPE exponent (n)	0.70632
ESR (experimental)/Ω	25.57
Double layer capacitance (at lowest frequency)/F	3.273×10^{-5}
Total device capacitance/F (from the model)	4.391×10^{-5}
Total device capacitance/F/g	3.0699
Experimental capacitance (1000 mV/s)/F/g	4.13

The efficacy of this exfoliation method is made readily apparent in the comparison shown in Figure 4, where a nearly two order of magnitude difference in energy and power densities is observed between a bulk graphite sample prepared and tested in the same manner as the LPEG samples. Several devices were prepared, and the relative variance in performance is shown to be small, indicating that while the nanoscale results can be somewhat chaotic, the macroscopic final product was relatively consistent in response.

When comparing current supercapacitor technologies gravimetrically, it becomes apparent that there is certainly growth and parameterization to be had through variation of the structure and composition of the electrodes used; however, the electrolyte used has a demonstrative effect on the performance of the prepared devices that can be combined with these efforts to produce even further advances in the technology. More complex fabrication methods [36,37] have produced supercapacitors with good energy densities in aqueous solutions, 78.8 Wh/kg and 73 Wh/kg, respectively, that exhibited lower power densities when compared with other supercapacitor configurations. In the past, it was shown that innovative electrode materials [28,38–40] could be used to tailor balanced performance in terms of both energy and power. In these cases, as is often seen when attempting to optimize performance, a tradeoff exists between the two (Table 2). These results almost mirror one another, one with 27 Wh/kg and 17 kW/kg and another with 14 Wh/kg and 25 kW/kg. Thus, current technologies are seemingly held back by varying degrees of cost, feasibility, and scalability when limited to aqueous electrolytes.

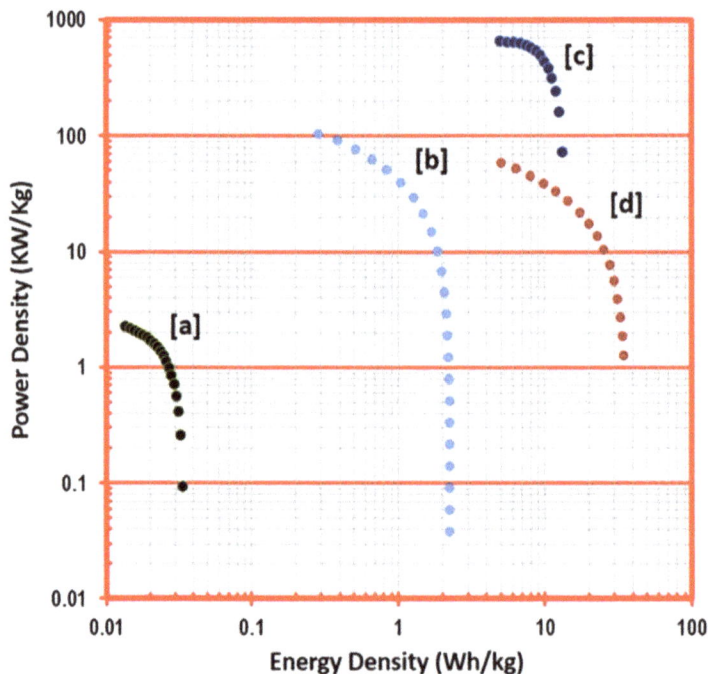

Figure 4. Ragone (power density vs energy density) performance plot of devices with graphite powder (**a**), LPEG with KOH as the electrolyte (**b**), 1-pyrene carboxylic-acid functionalized graphene with KOH as the electrolyte (**c**) [39], and LPEG with BMP-FAP as electrolyte devices (**d**).

Table 2. A comparison of a variety of graphene-based device electrochemical power and energy performance.

Source	Method	Measurement	Electrolyte	ED (Wh/kg)	PD (kW/kg)
[36]	CNG@NCH	Three-electrode	Aqueous (1 M LiOH)	78.8	8.4
[37]	RuO$_2$/PEDOT-G30	Three-electrode	Aqueous (0.5 M H$_2$SO$_4$)	73	1.1
[38]	Graphene w/Mg(OH)$_2$ template	Three-electrode & Two-electrode	Aqueous (6 M KOH)	≈27	≈17
[28]	Graphene Yarns	Two-electrode	Aqueous (1 M H$_2$SO$_4$)	14	25
[40]	NI$_3$S$_2$ & Co$_3$S$_4$ on RGO Hydrogel @Ni Foam	Asymmetrical Two-electrode	Aqueous (6 M KOH)	55.2	13
[29]	LRGONR	Two-electrode	Aqueous (2 M H$_2$SO$_4$)	15.1	≈10
			IL ([TEA][BF$_4$])	90	2.0 [a]
			IL ([BMIM][BF$_4$])	181.5	2.3 [a]
This work	LPEG	Two-electrode	IL ([BMP][FAP])	24.9	47.5

[a] Indicates simultaneous results rather than respective maximums.

To be able to further expand performance without complicating fabrication is an important possibility in need of exploration. Sahu et al. [29] demonstrate the viability of this approach quite well with their Ragone measurements for lacey-reduced graphene oxide nanoribbons (LRGONR) using 2M H$_2$SO$_4$, N,N,N,N-tetraethylammonium tetrafluoroborate (TEA-BF$_4$), and 1-butyl-1-methylimidazolium tetrafluoroborate (BMIM-BF$_4$). This work demonstrates the general electrochemical performance advances between 6M KOH, BMIM-PF$_6$, and BMP-FAP, as well as the energy and power density results for BMP-FAP. Sahu et al. demonstrate a particular jump in energy density, going from an aqueous maximum of 15.1 Wh/kg to 90 Wh/kg and 181.5 Wh/kg with each change in electrolyte.

4. Conclusions

The results of this work have shown strong improvements in general capacitive behavior from the addition of this ionic liquid and when compared to other supercapacitor electrode research, a stronger balanced performance in terms of energy and power is observed as well. The simple introduction of an ionic liquid into pre-existing device electrode architecture allows for an increase in energy density normally only achieved through more complex fabrication procedures or for even further optimization of heterostructures or tailored electrode materials. This allows for devices of various compositions and structures to be further improved by a change in electrolyte without a significant impact upon device reversibility or longevity and for simpler, scalable procedures to potentially demonstrate commercial viability due to increased yields or reduced production costs.

Supplementary Materials: The following are available online at http://www.mdpi.com/2079-9292/7/10/229/s1, Figure S1: (a) UV-VIS absorbance spectroscopy of LPEG, (b) XPS analysis survey scan, (c) O 1s scan, and (d) C 1s scan. Figure S2: (a) Cyclic voltammetric plot for bulk graphite powder in BMP-FAP, and (b) the resulting specific capacitance vs scan rate used. Figure S3: (a) Cyclic voltammetric plot of LPEG in 6M KOH aqueous electrolyte, and (b) the resulting specific capacitance vs scan rate used. (c) Cyclic voltammetric plot of LPEG in BMIM-PF6 ionic liquid electrolyte and (d) the resulting specific capacitance vs. scan rate used. Table S1: Specific capacitance (Csp) (in F/g) for all scan rates and voltage windows that were taken for LPEG.

Author Contributions: Conceptualization, S.T.; Methodology, J.H., M.W., A.W., and S.G.; Validation, J.R. and J.B.; Formal Analysis, J.H. and M.W.; Investigation, J.H., M.W., J.R., and J.B.; Resources, A.W., S.G., C.Y., and W.X.; Data Curation, M.W.; Writing-Original Draft Preparation, J.H. and M.W.; Writing-Review & Editing, S.T., L.S., and S.K.; Visualization, J.H. and M.W.; Supervision, S.T.; Project Administration, M.W. and J.H.; Funding Acquisition, S.T. and L.S.

Funding: J.H. was funded by NSF EAPSI program Award ID #1414819, and L.S. acknowledges the funding support through NSF of China (U1232131, U1532112, 11375198, 11574280).

Acknowledgments: M.W. acknowledges the DRA funding support through SIU Graduate School. MaSK [41] (Molecular Modeling and Simulation Kit http://ccmsi.us/mask) was used in the creation of the Figure 2d inset.

Conflicts of Interest: The authors declare no conflicts of interest.

References

1. Bard, A.J.; Faulkner, L.R. *Electrochemical Methods: Fundamentals and Applications*, 2nd ed.; John Wiley & Sons: Hoboken, NJ, USA, 2007.

2. Conway, B.E. *Electrochemical Supercapacitors: Scientific Fundamentals and Technological Applications*; Kluwer Acad.: New York, NY, USA, 2009; ISBN 978-1-4757-3058-6.

3. Goodenough, J.B.; Abruna, H.D.; Buchanan, M.V. *Basic Research Needs for Electrical Energy Storage*; Technical Report of the basic energy sciences workshop on electrical energy storage; DOESC (USDOE Office of Science (SC)): Washington, DC, USA, 2–4 April 2007. [CrossRef]

4. Rufer, A.; Barrade, P. A supercapacitor-based energy-storage system for elevators with soft commuted interface. *IEEE Trans. Ind. Appl.* **2002**, *38*, 1151–1159. [CrossRef]

5. Christen, T.; Carlen, M.W. Theory of ragone plots. *J. Power Sources* **2000**, *91*, 210–216. [CrossRef]

6. Borenstein, A.; Hanna, O.; Attias, R.; Luski, S.; Brousse, T.; Aurbach, D. Carbon-based composite materials for supercapacitor electrodes: A review. *J. Mater. Chem. A* **2017**, *5*, 12653–12672. [CrossRef]

7. Khan, A.H.; Ghosh, S.; Pradhan, B.; Dalui, A.; Shrestha, L.K.; Acharya, S.; Ariga, K. Two-dimensional (2d) nanomaterials towards electrochemical nanoarchitectonics in energy-related applications. *Bull. Chem. Soc. Jpn.* **2017**, *90*, 627–648. [CrossRef]

8. Wang, Y.; Mayorga-Martinez, C.C.; Pumera, M. Polyaniline/mosx supercapacitor by electrodeposition. *Bull. Chem. Soc. Jpn.* **2017**, *90*, 847–853. [CrossRef]

9. Sheberla, D.; Bachman, J.C.; Elias, J.S.; Sun, C.-J.; Shao-Horn, Y.; Dincă, M. Conductive mof electrodes for stable supercapacitors with high areal capacitance. *Nat. Mater.* **2016**, *16*, 220. [CrossRef] [PubMed]

10. Kim, J.; Kim, J.H.; Ariga, K. Redox-active polymers for energy storage nanoarchitectonics. *Joule* **2017**, *1*, 739–768. [CrossRef]

11. Frackowiak, E.; Béguin, F. Electrochemical storage of energy in carbon nanotubes and nanostructured carbons. *Carbon* **2002**, *40*, 1775–1787. [CrossRef]

12. Hu, S.; Rajamani, R.; Yu, X. Flexible solid-state paper based carbon nanotube supercapacitor. *Appl. Phys. Lett.* **2012**, *100*, 104103. [CrossRef]

13. Pandey, S.; Maiti, U.N.; Palanisamy, K.; Nikolaev, P.; Arepalli, S. Ultrasonicated double wall carbon nanotubes for enhanced electric double layer capacitance. *Appl. Phys. Lett.* **2014**, *104*, 233902. [CrossRef]

14. Rakesh, S.; Xianfeng, Z.; Saikat, T. Electrochemical double layer capacitor electrodes using aligned carbon nanotubes grown directly on metals. *Nanotechnology* **2009**, *20*, 395202.

15. Gao, Y.; Zhou, Y.S.; Xiong, W.; Jiang, L.J.; Mahjouri-samani, M.; Thirugnanam, P.; Huang, X.; Wang, M.M.; Jiang, L.; Lu, Y.F. Transparent, flexible, and solid-state supercapacitors based on graphene electrodes. *APL Mater.* **2013**, *1*, 012101. [CrossRef]

16. Mishra, A.K.; Ramaprabhu, S. Ultrahigh arsenic sorption using iron oxide-graphene nanocomposite supercapacitor assembly. *J. Appl. Phys.* **2012**, *112*, 104315. [CrossRef]

17. Wang, Y.; Shi, Z.; Huang, Y.; Ma, Y.; Wang, C.; Chen, M.; Chen, Y. Supercapacitor devices based on graphene materials. *J. Phys. Chem. C* **2009**, *113*, 13103–13107. [CrossRef]

18. Wu, Q.; Xu, Y.; Yao, Z.; Liu, A.; Shi, G. Supercapacitors based on flexible graphene/polyaniline nanofiber composite films. *ACS Nano* **2010**, *4*, 1963–1970. [CrossRef] [PubMed]

19. Yu, A.; Roes, I.; Davies, A.; Chen, Z. Ultrathin, transparent, and flexible graphene films for supercapacitor application. *Appl. Phys. Lett.* **2010**, *96*, 253105. [CrossRef]

20. Zang, X.; Li, P.; Chen, Q.; Wang, K.; Wei, J.; Wu, D.; Zhu, H. Evaluation of layer-by-layer graphene structures as supercapacitor electrode materials. *J. Appl. Phys.* **2014**, *115*, 024305. [CrossRef]

21. Zhang, L.L.; Zhou, R.; Zhao, X.S. Graphene-based materials as supercapacitor electrodes. *J. Mater. Chem.* **2010**, *20*, 5983–5992. [CrossRef]

22. Prabaharan, S.R.S.; Vimala, R.; Zainal, Z. Nanostructured mesoporous carbon as electrodes for supercapacitors. *J. Power Sources* **2006**, *161*, 730–736. [CrossRef]

23. Al-Asadi, A.S.; Henley, L.A.; Wasala, M.; Muchharla, B.; Perea-Lopez, N.; Carozo, V.; Lin, Z.; Terrones, M.; Mondal, K.; Kordas, K.; et al. Aligned carbon nanotube/zinc oxide nanowire hybrids as high performance electrodes for supercapacitor applications. *J. Appl. Phys.* **2017**, *121*, 124303. [CrossRef]

24. Gong, W.; Fugetsu, B.; Wang, Z.; Sakata, I.; Su, L.; Zhang, X.; Ogata, H.; Li, M.; Wang, C.; Li, J.; et al. Carbon nanotubes and manganese oxide hybrid nanostructures as high performance fiber supercapacitors. *Commun. Chem.* **2018**, *1*, 16. [CrossRef]

25. Cakici, M.; Kakarla, R.R.; Alonso-Marroquin, F. Advanced electrochemical energy storage supercapacitors based on the flexible carbon fiber fabric-coated with uniform coral-like mno2 structured electrodes. *Chem. Eng. J.* **2017**, *309*, 151–158. [CrossRef]

26. Fletcher, S.I.; Sillars, F.B.; Hudson, N.E.; Hall, P.J. Physical properties of selected ionic liquids for use as electrolytes and other industrial applications. *J.Chem. Eng. Data* **2010**, *55*, 778–782. [CrossRef]

27. Ignat'ev, N.V.; Welz-Biermann, U.; Kucheryna, A.; Bissky, G.; Willner, H. New ionic liquids with tris(perfluoroalkyl)trifluorophosphate (fap) anions. *J. Fluorine Chem.* **2005**, *126*, 1150–1159. [CrossRef]

28. Aboutalebi, S.H.; Jalili, R.; Esrafilzadeh, D.; Salari, M.; Gholamvand, Z.; Aminorroaya Yamini, S.; Konstantinov, K.; Shepherd, R.L.; Chen, J.; Moulton, S.E.; et al. High-performance multifunctional graphene yarns: Toward wearable all-carbon energy storage textiles. *ACS Nano* **2014**, *8*, 2456–2466. [CrossRef] [PubMed]

29. Sahu, V.; Shekhar, S.; Sharma, R.K.; Singh, G. Ultrahigh performance supercapacitor from lacey reduced graphene oxide nanoribbons. *ACS Appl. Mater. Interfaces* **2015**, *7*, 3110–3116. [CrossRef] [PubMed]

30. Albrektsen, O.; Eriksen, R.L.; Novikov, S.M.; Schall, D.; Karl, M.; Bozhevolnyi, S.I.; Simonsen, A.C. High resolution imaging of few-layer graphene. *J. Appl. Phys.* **2012**, *111*, 064305. [CrossRef]

31. Wu, J.X.; Xu, H.; Zhang, J. Raman spectroscopy of graphene. *Acta Chim. Sin.* **2014**, *72*, 301–318. [CrossRef]

32. Pensado, A.S.; Malberg, F.; Gomes, M.F.C.; Pádua, A.A.H.; Fernández, J.; Kirchner, B. Interactions and structure of ionic liquids on graphene and carbon nanotubes surfaces. *RSC Adv.* **2014**, *4*, 18017–18024. [CrossRef]

33. Jorcin, J.-B.; Orazem, M.E.; Pébère, N.; Tribollet, B. Cpe analysis by local electrochemical impedance spectroscopy. *Electrochim. Acta* **2006**, *51*, 1473–1479. [CrossRef]

34. Kötz, R.; Hahn, M.; Gallay, R. Temperature behavior and impedance fundamentals of supercapacitors. *J. Power Sources* **2006**, *154*, 550–555. [CrossRef]

35. Macdonald, J.R. Impedance spectroscopy. *Ann. Biomed. Eng.* **1992**, *20*, 289–305. [CrossRef] [PubMed]

36. Qu, L.; Zhao, Y.; Khan, A.M.; Han, C.; Hercule, K.M.; Yan, M.; Liu, X.; Chen, W.; Wang, D.; Cai, Z.; et al. Interwoven three-dimensional architecture of cobalt oxide nanobrush-graphene@nixco2x(oh)6x for high-performance supercapacitors. *Nano Lett.* **2015**, *15*, 2037–2044. [CrossRef] [PubMed]

37. Cho, S.; Kim, M.; Jang, J. Screen-printable and flexible ruo2 nanoparticle-decorated pedot:Pss/graphene nanocomposite with enhanced electrical and electrochemical performances for high-capacity supercapacitor. *ACS Appl. Mater. Interfaces* **2015**, *7*, 10213–10227. [CrossRef] [PubMed]

38. Yan, J.; Wang, Q.; Wei, T.; Jiang, L.; Zhang, M.; Jing, X.; Fan, Z. Template-assisted low temperature synthesis of functionalized graphene for ultrahigh volumetric performance supercapacitors. *ACS Nano* **2014**, *8*, 4720–4729. [CrossRef] [PubMed]

39. Ghosh, S.; An, X.; Shah, R.; Rawat, D.; Dave, B.; Kar, S.; Talapatra, S. Effect of 1- pyrene carboxylic-acid functionalization of graphene on its capacitive energy storage. *J. Phys. Chem. C* **2012**, *116*, 20688–20693. [CrossRef]

40. Ghosh, D.; Das, C.K. Hydrothermal growth of hierarchical ni3s2 and co3s4 on a reduced graphene oxide hydrogel@ni foam: A high-energy-density aqueous asymmetric supercapacitor. *ACS Appl. Mater. Interfaces* **2015**, *7*, 1122–1131. [CrossRef] [PubMed]

41. Podolyan, Y.; Leszczynski, J. Mask: A visualization tool for teaching and research in computational chemistry. *Int. J. Quantum Chem.* **2009**, *109*, 8–16. [CrossRef]

© 2018 by the authors. Licensee MDPI, Basel, Switzerland. This article is an open access article distributed under the terms and conditions of the Creative Commons Attribution (CC BY) license (http://creativecommons.org/licenses/by/4.0/).

![electronics logo] *electronics*

MDPI

Article

Multichannel and Multistate All-Optical Switch Using Quantum-Dot and Sample-Grating Semiconductor Optical Amplifier

Omar Qasaimeh

Department of Electrical Engineering, Jordan University of Science and Technology, P.O. Box 3030, Irbid 22110, Jordan; qasaimeh@just.edu.jo; Tel.: +962-2-7201-000

Received: 27 July 2018; Accepted: 22 August 2018; Published: 29 August 2018

Abstract: A novel type of multichannel and multistate all-optical switch using a single sample-grating quantum-dot-distributed feedback semiconductor optical amplifier has been proposed and theoretically demonstrated. The multichannel device, which operates below threshold, utilizes cross-gain modulation and the sample-grating technique. The multichannel outputs are strongly coupled and are utilized to get multistability at several wavelength channels. Three logic states can be obtained when the inputs are properly detuned to the sample-grating comb modes. The three logic states, which exhibit reasonable gain, are separated by wide hysteresis width and can be tuned to a different wavelength channels. The device characteristics are very useful for building all-optical logic gates, flip-flops, and decision circuits.

Keywords: quantum dot; sample grating; cross-gain modulation; bistability; distributed Bragg; semiconductor optical amplifier

1. Introduction

Multistate optical switches that can operate at multi-wavelength channels are attractive for parallel all-optical processing in modern photonics and wavelength-division multiplexing systems. Multichannel all-optical processors reduce the cost, offer reliable routing and ultrafast processing, and enhance the capacity/density of wavelength-routing in optical networks [1]. Distributed-feedback semiconductor optical amplifiers (DFB-SOAs) that display strong nonlinear phenomena and adjustable optical bistability can be adopted to build all-optical switches, photonic logic gates, and all-optical flip-flops [2,3]. The functionality as well as the nonlinear effects in DFB-SOAs can be expanded by incorporating sample grating and enhanced by incorporating quantum-dot nanostructures in the active region. Conventional DFB-SOAs, which can be integrated with other optoelectronics devices, display cross-gain optical bistability when the device is properly designed [2]. The optical bistability of the device arises from mutual changes in the modal gain and the refractive index of the active region when the device operates slightly below threshold and when the input wavelength is properly detuned to the device modes. Applying an optical input to the device increases the internal power and changes the refractive index of the active region, which shifts the Bragg resonance of the device and consequently causes optical bistability in the input–output power characteristics of the device.

Sample Bragg gratings, which can be precisely fabricated with high quality and have recently been utilized to enhance many active and passive devices, have drawn significant attraction in recent years for use in dense wavelength-division multiplexing, long-haul optical-communication systems and optical signal processing [4–8]. Introducing sample grating to DFB-SOAs results in ultra-narrowband optical comb filtering, which may be utilized to increase the functionality of the amplifier and permit multichannel probing and switching. The multistable characteristics of sample grating DFB-SOAs are extensively desired for parallel processing. The optical characteristics of these channels can be controlled by cross-gain modulation (XGM) and can be employed to realize optical memories [9].

Semiconductor optical amplifiers made of quantum-dot (QD) nanostructures have recently demonstrated excellent optical performance compared with higher-order structures. The discrete energy states of QDs result in many unique features such as sharp optical transitions, high differential gain, low temperature sensitivity, and enhanced optical nonlinearities. QD-SOAs have also displayed enhanced ultrafast nonlinear optical response on very short time scales [10–12]. The enhanced nonlinear response can be utilized to realize efficient microwave and millimeter-wave signal generation, all-optical signal processing, all-optical wavelength conversion, and add–drop functionalities.

Multistability has been obtained in a single-mode laser subject to feedback through monolithically integrated phase-tuning and amplifier sections [13] and in two coupled active microrings [14]. The latter utilizes gain saturation, intrinsic feedback, and optical feedbacks between the two active microrings to obtain different types of multistability. Cross-gain modulation is simple to realize in SOAs and has shown impressive performance for bit rates up to 40 Gb/s. It has also been utilized to get multistability in dual-mode DFB semiconductor optical amplifiers [2]. To our knowledge, there is no work in the literature that utilizes sample gating and cross-gain modulation to get multistability in DFB-SOAs. In this paper, a novel type of multi-wavelength and multistate all-optical switch has been proposed. The device, which utilizes sample-grating and cross-gain modulation in DFB-SOA, exhibits multiple modes where the control input is tuned to one of these modes and the data signal is tuned to another mode. More than one control signal can be used to perform switching. The proposed device is very useful for constructing new controllable all-optical switches, all-optical flip-flops, and single-memory units.

2. Theory

The investigated device is a monolithically integrated sample-grating DFB-SOA having 5 QD layers in the active region. The device, which consists of multiple pairs of gain and DFB sections, operates in the transmission mode as shown in Figure 1. Using sample grating introduces multiple degenerate comb modes that exhibit large optical gain. Multiple optical inputs (for example, control signal and data signal) are injected into the device. Due to the nonlinearity of the device, the outputs would be coupled and dependent on the device/input parameters. The field amplitude of the control signal is denoted by A^C and its wavelength by λ_C. Similarly, the field amplitude of the data signal is denoted by A^D and its wavelength by λ_D. The control and the data signals interact via cross-gain modulation that can be modeled by using coupled-mode equations. The forward and backward waves for the control and data signals are governed by [2]:

$$\frac{\partial A^j}{\partial z} + \frac{1}{v_g}\frac{\partial A^j}{\partial t} = i\left[\Delta\beta^j A^j + \kappa B^j\right] \tag{1}$$

$$\frac{\partial B^j}{\partial z} - \frac{1}{v_g}\frac{\partial B^j}{\partial t} = -i\left[\Delta\beta^j B^j + \kappa A^j\right] \tag{2}$$

where the superscript $j = C$ indicates control signal and superscript $j = D$ indicates data signal, $i = \sqrt{-1}$, t is time, v_g is the group velocity, and z is the distance. $\Delta\beta^j = \beta^j - \beta_B$ is the detuning of the wavenumber β^j from the Bragg wavenumber β_B and k is the coupling coefficient. The boundary conditions at $z = 0$ are given by $A^j(0,t) = A^j_{in}$. Similarly, the boundary conditions at $z = L$, where L is the length of the device, are given by $B^j(L,t) = 0$; moreover, $A^j_{tr}(L,t) = A^j(L,t)$, where A^j_{in} and A^j_{tr} are the slowly varying amplitudes of the incident and transmitted waves, respectively.

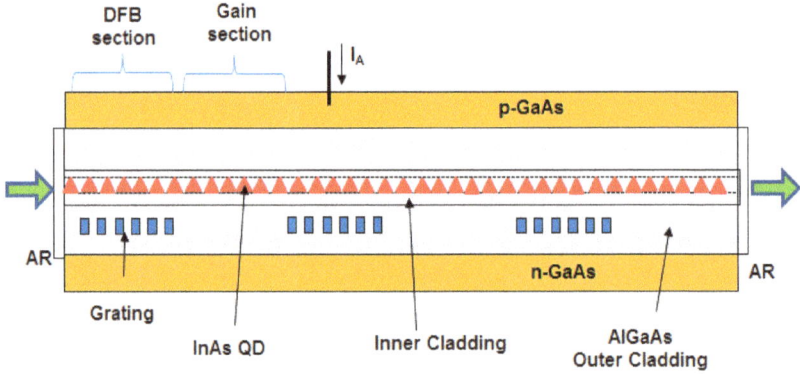

Figure 1. Schematic of sample-grating quantum-dot-distributed Bragg semiconductor optical amplifier.

By splitting the length of the device (L) into equal uniform sections, the spatial-step size is equal to l where $l = L/N_L$ and N_L is the number of subsections. Each subsection will be labeled by $\varepsilon = 1, 2, \cdots, N_L$. With the help of the transfer-matrix method, one can write the solution of the coupled-mode equations as:

$$\begin{bmatrix} A^j_{\varepsilon+1} \\ B^j_{\varepsilon+1} \end{bmatrix} = M^j_\varepsilon \begin{bmatrix} A^j_\varepsilon \\ B^j_\varepsilon \end{bmatrix} \tag{3}$$

The matrix M^j_ε is given by [2]:

$$M^j_\varepsilon = \frac{1}{1 - r^2_{j\varepsilon}} \times \begin{bmatrix} e^{i\gamma_{j\varepsilon}l} - r^2_{j\varepsilon}e^{-i\gamma_{j\varepsilon}l} & -r_{j\varepsilon}\left(e^{i\gamma_{j\varepsilon}l} - e^{-i\gamma_{j\varepsilon}l}\right) \\ r_{j\varepsilon}\left(e^{i\gamma_{j\varepsilon}l} - e^{-i\gamma_{j\varepsilon}l}\right) & e^{-i\gamma_{j\varepsilon}l} - r^2_{j\varepsilon}e^{i\gamma_{j\varepsilon}l} \end{bmatrix} \tag{4}$$

where

$$\gamma_{j\varepsilon} = \sqrt{\left(\Delta\beta^j_\varepsilon\right)^2 - \kappa^2} \tag{5}$$

and

$$r_{j\varepsilon} = \frac{-\kappa}{\gamma_{j\varepsilon} + \Delta\beta^j_\varepsilon} = \frac{\gamma_{j\varepsilon} - \Delta\beta^j_\varepsilon}{\kappa} \tag{6}$$

The parameter $\Delta\beta^j_\varepsilon$ is given by

$$\Delta\beta^j_\varepsilon = \delta_j - i\frac{g^j_\varepsilon}{2}(1 - i\alpha_H) + i\frac{\alpha_{\text{int}}}{2} \tag{7}$$

where δ_j is the initial detuning, α_H is the linewidth enhancement factor, α_{int} is the internal loss of the structure, and g^j_ε is the material gain in the subsection ε. This includes the spatial dependence of the material gain.

In a separate-confinement heterostructure that has inner- and outer-cladding layers and a QD active layer, the energy-band diagram of the device consists of multiple energy states in the conduction and the valence bands. Experimental data have shown that 1.3 µm InAs/GaAs QD exhibits three energy states in the conduction band and eight energy states in the valence band plus a wetting layer state [11,12]. The separations of the electron and hole energy states are 60 and 10 meV, respectively. For simplicity, $k = 0$ denotes the ground state, $k = 1$ denotes the next higher energy state, and $k = w$ denotes the wetting layer state. The energy diagram in the conduction band is shown in Figure 2.

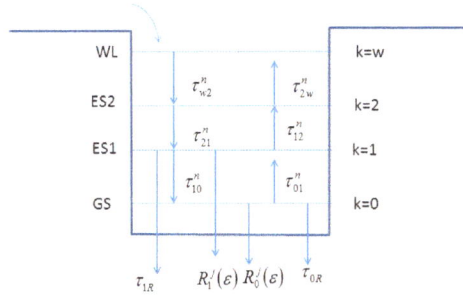

Figure 2. Energy-band diagram of the conduction band of the device.

Carriers injected to the active layer see and interact with the QD energy states and their dynamics are described by the rate equations. Carrier transport between the dot states is determined by capture/escape rates, spontaneous-emission rate, and stimulated-emission rate. The rate equation for electrons in the k-th energy state for subsection ε will be written as [15]:

$$\frac{\partial f_k^n(\varepsilon)}{\partial t} = \left(R_{k+1,i}^{nc}(\varepsilon) - R_{k,k+1}^{ne}(\varepsilon)\right) - \left(R_{k,k-1}^{nc}(\varepsilon) - R_{k-1,k}^{ne}(\varepsilon)\right) - \frac{f_k^n(\varepsilon)f_k^p(\varepsilon)}{\tau_{kR}} - \sum_j R_k^j(\varepsilon) \tag{8}$$

where $f_k^n(\varepsilon)$ and $f_k^p(\varepsilon)$ are, respectively, the occupation probability for the electrons and holes in the k-th state, and τ_{kR} is the spontaneous radiative lifetime in k-th state. The electron capture rate is given by

$$R_{k+1,k}^{nc}(\varepsilon) = \frac{(1 - f_k^n(\varepsilon))f_{k+1}^n(\varepsilon)}{\tau_{k+1,k}^n} \tag{9}$$

The electron emission rate is

$$R_{k,k+1}^{ne}(\varepsilon) = \frac{f_k^n(\varepsilon)(1 - f_{k+1}^n(\varepsilon))}{\tau_{k,k+1}^n} \tag{10}$$

where $\tau_{k+1,k}^n$ is the electron capture lifetime and $\tau_{k,k+1}^n$ is the electron escape lifetime. $R_k^j(\varepsilon)$ in Equation (8) is the stimulation emission rate, which is given by

$$R_k^j(\varepsilon) = \frac{v_g}{N_Q} \sum_{k_p} a_{kk_p}^j (f_k^n(\varepsilon) + f_{k_p}^p(\varepsilon) - 1)\left(\left|A_\varepsilon^j\right|^2 + \left|B_\varepsilon^j\right|^2\right) \tag{11}$$

where N_Q is the dot-volume density and a_{kk_p} is the material gain coefficient of the QD layer, which is a Gaussian function and is function of the transition-matrix elements. For the ground state ($k = 0$), the second term in the right-hand side of Equation (8) is equal to $\left(R_{k,k-1}^{nc}(\varepsilon) - R_{k-1,k}^{ne}(\varepsilon)\right) = 0$. While for the wetting layer state ($k = w$), the first term in the right-hand side of Equation (8) is equal to $\left(R_{k+1,k}^{nc}(\varepsilon) - R_{k,k+1}^{ne}(\varepsilon)\right) = \frac{I_A}{V\tau_{kR}}$, where I_A is the normalized applied current. Similar-rate equations can be written for the hole states. The relation between the electron and the hole concentration is governed by the charge-neutrality equation, which is given by

$$\sum_{k=0}^{M_n} N_k f_k^n(\varepsilon) + N_w = \sum_{k_p=0}^{M_p} N_{k_p} f_{k_p}^p(\varepsilon) + P_w \tag{12}$$

where N_k is the volume density of the k-th state, N_w and P_w are the electron and the hole concentration of the wetting layer, and M_n and M_p are the number of electron and hole states, respectively. The modal gain of the active layer is

$$g_\epsilon^j = \sum_{k=0}^{M_n} \sum_{k_p=0}^{M_p} a_{kk_p} \left(f_k^n(\epsilon) + f_{k_p}^p(\epsilon) - 1 \right) \tag{13}$$

where a_{kk_p} is the material gain coefficient of the active layer which is given by [15]:

$$a_{kk_p} = g_{kk_p}^{max} \frac{C}{\sigma_{kk_p}} \frac{\hbar\omega_{kk_p}^{max}}{\hbar\omega} Exp \left(\frac{-(\hbar\omega - \hbar\omega_{kk_p}^{max})^2}{2\sigma_{kk_p}^2} \right) \tag{14}$$

where σ_{kk_p} is the inhomogeneous line broadening, $g_{kk_p}^{max}$ is the maximum gain coefficient for the k–k_p transition, $\hbar\omega$ is the photon energy of the amplifier, $\hbar\omega_{kk_p}^{max}$ is the energy corresponding to the gain peak of the k–k_p transition, and C is a constant extracted from the transition-matrix elements, which takes into account the selection rule and homogeneous line broadening [16]. In the following analysis, the homogeneous line broadening is not included since the bandwidth of the device is extremely narrow.

3. Results and Discussions

The device consists of 3 periods of sample-grating DFB and gain sections as shown in Figure 1. The length of the DFB section is 100 μm and the length of the gain section is 470 μm. For 3 periods, the total length of the device is 1710 μm. The device parameters, obtained from the literature, are given in Table 1 [15–17]. The Bragg wavelength is chosen to be equal to the QD ground-state wavelength and the device operates at 0.98 of the threshold conditions. To include the spatial dependence of the gain, the device is divided into $N_L = 350$ segments, the rate equations are simultaneously solved in each section, and the boundary conditions are applied. The input–output characteristics of the device are evaluated in a reverse manner to resolve the device nonlinearity (i.e., the inputs are evaluated when the output is known). The procedure is as follows: the output powers for the two signals is varied by running two do loops (one for the data output and the other for the control output) and the corresponding inputs are evaluated. The input powers are saved in two matrices. The outputs that correspond to fixed inputs are then extracted. In the following analysis, we define the wavelength shift as the shift between the input wavelength and the Bragg wavelength i.e., $\partial\lambda_j = \lambda_j - \lambda_B$. The input power of the control and data signals is denoted by P_{in}^C and P_{in}^D, respectively.

The optical-transmission spectrum of the device displays multiple peaks corresponding to the multiple modes of the sample-grating structure. The optical gain of the device for single input, i.e., no XGM, is shown in the inset of Figure 3. The first 3 dual modes are shown and labeled by $1^\pm$, $2^\pm$, and $3^\pm$ in the inset of Figure 3. As evident, the optical-gain spectrum exhibits comb peaks separated by $\Delta\lambda = 1.5$ Å between closely adjacent modes like $1^\pm$ and $2^\pm$, and $\Delta\lambda = 4.8$ Å between next adjacent modes like 1^- and 1^+. The unsaturated optical gain is about 50 dB. For input power equal to 10 μW, the comb peaks vary between 17 dB to 18 dB, and, for higher input power, the multiple channels show approximately equal response. Figure 3 shows a zoom in for the optical spectrum for mode 1^+ and 2^+ for 8 μW and 20 μW input power. As shown, the peak of mode 1^+ is 0.3 dB higher gain than mode 2^+ for $P_{in} = 20$ μW, and this gain variation reduces when P_{in} increases. As evident, increasing the input power (P_{in}) shifts the mode spectrum to higher wavelength. For example, when the input power is very small compared with the input saturation power, mode 1^+ is peaked at $\partial\lambda = 0.2$ Å (i.e., the mode unsaturated wavelength shift is $\partial\lambda^{us} = 0.2$ Å). When $P_{in} = 20$ μW, the wavelength shift that corresponds to the mode peak is increased to $\partial\lambda = 0.4$ Å. According to Figure 3, the device modes exhibit spectral bistability at the higher-wavelength side of each mode spectrum. Clockwise spectral hysteresis loops are obtained at specific wavelengths for all comb modes. When P_{in} increases, the optical gain peak decreases and the spectral hysteresis width increases as well. Operating the device at the wavelengths that correspond to the bistable regions will be useful for multichannel optical switching. We expect that enhanced bistable characteristics will be obtained when cross-gain modulation is presented.

Table 1. Parameters used for the investigated device.

Parameters	Values
Length of distributed-feedback (DFB) section	100 μm
Length of Gain section	470 μm
Energy separations for electron states	60 meV
Energy separations for hole states	10 meV
GS electron-relaxation lifetime (τ_{10}^{n})	8 ps
Higher-energy electron-relaxation lifetime ($\tau_{k+1,k}^{n}$)	2 ps
Hole relaxation time ($\tau_{k+1,k}^{p}$)	0.1 ps
Recombination lifetime (τ_{kR})	0.4 ns
Inhomogeneous line broadening (σ_{kk_p})	30 meV
Quantum-dot (QD) volume density (N_Q)	2.5×10^{17} cm^3
Coupling coefficient (κ)	90 cm^{-1}
Linewidth enhancement factor (α_H)	3
GS gain coefficient (g_{00}^{max})	14 cm^{-1}
First ES gain coefficient (g_{11}^{max})	20 cm^{-1}
Second ES gain coefficient (g_{22}^{max})	10 cm^{-1}

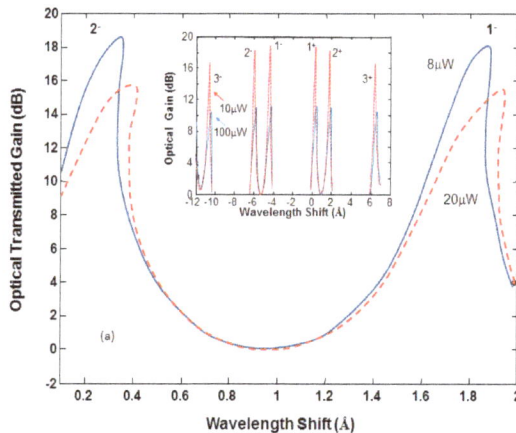

Figure 3. Optical-transmission gain as a function of wavelength shift for single-input signal. The inset shows a wider spectral range for input power equal to 8 μW and 20 μW.

In the following analysis, the bistable characteristics of the device, governed by cross-gain modulation, are studied when multiple input signals are coupled to the comb modes. The data signal is detuned to mode 1^+ and has its input power fixed at P_{in}^{D} = 8 μW. The control signal, which has an input power of P_{in}^{C} = 20 μW, is detuned to the other comb modes as shown in Figure 4. In Figure 4, we detuned the data wavelength to $\partial\lambda_D$ = 0.36 Å, which places mode 1^+ at the upper wavelength edge of its bistable region when P_{in}^{C} = 0. The optical gain of the control signal as a function of its wavelength shift $\partial\lambda_C$ is shown in Figure 4. The figure shows the control spectrum for modes 1^- and 2^- only. According to Figure 4 and as a result of XGM, the data signal significantly modifies the bistable characteristics of the control signal. When P_{in}^{D} = 0 (i.e., without XGM), the control bistability region is located at the upper wavelength side of each mode, as shown by the solid line in Figure 4, and when P_{in}^{D} = 8 μW (i.e., with XGM), the control signal exhibits two wider regions of bistabilities; one is located at the upper wavelength side and the other is located at the lower wavelength side, as shown by the dash line in Figure 4. This is attributed to an overlap between the control- and data-bistability regions and due to coupling between the control and data signals. Under the effect of XGM, we find that the bistability characteristics of mode 1^- are different from those of mode 2^-. In the following analysis, we will show that the bistable characteristics of mode 1^- are approximately similar to those of mode 2^+. This is attributed to the fact that modes 1^- and 2^+, which are separated by one spectral period, exhibit approximately similar spectral gain and similar power dependence. The optical gain of the data signal is shown in Figure 5 as a function of $\partial\lambda_C$. As evident, the optical gain of the data signal as well as its bistable regions can be controlled by detuning $\partial\lambda_C$; detuning $\partial\lambda_C$ to mode 1^- and 2^- is only shown in the figure. We find that detuning the control wavelength to mode 1^- provides different hysteresis width and shape compared with mode 2^-. Modes that have similar power dependence exhibit similar bistable characteristics. According to Figure 5, 16 dB contrast ratio can be obtained by operating the device at $\partial\lambda_C = -4.1$ Å or -5.6 Å.

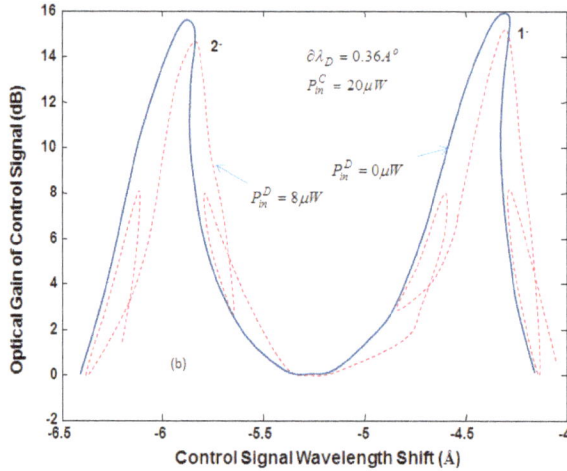

Figure 4. Optical gain of control signal as a function of control wavelength shift at fixed input powers.

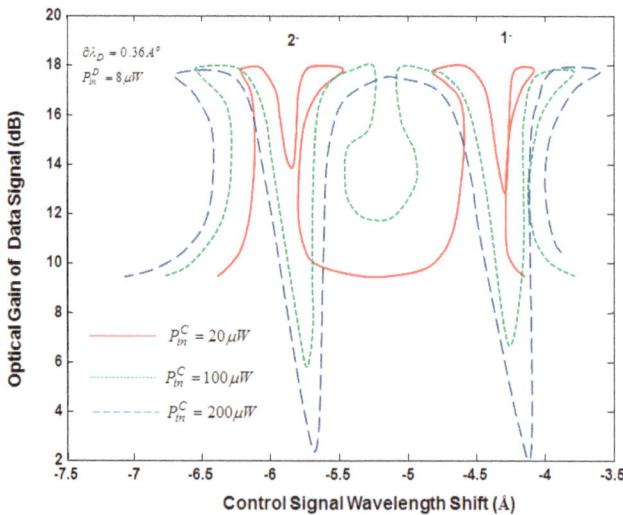

Figure 5. Optical gain of data signal as a function of control-wavelength shift at fixed input.

The shape of the spectral bistability, as well as the hysteresis width, can be adjusted by changing input parameters. The optical gain of the data signal is shown in Figure 5 as a function of $\partial\lambda_C$ for different control-input power (P_{in}^C). A dip in the data gain, which occurred at $\partial\lambda_C = -4.1$ Å or -5.6 Å when P_{in}^C is increased, was due to gain saturation caused by large control gain. The dip split the mode into two regions; upper bistable region and lower bistable region for $P_{in}^C < 100$ µW. This feature was obtained when the control and data wavelengths are properly detuned by 0.4 Å and 0.16 Å above the mode unsaturated wavelength. According to Figure 5, the upper bistable region of all modes is shifted to higher wavelength and the lower bistable region is shifted to lower wavelength when P_{in}^C is increased. When P_{in}^C reaches 200 µW or more, the lower wavelength side of mode 1^- and the upper wavelength side of mode 2^- merges and becomes stable. According to Figures 4 and 5, bistability is limited to a narrow wavelength range.

The hysteresis width and shape are power-dependent. This phenomenon can be utilized to get multiple optical bistability if the control and data inputs are properly tuned to the comb modes. The power multistable characteristics for $\partial\lambda_D = +0.36$ Å are shown in Figure 6, when the control wavelength is tuned to modes 1^-, 2^-, 2^+ and 3^+, i.e., $\partial\lambda_C$ is tuned to -5.6 Å, -4.1 Å, $+2.15$ Å and $+6.77$ Å. This corresponds to 0.4 Å wavelength shift above the modes' unsaturated wavelengths. The data response may be classified into 3 states: the 9 dB state, the ~18 dB state, and the ~1 dB state. The three states are surrounded by bistability regions (with wide hysteresis widths) that increase the noise margin between these states. The three states are useful for setting and resetting all-optical flip-flops built using this device. The characteristics of the device when the control wavelength is tuned to mode 1^- and 2^+ look very similar since these modes approximately exhibit similar power dependence. A three-state logic gate can be implemented from the output of the data signal. Figure 6 indicates that multichannel control inputs can be used to adjust the gain and bistability of another data channel. This feature is very attractive for building all-optical logic gates and flip-flops. Figure 7 shows the characteristics of the device when the control wavelength is tuned to mode 1^- (i.e., $\partial\lambda_C = -4.1$ Å) and data wavelength is detuned to different comb modes (i.e., the data wavelength is tuned by 0.16 Å above the mode unsaturated wavelength). As evidence, one control channel can be used to adjacent the bistability of multiple data channels. The tristability is also limited to a very narrow wavelength range, which can be considered as the main drawback of the switch.

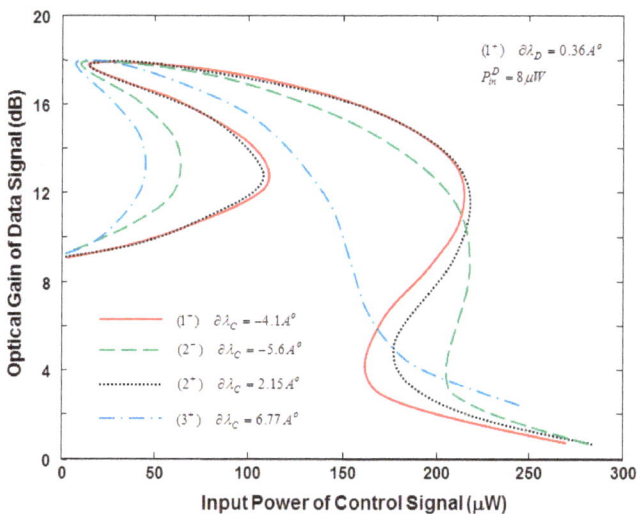

Figure 6. Optical gain of data signal as a function of control-signal input power when the control-signal wavelength is tuned to different comb modes.

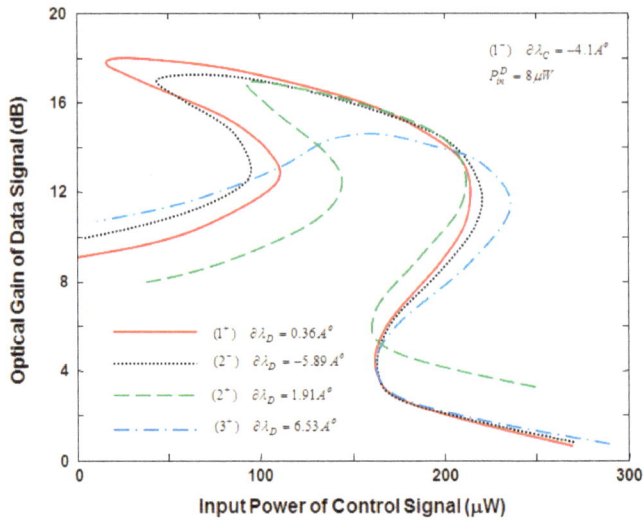

Figure 7. Optical gain of data signal as a function of control-signal input power when the control signal wavelength is fixed and the data-signal wavelength is tuned to different comb modes.

4. Conclusions

A novel type of all-optical switch using a single sample-grating quantum-dot-distributed Bragg semiconductor optical amplifier is proposed. Cross-gain modulation is utilized to switch the state of a data signal by one or more control channels. It has been shown that multistability of a data channel can be adjusted by multichannel control inputs when the input signals are properly tuned to the sample-grating comb modes. Switching between three logic states with a contrast ratio of 8 dB and 18 dB has been demonstrated. A relatively wide noise margin/ hysteresis width between the logic states is obtained. The device characteristics are very useful for building all-optical logic gates, flip-flops, and design circuits.

Funding: This research received no external funding.

Acknowledgments: Part of the open-access fees is paid by the deanship of research at Jordan University of Science and Technology.

Conflicts of Interest: The authors declare no conflict of interest.

References

1. Huang, Z.; Cao, T.; Chen, L.; Yu, Y.; Zhang, X. Monolithic Integrated Chip with SOA and Tunable DI for Multichannel All-Optical Signal Processing. *IEEE Photonics J.* **2018**, *10*, 1–9. [CrossRef]
2. Qasaimeh, O. All-optical multistability using cross-gain modulation in quantum-dot distributed feedback semiconductor optical amplifier. *Opt. Quantum Electron.* **2018**, *50*, 1–9. [CrossRef]
3. Huybrechts, K.; D'Oosterlinck, W.; Morthier, G.; Baets, R. Proposal for an All-Optical Flip-Flop Using a Single Distributed Feedback Laser Diode. *IEEE Photonics Technol. Lett.* **2008**, *20*, 18–20. [CrossRef]
4. Dhoore, S.; Li, L.; Abbasi, A.; Roelkens, G.; Morthier, G. Demonstration of a Discretely Tunable III-V-on-Silicon Sampled Grating DFB Laser. *IEEE Photonics Technol. Lett.* **2016**, *28*, 2343–2346. [CrossRef]
5. Akrout, A.; Dridi, K.; Abdul-Majid, S.; Seregelyi, J.; Hall, T.J. Numerical Study of Dual Mode Generation Using a Sampled-Grating High-Order Quantum-Dot Based Laterally-Coupled DFB Laser. *IEEE J. Quantum Electron.* **2013**, *49*, 821–828. [CrossRef]

6. Liu, S.; Shi, Y.; Hao, L.; Xiao, R.; Chen, X. Experimental Demonstration of the Anti-Symmetric Sampled Bragg Grating. *IEEE Photonics Technol. Lett.* **2017**, *29*, 353–356. [CrossRef]

7. Li, L.; Shi, Y.; Zhang, Y.; Zou, H.; Shen, J.; Chen, X. Study on a DFB Laser Diode Based on Sampled Grating Technique for Suppression of the Zeroth Order Resonance. *IEEE Photonics J.* **2017**, *9*, 1–9. [CrossRef]

8. Happach, M.; Felipe, D.; Friedhoff, V.; Kleinert, M.; Zawadzki, C.; Rehbein, W.; Brinker, W.; Möhrle, M.; Keil, N.; Hofmann, W.; et al. Temperature-Tolerant Wavelength-Setting and -Stabilization in a Polymer-Based Tunable DBR Laser. *J. Lightwave Technol.* **2017**, *35*, 1797–1802. [CrossRef]

9. Vagionas, C.; Fitsios, D.; Vyrsokinos, K.; Kanellos, G.; Miliou, A.; Pleros, N. XPM- and XGM-Based Optical RAM Memories: Frequency and Time Domain Theoretical Analysis. *IEEE J. Quantum Electron.* **2014**, *50*, 683–689. [CrossRef]

10. Hurtado, A.; Raghunathan, R.; Henning, I.; Adams, M.; Lester, L. Simultaneous Microwave- and Millimeter-Wave Signal Generation With a 1310-nm Quantum-Dot-Distributed Feedback Laser. *IEEE J. Sel. Top. Quantum Electron.* **2015**, *21*, 1801207. [CrossRef]

11. Ababneh, J.; Qasaimeh, O. Simple model for quantum-dot semiconductor optical amplifiers using artificial neural networks. *IEEE Trans. Electron Devices* **2006**, *53*, 1543–1550. [CrossRef]

12. Qasaimeh, O. Linewidth Enhancement Factor of Quantum Dot Lasers. *Opt. Quantum Electron.* **2005**, *37*, 495–507. [CrossRef]

13. Loose, A.; Goswami, B.; Wünsche, H.; Henneberger, F. Tristability of a semiconductor laser due to time delayed optical feedback. *Phys. Rev. E* **2009**, *79*, 036211. [CrossRef] [PubMed]

14. Zhang, Q.; Qin, C.; Chen, K.; Xiong, M.; Zhang, X. Novel optical multibistability and multistability characteristics of coupled active microrings. *IEEE J. Quantum Electron.* **2013**, *49*, 365–374. [CrossRef]

15. Qasaimeh, O. Novel closed-form solution for spin-polarization in quantum dot VCSEL. *Opt. Commun.* **2015**, *350*, 83–89. [CrossRef]

16. Gioannini, M.; Montrosset, I. Numerical analysis of the frequency chirp in quantum-dot semiconductor lasers. *IEEE J. Quantum Electron.* **2007**, *43*, 941–949. [CrossRef]

17. Tatebayashi, J.; Ishida, M.; Hatori, N.; Ebe, H.; Sudou, H.; Kuramata, A.; Sugawara, M.; Arakawa, Y. Lasing at 1.28 /spl mu/m of InAs-GaAs quantum dots with AlGaAs cladding layer grown by metal-organic chemical vapor deposition. *IEEE J. Sel. Top. Quantum Electron.* **2005**, *11*, 1027–1034. [CrossRef]

© 2018 by the author. Licensee MDPI, Basel, Switzerland. This article is an open access article distributed under the terms and conditions of the Creative Commons Attribution (CC BY) license (http://creativecommons.org/licenses/by/4.0/).

![electronics logo] *electronics*

MDPI

Article

Determination of Complex Conductivity of Thin Strips with a Transmission Method

Morteza Shahpari

School of Engineering & Built Environment, Gold Coast Campus, Griffith University,
Southport, QLD 4215, Australia; morteza.shahpari@ieee.org

Received: 6 December 2018; Accepted: 19 December 2018; Published: 24 December 2018

Abstract: Induced modes due to discontinuities inside the waveguide are dependent on the shape and material properties of the discontinuity. Reflection and transmission coefficients provide useful information about material properties of discontinuities inside the waveguide. A novel non-resonant procedure to measure the complex conductivity of narrow strips is proposed in this paper. The sample is placed inside a rectangular waveguide which is excited by its fundamental mode. Reflection and transmission coefficients are calculated by the assistance of the Green's functions and enforcing the boundary conditions. We show that resistivity only impacts one of the terms in the reflection coefficient. The competency of the method is demonstrated with a comparison of theoretic results and full wave modelling of method of moments and finite element methods.

Keywords: conductivity; 2D material; Green's function; reflection transmission method; variational form

---

1. Introduction

Developing new methods to measure various material parameters are of prominent importance as they enable one to perform precise measurements. Conventional methods to measure the conductivity of materials at microwave frequencies are the cavity, reflection and transmission methods [1,2]. Cavity methods [3–12] are generally based on the pertubation theory, where the sample under test (SUT) is placed inside a cavity and perturbs the natural modes of the cavity. Perturbations result in the shifts in the resonant frequency f_r and quality factor Q. Material properties are extracted from the changes in f_r and Q. In reflection methods [13–16], SUT is placed as the termination of the transmission line, and the material is characterized through investigation of the reflection coefficient. Transmission methods [1,17–22] are based on placing the object inside the transmission line, and both transmission and reflection coefficients are used. Cavity methods [3–11] are inherently narrowband, while reflection [13,15,16] and transimssion methods [18–21] are broadband methods. However, cavity methods are often considered as more accurate methods for material characterizations. Another method for measuring the permittivity and dielectric properties is recently proposed by Geyi and colleagues [23–25]. They place the unknown material in the near field of the antenna and use the variations in the reflection coefficient to find the dielectric constant. The limitation of the method is that SUT has to be electrically small (much smaller than wavelength λ) so fields can be approximated in the antenna near zone. A mode matching technique based on the transmission methods is reported in [26].

In this paper, a method to measure the complex conductivity of the graphene and thin film materials is proposed. The method is based on the standard reflection/transmission methods, which are inherently broadband measurement methods. The whole aperture of the waveguide had to be covered with the sample under test in previous transmission methods (e.g., [20]) for surface conductivity measurement. As far as the authors are aware, this is the first time an analytic method for the conductivity of a thin strip is proposed, and it should be useful to measure the performance of materials that are synthesized in a strip shape. An example of such materials is graphene produced by reduction of the graphene oxide by laser [27,28] (although the reduction of graphene oxide hardly produces one-atom thick 2D layers of graphene, the product is still so thin compared to the wavelength of microwave frequencies that we can consider it as infinitely thin). The geometry of the problem in Bogle et al. [26] is similar to this contribution. The difference between this work and [26] is that a mode matching technique with optimisation solver was used to find the unknowns. On the other hand, Green's functions are used here and variational formulations are provided that are approximated with some choice of basis functions.

The outline of the paper is as follows: initially, Section 2 provides rigorous field theory for a conductive strip inside a rectangular waveguide. Simplified results for the impedance of the conductive strip is discussed in Section 3 for uniform and cosinus hyperbolic distributions, and their corresponding fields on the aperture are reported in Section 4. Numerical results from the analytic theory are validated in Section 5 by modelling similar structure using full-wave commercial packages. Step by step procedure to measure the surface conductivity is also proposed (four steps). A time convention of the $e^{j\omega t}$ is assumed throughout the paper and vector quantities are presented with bold symbols.

2. Theory

This study follows the procedure described by Collin ([29] Section 8.5) to find a reflection coefficient due to a narrow strip in a rectangular waveguide. The difference is that we assume a finite conductivity for the strip, while Collin assumed a perfect electric conductor (PEC) strip. We consider a waveguide with a cross section $a \times b$ in the $x - y$ plane as shown in Figure 1. An infinitely thin conductive strip with width $2t$ and conductivity of σ is placed in $x = x_0$ and $z = 0$, and is stretched from $y = 0$ to $y = b$. It is assumed that a tranverse electric mode TE_{10} mode travels from $-z$. Electric and magnetic fields due to this mode are denoted by E^i and H^i. Presence of the strip makes discontinuity in the waveguide, induces currents J on the strip, and scatters the wave in both directions. One way to represent the scattered fields is to use Green's functions which satisfy the waveguide boundary value problem. One can write Green's function for the scattered electric field due to a current $J = J(x')\hat{y}$ inside rectangular waveguide as ([29], Section 5.6):

$$G_e = -\frac{j\omega\mu_0}{a} \sum_{n=1}^{\infty} \frac{1}{\gamma_n} \sin\frac{n\pi x}{a} \sin\frac{n\pi x'}{a} e^{-\gamma_n|z|}\hat{y}, \tag{1}$$

where γ_n is

$$\gamma_n = j\sqrt{\omega^2\mu\epsilon - \frac{n^2\pi^2}{a^2}} \tag{2}$$

and defined as the complex propagation constant of mode n in the waveguide. The scattered electric field E^s is

$$E^s(x,z) = \int_S G_e(x,x') J(x') \, dx' \tag{3}$$

and the total electric field inside the waveguide is $E = E^i + E^s$. Here, we assume the TE_{10} mode with the incident field

$$E^i = \sin\frac{\pi x}{a}e^{-\gamma_1 z}\hat{y}, \tag{4}$$

$$H^i = -\frac{\gamma_1}{j\omega\mu_0} \sin\frac{\pi x}{a}e^{-\gamma_1 z}\hat{x} - \frac{\pi}{j\omega\mu_0 a} \cos\frac{\pi x}{a}e^{-\gamma_1 z}\hat{z}. \tag{5}$$

Because the strip is made of conductive material, so-called standard impedance boundary conditions (SIBC) ([30], Section 2.4) are to be satisfied on the strip:

$$\hat{n} \times (E^+ + E^-) = \bar{\bar{\eta}} \cdot \hat{n} \times \hat{n} \times (H^+ - H^-), \tag{6}$$

where $\bar{\bar{\eta}}$ is the condition tensor of the sheet. In the following, we assume an isotropic non-magnetic sheet with conductivity of σ_s. Therefore, $\bar{\bar{\eta}} = (2\sigma_s)^{-1}\bar{\bar{I}}$, where $\bar{\bar{I}}$ is the identity tensor. Plus and minus superscripts denote the field on $z = 0^+$ or $z = 0^-$, respectively. The tangential component of E has to be continuous at the junction. Therefore:

$$E^+ = E^- = \sin\frac{\pi x}{a}\,\hat{y} + \int G_e(x,x')J(x')\,dx'\,\hat{y}. \tag{7}$$

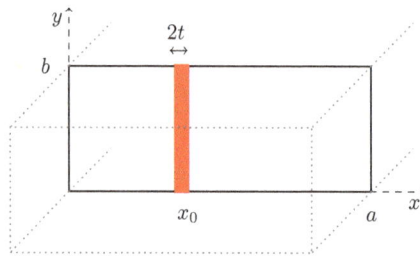

Figure 1. Geometry of the problem; thin strip of width $2t$ inside the rectangular waveguide.

We also find H^s by inserting Equation (1) in Equation (3), and $\nabla \times E^s = -j\omega\mu_0 H^s$. Tangential components of H^s on each side of the boundary are:

$$H_t^{s+} = \frac{1}{a}\sum_{n=1}^{\infty}\sin\frac{n\pi x}{a}\int\sin\frac{n\pi x'}{a}J(x')\,dx'\,\hat{x}, \tag{8}$$

$$H_t^{s-} = -\frac{1}{a}\sum_{n=1}^{\infty}\sin\frac{n\pi x}{a}\int\sin\frac{n\pi x'}{a}J(x')\,dx'\,\hat{x}. \tag{9}$$

The difference in the sign is due to the fact that one mode is propagating to $z > 0$ while the other propagates to $z < 0$.

We juxtapose components into Equation (6):

$$-2\sin\frac{\pi x}{a} - 2\frac{j\omega\mu_0}{a}\sum_{n=1}^{\infty}\frac{1}{\gamma_n}\sin\frac{n\pi x}{a}\int\sin\frac{n\pi x'}{a}J(x')\,dx' = -\frac{1}{2\sigma_s}\frac{2}{a}\sum_{n=1}^{\infty}\sin\frac{n\pi x}{a}\int\sin\frac{n\pi x'}{a}J(x')\,dx' \tag{10}$$

The coefficient of the non-evanescent part of the scattered E^s traveling in the $z < 0$ would be the reflection coefficient Γ:

$$\Gamma = -\frac{j\omega\mu_0}{a\gamma_1}\int\sin\frac{\pi x'}{a}J(x')\,dx'. \tag{11}$$

The transmission coefficient T is also equal to sum of incident wave and non-evanescent part of scattered wave traveling towards $z > 0$:

$$T = 1 - \frac{j\omega\mu_0}{a\gamma_1}\int\sin\frac{\pi x'}{a}J(x')\,dx. \tag{12}$$

From Equation (11) and Equation (12), one finds $T = \Gamma + 1$ relationship. Therefore, we can consider the strip discontinuity as a shunt element across a transmission line (see Figure 2). One can also show that $Z_{shunt} = -\dfrac{\Gamma + 1}{2\Gamma}$. As a result, we find expressions for $(\Gamma + 1)/(2\Gamma)$ in some variational form.

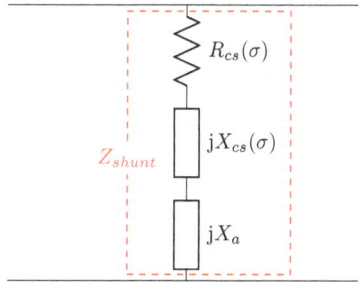

Figure 2. Equivalent circuit for a conductive strip inside waveguide.

We can re-arrange Equation (10) by taking the $n = 1$ term out of the series, and also moving all terms with $n \geq 2$ to the right-hand side:

$$\sin\frac{\pi x}{a} - \frac{j\omega\mu_0}{a\gamma_1}\sin\frac{\pi x}{a}\int\sin\frac{\pi x'}{a}J(x')\,dx'$$
$$= \frac{1}{2a\sigma_s}\sum_{n=1}^{\infty}\sin\frac{n\pi x}{a}\int\sin\frac{n\pi x'}{a}J(x')\,dx' + \sum_{n=2}^{\infty}\frac{j\omega\mu_0}{a\gamma_n}\sin\frac{n\pi x}{a}\int\sin\frac{n\pi x'}{a}J(x')\,dx'. \tag{13}$$

The left-hand side of (13) can be factorized as $(1+\Gamma)\sin\frac{\pi x}{a}$. Following the Collin's procedure, we arrive at $(\Gamma + 1)/2\Gamma$ form by multiplying both sides by $J(x)$, integrating over x, and dividing both sides by 2Γ. Therefore, the total shunt impedance is calculated as:

$$Z_{shunt} = \frac{a\gamma_1}{j2\omega\mu_0}\frac{1}{\left[\int\sin\frac{\pi x}{a}J(x)\,dx\right]^2}\left\{\frac{1}{2a\sigma_s}\sum_{n=1}^{\infty}\int\int\sin\frac{n\pi x}{a}\sin\frac{n\pi x'}{a}J(x')J(x)\,dx'\,dx \right.$$
$$\left. + \sum_{n=2}^{\infty}\frac{j\omega\mu_0}{a\gamma_n}\int\int\sin\frac{n\pi x}{a}\sin\frac{n\pi x'}{a}J(x')J(x)\,dx\,dx'\right\}. \tag{14}$$

By direct comparison with the analytic case solved by Collin, we can identify the terms corresponding to the equivalent circuit proposed in Figure 2. Interestingly, the second term in (14) is identical to the result derived by Collin for a lossless strip inside a waveguide which is called jX_a. On the other hand, Z_{cs} is the contribution due to the material properties of the conductive strip (R_{cs} and X_{cs} refer to the real and imaginary parts of Z_{cs}). Thus,

$$Z_{shunt} = Z_{cs} + jX_a, \tag{15}$$

where

$$Z_{cs} = \frac{\gamma_1}{j4\omega\mu_0\sigma_s}\frac{\displaystyle\sum_{n=1}^{\infty}\int\int\sin\frac{n\pi x}{a}\sin\frac{n\pi x'}{a}J(x')J(x)\,dx'\,dx}{\left[\int\sin\frac{\pi x}{a}J(x)\,dx\right]^2}, \tag{16}$$

$$jX_a = \frac{\gamma_1}{2}\frac{\displaystyle\sum_{n=2}^{\infty}\frac{1}{\gamma_n}\int\int\sin\frac{n\pi x}{a}\sin\frac{n\pi x'}{a}J(x)J(x')\,dx\,dx'}{\left[\int J(x)\sin\frac{\pi x}{a}\,dx\right]^2}. \tag{17}$$

It should be noted that, for a waveguide with lossless walls, γ_1 is pure imaginary while γ_n with $n \geq 2$ are pure real numbers. This is due to the fact that the driving frequency f is chosen so that it is only above the cut-off frequency of the first mode and below the cut-off frequencies of all higher order modes. As a result, we see that jX_a is always pure imaginary. On the other hand, Z_{cs} is a complex number in general, due to complex conductivity σ_s of the conductive strip.

3. Current Distribution on the Strip

The formulations for Z_{cs} and X_a in Equation (16) and Equation (17) are in the variational forms. This implies that choosing a testing function for $J(x)$ does not make a huge influence on the results. Choosing an appropriate testing function, which is closer to the true current distribution, would undoubtedly improve the accuracy of the results. Nevertheless, trying different testing functions also assists with examining the numerical sensitivity to the choice of the testing functions.

3.1. Uniform Current

If the strip is reasonably thin $t \ll a$, a rough approximation is to assume $J(x)$ as uniform over the strip [29]

$$jX = \frac{\gamma_1}{2} \frac{\sum\limits_{n=2}^{\infty} \frac{1}{n^2 \gamma_n} \left[\sin \frac{n\pi x_0}{a} \sin \frac{n\pi t}{a}\right]^2}{\left[\sin \frac{\pi x_0}{a} \sin \frac{\pi t}{a}\right]^2}. \tag{18}$$

Collin also simplifies (18) for the case when the strip is exactly in the middle of the waveguide $(x_0 = a/2)$

$$jX = \frac{\gamma_1}{2} \csc^2 \frac{\pi t}{a} \sum\limits_{n=3,5,\dots}^{\infty} \frac{1}{n^2 \gamma_n} \sin^2 \frac{n\pi t}{a}. \tag{19}$$

Similarly, we also find expressions for Z_{cs}:

$$Z_{cs} = \frac{\gamma_1}{j4\omega\mu_0\sigma_s} \frac{\sum\limits_{n=1}^{\infty} \frac{1}{n^2} \left[\sin \frac{n\pi x_0}{a} \sin \frac{n\pi t}{a}\right]^2}{\left[\sin \frac{\pi x_0}{a} \sin \frac{\pi t}{a}\right]^2}, \tag{20}$$

and, for the case of the centered strip, we have:

$$Z_{cs} = \frac{\gamma_1}{j4\omega\mu_0\sigma_s} \csc^2 \frac{\pi t}{a} \sum\limits_{n=1,3,\dots}^{\infty} \frac{1}{n^2} \sin^2 \frac{n\pi t}{a}. \tag{21}$$

This can be further simplified to (see Appendix A) [31]:

$$Z_{cs} = \frac{\gamma_1}{j4\omega\mu_0\sigma_s} \frac{\pi^2 t}{4a} \csc^2 \frac{\pi t}{a}. \tag{22}$$

3.2. Hyperbolic Cosine Distribution

A second current distribution of $J(x) = J_0 \cosh[(x - x_0)b/t]\hat{y}$ was also studied for this problem. The motivation for such choice of testing function is to analytically model the singularity on the **H** fields, especially H_z component on the edges of the strip $(x \rightarrow x_0 \pm t)$. This effect significantly increases

J near the edges (see Figure 3 for currents on a PEC strip). By the assumption on the current $J(x)$ as hyper cosine distribution, we get the Z_{cs} and jX_a as

$$Z_{cs} = \frac{\gamma_1}{j4\omega\mu_0\sigma_s} \frac{\sum_{n=1}^{\infty} I_n^2}{I_1^2}, \tag{23}$$

$$jX_a = \frac{\gamma_1}{2} \frac{\sum_{n=2}^{\infty} \frac{1}{\gamma_n} I_n^2}{I_1^2}, \tag{24}$$

where

$$I_n = \int_{x_-}^{x_+} J(x)\sin\frac{n\pi x}{a}dx = \frac{A\left[\sin\frac{n\pi x_+}{a} + \sin\frac{n\pi x_-}{a}\right] - B\left[\cos\frac{n\pi x_+}{a} - \cos\frac{n\pi x_-}{a}\right]}{2\exp(b)[(n\pi t)^2 + (ab)^2]} \tag{25}$$

b is the scaling factor in the cosh basis function, and A, B and $x_{\pm}$ are:

$$A = a^2 bt(\exp(2b) - 1), \tag{26}$$
$$B = \pi a t^2(\exp(2b) + 1), \tag{27}$$
$$x_{\pm} = x_0 \pm t. \tag{28}$$

It should be noted that I_1 is found by setting $n = 1$ in (25).

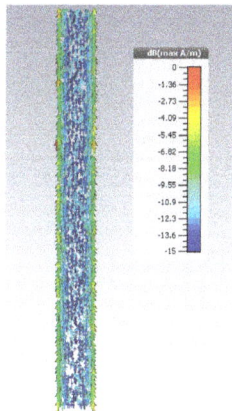

Figure 3. Currents on the conductive strip (x–y plane).

4. Field on the Conductive Strip

It is instrumental to study the E and H fields on the discontinuity since they provide deeper insights into the problem. One can also check whether appropriate boundary conditions (here SIBC) were satisfied or not.

In the previous section, we have assumed that induced current is in the form of $J = J_0\hat{y}$ or $J(x) = J_0\cosh[(x - x_0)/t]\hat{y}$, where J_0 is a complex number. One finds J_0 from (11) after computing the LHS from the results of Section 3. The scattered field E_x^s is then found by using (3):

$$E_x^s = \sum_n E_n^s \sin\frac{n\pi x}{a}, \tag{29}$$

where E_n^s is in the following form for uniform and cosh distributions, respectively:

$$E_n^s = \frac{\gamma_1 \Gamma}{n \gamma_n} \frac{\left[\cos \frac{n\pi}{a}(x_0 - t) - \cos \frac{n\pi}{a}(x_0 + t) \right]}{\left[\cos \frac{\pi}{a}(x_0 - t) - \cos \frac{\pi}{a}(x_0 + t) \right]}, \tag{30}$$

$$E_n^s = \frac{\gamma_1 \Gamma}{n \gamma_n} \frac{I_n}{I_1}. \tag{31}$$

5. Numerical Results

Intuitive tests can be readily performed to validate the rationality of results. If the strip is assumed as PEC, then $\sigma_s \to \infty$, Therefore, Equation (14) would also reduces to Equation (17). On the other hand, Z_{cs} goes to infinity if we presume that the strip in the waveguide aperture has very low conductivity ($\sigma_s \to 0$). This is equivalent to replacing Z_{cs} with an open circuit in the equivalent circuit (see Figure 2). Therefore, no reflections occur at zero conductivity, which resembles no discontinuity on the aperture.

To show the competency of the method, we compare our analytic method with modelling results. A general purpose programming tool [32] was used to find analytic results from Equations (18) and (22) or Equations (23) and (24). We used two commercial electromagnetic packages with a finite element method (FEM) solver [33] and method of moments (MoM) solver [34] to simulate the structures. Two different values are chosen for the complex conductivity of the strips as $0.01 - j0.01 \, \text{S m}^{-1}$ and $0.001 - j0.001 \, \text{S m}^{-1}$ which are chosen close to conductivity of the graphene at X-band.

5.1. Reflection and Transmission Coefficients

Figures 4 and 5 illustrate the magnitude and phase of the reflection and transmission coefficients caused by the conductive strip discontinuity. In the modelling, we de-embedded the excitation ports to the plane of the discontinuity, in order to achieve the correct phase for Γ and T.

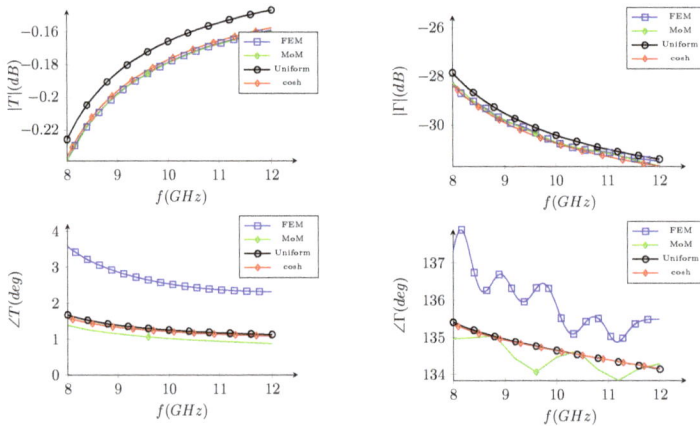

Figure 4. Magnitude and phase of the reflection Γ and transmission coefficient T from a strip with $\sigma = 0.001 - j0.001 \, \text{S m}^{-1}$.

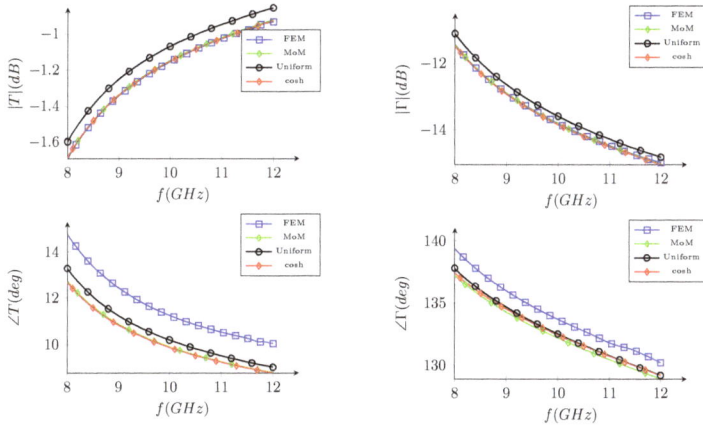

Figure 5. Magnitude and phase of the reflection Γ and transmission coefficient T from a strip with $\sigma = 0.01 - \text{j}0.01 \, \text{S} \, \text{m}^{-1}$.

A good agreement between analytic and simulation results are obtained for both current distribution (see Figures 4 and 5). However, it is evident that the Γ and T from cosh assumption are much closer to full wave modelling results. This is best illustrated on the magnitude of T and Γ coefficients. It is also observed that MoM simulations lie closer to the theoretical calculations.

Comparing Figures 4 and 5, we see a relatively drop in T and a jump in the Γ with an increase in surface conductivity σ_s. This is expected as the higher σ_s causes more reflection and reduces the transmission of the wave through the waveguide.

5.2. Fields on the Aperture

A perture fields on the strip discontinuity are examined in this subsection. Fields depicted in Figures 6 and 7 are the total fields (incident+scattered) for a strip with 1 mm width. The frequency is set to 10 GHz. The fields from both current approximations are close to the simulation results when $|x - x_0| > 2t$. Particularly, E_x and H_z components by our approximations and full-wave solver are almost identical even on the conductive strip. H_x component by cosh approximation disagrees slightly with results from other methods that are due to the choice of the basis function; although such a disagreement is only observed on the conductive strip and everywhere else, they are in total agreement. The singular like behaviour of the H_z near the edges of the strip is to make a closed loop of H field around the strip (also see Figure 3).

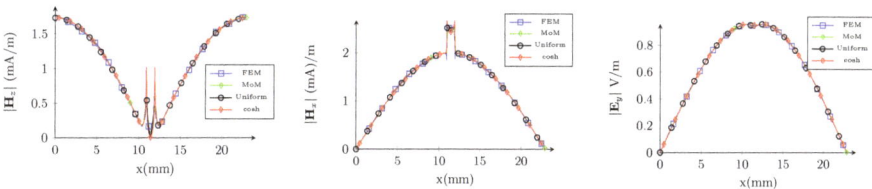

Figure 6. Fields on the aperture of the waveguide and discontinuity of conductive strip. E field at $z = 0$ and H fields at $z = 0^+$ with conductivity of strip set to $\sigma = 0.001 - \text{j}0.001 \, \text{S} \, \text{m}^{-1}$.

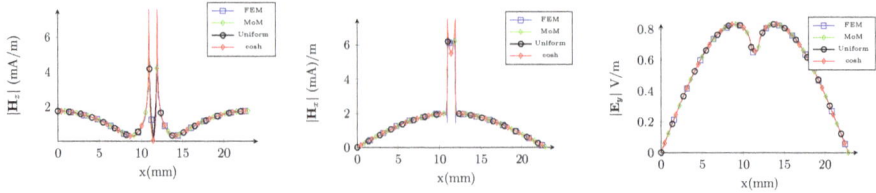

Figure 7. Fields on the aperture of the waveguide and discontinuity of conductive strip. *E* field at $z = 0$ and *H* fields at $z = 0^+$ with conductivity of strip set to $\sigma = 0.01 - j0.01 \, \mathrm{S \, m^{-1}}$.

It is interesting to see that there is a drop in E_y and a significant jump in H_x field to satisfy the imposed boundary conditions. To understand this phenomena, we start by recalling that the ratio of E_y and H_x at distances far from the discontinuity is set by the characteristic impedance of the waveguide $Z_0 = \frac{j\omega\mu}{\gamma_1}$. For a WR90 waveguide at 10 GHz, Z_0 is around 499 Ω. However, the conductivity of the strip dictates the ratios of the transverse components of *E* and *H* by enforcing (6). In Figures 6 and 7, $\frac{1}{|2\sigma_s|}$ is 353.53 Ω and 35.353 Ω, respectively. Therefore, comparing Figure 7 with Figure 6, a significant drop in *E* and also a larger jump in H_x are needed to satisfy the boundary conditions in Figure 7.

5.3. Conductivity Estimation

In the following, we demonstrate how one can use the analytic results of Section 3 to measure the real and imaginary part of the conductivity of a thin strip. The measurement apparatus should basically include a vector network analyser and waveguides. The diagram in Figure 8 shows the required steps in a typical apparatus. For the uniform current approximation, Δ is found from

$$\Delta = \frac{\gamma_1}{j4\omega\mu_0} \csc^2 \frac{\pi t}{a} \sum_{n=1,3,\dots}^{\infty} \frac{1}{n^2} \sin^2 \frac{n\pi t}{a}, \tag{32}$$

while Δ, for cosh distribution, is:

$$\Delta = \frac{\gamma_1}{j4\omega\mu_0 I_1^2} \sum_{n=1}^{\infty} I_n^2. \tag{33}$$

We used the modelling results of the full-wave simulators (FEM and MoM) as the input to examine the measurement procedure. *S*-parameters from the modelling packages are fed into the Matlab® code to estimate the values of the conductivity based on the presented theory with uniform and cosh approximations. Figures 9 and 10 illustrate the computed conductivity by our method for a conductive strip of $0.001 - j0.001 \, \mathrm{S \, m^{-1}}$ and $0.01 - j0.01 \, \mathrm{S \, m^{-1}}$, respectively.

Generally, the theory presented in this paper is valid for the various positions of the conductive strip inside the waveguide and along its long side. When the strip is placed at the center of the guide, the maximum reflection from the strip occurs. The interaction with the fundamental mode is stronger with the conductive strip in the middle, which improves the dynamic range of the measurement method. Further investigations are needed to explore the accuracy and sensitivity of this method to various parameters in the apparatus.

To examine the method, conductivity of a material with $\angle\sigma_s = -45$ deg is swept over a range of $0.0001 \, \mathrm{S \, m^{-1}}$ to $10{,}000 \, \mathrm{S \, m^{-1}}$. Reflection and transmission coefficients are plotted in Figure 11 at 10 GHz for a strip with a width of $2t = 1$ mm, which is placed at the centre of the waveguide. Both Γ and T change reasonably as long as the $10^{-4} \ll \sigma_s \ll 10^2 \, \mathrm{S \, m^{-1}}$. On the other hand, they have minimal changes if the surface conductivity is out of the specified range (it is hard to measure the $S_{11} \ll -30$ dB with current status quo). This method would be most instrumental to measure the surface conductivity of the materials in the above range.

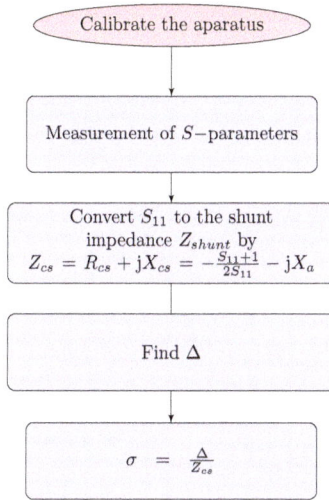

Figure 8. The procedure to measure the conductivity by the proposed method.

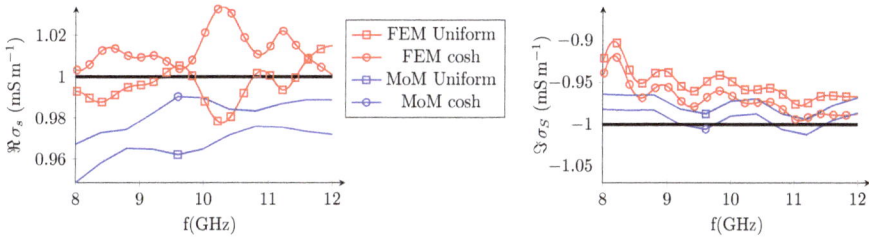

Figure 9. Estimated conductivity for a strip with surface conductivity of $(0.001 - j0.001)\,\mathrm{S\,m^{-1}}$. The solid black line shows the expected value; (top) real part (bottom) imaginary part of surface conductivity.

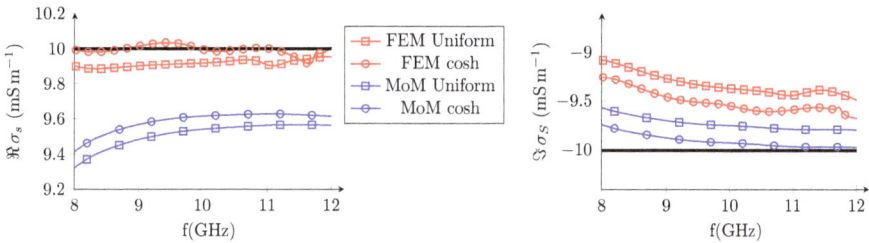

Figure 10. Estimated conductivity for a strip with surface conductivity of $(0.01 - j0.01)\,\mathrm{S\,m^{-1}}$. The solid black line shows the expected value; (top) real part; (bottom) imaginary part of surface conductivity.

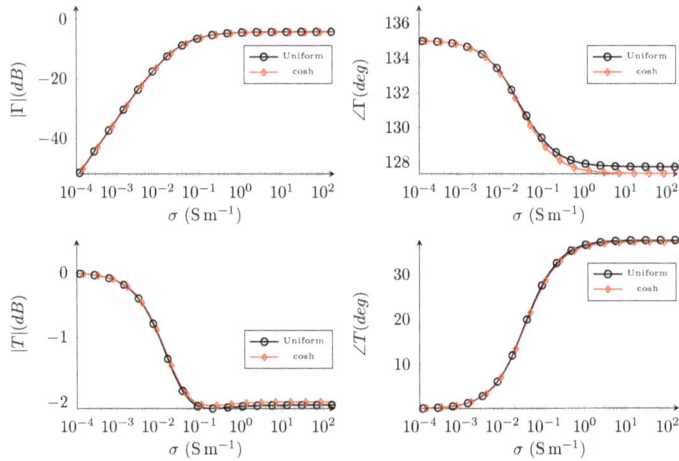

Figure 11. Variation of Γ and T coefficients with sweeping conductivity.

6. Conclusions

In this paper, a method to measure the conductivity of thin layers is proposed which is based on the reflection and transmission of the TE modes in a rectangular waveguide. Other transmission–reflection methods need sample under test to cover the aperture of the waveguide; however, our method needs SUT to only cover a small portion of the cross section. An equivalent circuit of the problem is proposed which is of assistance for intuitive understanding. We provide analytic formulas for the reflection Γ and transmission T coefficients, and derive terms related to each component in the equivalent circuit. Distribution of the fields and currents over the SUT are also reported. The reflection and transmission coefficients from a resistive sheet are compared with S-parameters from the commercial software with FEM and MoM solvers. A reasonable agreement is observed for Γ and T coefficients, as well as fields and currents over the aperture.

Funding: This work was partly supported by the Australian Research Council grant DP130102098.

Acknowledgments: The author is grateful to Prof. David Thiel of Griffith University for pointing out this problem as well as fruitful discussions. The manuscript was also greatly benefited by suggestions of Trevor Bird of Antengenuity, as well as his simplified closed form formula for $\sum_{n=1,3,\dots} \sin^2 nx/n^2$. Finally, the author thanks Mohammad Albooyeh of UC Irvine and Dr. Mehdi Sadatgol of MTU Houghton for quickly reviewing a draft of the manuscript.

Conflicts of Interest: The author declares no conflict of interest.

Appendix A. The Simplification of Series in Equation (21)

In Equation (21), we have a series in the form of:

$$S_2(x) = \sum_{n=1,3,\dots}^{\infty} \frac{1}{n^2} \sin^2 nx, \tag{A1}$$

where $x = (\pi t/a)$. This series can be slow to converge under some conditions and it is useful to obtain a closed form solution. It is recognized that this series is the difference of two infinite series containing all terms. Thus:

$$S_2(x) = \sum_{n=1,2,\dots}^{\infty} \frac{1}{n^2} \sin^2 nx - \frac{1}{4} \sum_{k=1,2,\dots}^{\infty} \frac{1}{n^2} \sin^2 2kx = S_\infty(x) - \frac{1}{4} S_\infty(2x), \tag{A2}$$

where $S_\infty(x)$ is:

$$S_\infty(x) = \sum_{n=1,2,\dots}^{\infty} \frac{1}{n^2} \sin^2 nx. \tag{A3}$$

Using $2\sin^2 z = 1 - \cos 2z$, we find $S_\infty(x)$:

$$S_\infty(x) = \frac{1}{2} \sum_{n=1,2,\dots}^{\infty} \frac{1}{n^2} - \frac{1}{2} \sum_{n=1,2,\dots}^{\infty} \frac{1}{n^2} \cos 2nx, \tag{A4}$$

where the first series is known as the Euler's result [35]

$$\sum_{n=1,2,\dots}^{\infty} \frac{1}{n^2} = \frac{\pi^2}{6}. \tag{A5}$$

The second series is looked up from the tables ([36], Section 1.443) with a change of variable of $x \to 2x$ with a restriction of $0 \le x \le \pi/2$:

$$\sum_{n=1,2,\dots}^{\infty} \frac{1}{n^2} \cos 2nx = \frac{\pi^2}{6} - \pi x + x^2. \tag{A6}$$

Therefore:

$$S_\infty = \frac{1}{2} \left(\pi x - x^2 \right). \tag{A7}$$

Finally, using Equation (A7) twice in Equation (A2) results in

$$S_2(x) = \sum_{n=1,3,\dots}^{\infty} \frac{1}{n^2} \sin^2 nx = \frac{\pi x}{4}, \tag{A8}$$

where $0 \le x \le \pi/2$.

References

1. Baker-Jarvis, J. *Transmission/Reflection and Short-Circuit Line Permittivity Measurements*; Technical Report, NIST Technical Note 1341; National Institute of Standards and Technology: Boulder, CO, USA, 1990.
2. Chen, L.F.; Ong, C.K.; Neo, C.P.; Varadan, V.V.; Varadan, V.K. *Microwave Electronics: Measurement and Materials Characterization*; John Wiley Sons: Hoboken, NJ, USA, 2004.
3. Kobayashi, Y.; Katoh, M. Microwave Measurement of Dielectric Properties of Low-Loss Materials by the Dielectric Rod Resonator Method. *IEEE Trans. Microw. Theory Tech.* **1985**, *33*, 586–592. [CrossRef]
4. Champlin, K.S.; Krongard, R.R. The Measurement of Conductivity and Permittivity of Semiconductor Spheres by an Extension of the Cavity Perturbation Method. *IRE Trans. Microw. Theory Tech.* **1961**, MTT-9, 545–551. [CrossRef]
5. Courtney, W. Analysis and Evaluation of a Method of Measuring the Complex Permittivity and Permeability Microwave Insulators. *IEEE Trans. Microw. Theory Tech.* **1970**, *18*, 476–485. [CrossRef]
6. Krupka, J.; Judek, J.; Jastrzębski, C.; Ciuk, T.; Wosik, J.; Zdrojek, M. Microwave complex conductivity of the YBCO thin films as a function of static external magnetic field. *Appl. Phys. Lett.* **2014**, *104*, 102603. [CrossRef]
7. Le Floch, J.M.; Fan, Y.; Humbert, G.; Shan, Q.; Férachou, D.; Bara-Maillet, R.; Aubourg, M.; Hartnett, J.G.; Madrangeas, V.; Cros, D.; et al. Invited article: Dielectric material characterization techniques and designs of high-Q resonators for applications from micro to millimeter-waves frequencies applicable at room and cryogenic temperatures. *Rev. Sci. Instrum.* **2014**, *85*, 031301. [CrossRef] [PubMed]

8. Krupka, J.; Strupinski, W.; Kwietniewski, N. Microwave Conductivity of Very Thin Graphene and Metal Films. *J. Nanosci. Nanotechnol.* **2011**, *11*, 3358–3362. [CrossRef] [PubMed]

9. Hao, L.; Gallop, J.; Goniszewski, S.; Shaforost, O.; Klein, N.; Yakimova, R. Non-contact method for measurement of the microwave conductivity of graphene. *Appl. Phys. Lett.* **2013**, *103*, 123103. [CrossRef]

10. Ilić, A.Ž.; Budimir, D. Electromagnetic analysis of graphene based tunable waveguide resonators. *Microw. Opt. Technol. Lett.* **2014**, *56*, 2385–2388. [CrossRef]

11. Obrzut, J.; Emiroglu, C.; Kirillov, O.; Yang, Y.; Elmquist, R.E. Surface conductance of graphene from non-contact resonant cavity. *Measurement* **2016**, *87*, 146–151. [CrossRef]

12. Kato, Y.; Horibe, M. New Permittivity Measurement Methods Using Resonant Phenomena for High-Permittivity Materials. *IEEE Trans. Instrum. Meas.* **2017**, *66*, 1191–1200. [CrossRef]

13. Nozaki, R.; Bose, T. Broadband complex permittivity measurements by time-domain spectroscopy. *IEEE Trans. Instrum. Meas.* **1990**, *39*, 945–951. [CrossRef]

14. Booth, J.C.; Wu, D.H.; Anlage, S.M. A broadband method for the measurement of the surface impedance of thin films at microwave frequencies. *Rev. Sci. Instrum.* **1994**, *65*, 2082–2090. [CrossRef]

15. Rzepecka, M.A.; Stuchly, S.S. A Lumped Capacitance Method for the Measurement of the Permittivity and Conductivity in the Frequency and Time Domain-A Further Analysis. *IEEE Trans. Instrum. Meas.* **1975**, *24*, 27–32. [CrossRef]

16. Nag, B.; Roy, S.; Chatterji, C. Microwave measurement of conductivity and dielectric constant of semiconductors. *Proc. IEEE* **1963**, *51*, 962. [CrossRef]

17. Abdulnour, J.; Akyel, C.; Wu, K. A generic approach for permittivity measurement of dielectric materials using a discontinuity in a rectangular waveguide or a microstrip line. *IEEE Trans. Microw. Theory Tech.* **1995**, *43*, 1060–1066. [CrossRef]

18. Hong, Y.K.; Lee, C.Y.; Jeong, C.K.; Lee, D.E.; Kim, K.; Joo, J. Method and apparatus to measure electromagnetic interference shielding efficiency and its shielding characteristics in broadband frequency ranges. *Rev. Sci. Instrum.* **2003**, *74*, 1098–1102. [CrossRef]

19. Wei, X.C.; Xu, Y.L.; Meng, N.; Xu, Y.; Hakro, A.; Dai, G.L.; Hao, R.; Li, E.P. A non-contact graphene surface scattering rate characterization method at microwave frequency by combining Raman spectroscopy and coaxial connectors measurement. *Carbon N. Y.* **2014**, *77*, 53–58. [CrossRef]

20. Gomez-Diaz, J.S.; Perruisseau-Carrier, J.; Sharma, P.; Ionescu, A. Non-contact characterization of graphene surface impedance at micro and millimeter waves. *J. Appl. Phys.* **2012**, *111*, 114908. [CrossRef]

21. Rostamnejadi, A. Microwave properties of La0.8Ag0.2MnO3 nanoparticles. *Appl. Phys. A* **2016**, *122*, 966. [CrossRef]

22. Hassan, A.M.; Obrzut, J.; Garboczi, E.J. A Q-Band Free-Space Characterization of Carbon Nanotube Composites. *IEEE Trans. Microw. Theory Tech.* **2016**, *64*, 3807–3819. [CrossRef]

23. Han, J.; Geyi, W. A New Method for Measuring the Properties of Dielectric Materials. *IEEE Antennas Wirel. Propag. Lett.* **2013**, *12*, 425–428. [CrossRef]

24. Jiang, J.; Geyi, W. Development of a new prototype system for measuring the permittivity of dielectric materials. *J. Eng.* **2014**, *2014*, 302–304. [CrossRef]

25. Wang, X.; Geyi, W. Design of a Wideband System for Measuring Dielectric Properties. *IEEE Trans. Instrum. Meas.* **2017**, *66*, 69–76. [CrossRef]

26. Bogle, A.; Havrilla, M.; Nyquis, D.; Kempel, L.; Rothwell, E. Electromagnetic Material Characterization using a Partially-Filled Rectangular Waveguide. *J. Electromagn. Waves Appl.* **2005**, *19*, 1291–1306. [CrossRef]

27. Thiel, D.V.; Li, Q.; Li, X.; Gu, M. Laser induced carbon nano-structures for planar antenna fabrication at microwave frequencies. In Proceedings of the 2014 IEEE Antennas and Propagation Society International Symposium (APSURSI), Memphis, TN, USA, 6–11 July 2014; Volume 3, pp. 898–899.

28. Wang, W.; Chakrabarti, S.; Chen, Z.; Yan, Z.; Tade, M.O.; Zou, J.; Li, Q. A novel bottom-up solvothermal synthesis of carbon nanosheets. *J. Mater. Chem. A* **2014**, *2*, 2390. [CrossRef]

29. Collin, R.E. *Field Theory of Guided Waves*, 2nd ed.; IEEE-Press: Piscataway, New Jersey, USA, 1991.

30. Senior, T.B.A.; Volakis, J.L. *Approximate Boundary Conditions in Electromagnetics*; No. 41; IET: London, UK, 1995.

31. Bird, T. Antengenuity, Eastwood NSW, Australia. Personal communication, 2017.

32. Mathworks Inc. Available online: www.MATHWORKS.com (accessed on 6 December 2018).

33. CST Microwave Studio. Available online: www.cst.com (accessed on 6 December 2018).

34. FEKO. EM Software & Systems. Available online: www.feko.info (accessed on 6 December 2018).
35. Kline, M. Euler and Infinite Series. *Math. Mag.* **1983**, *56*, 307–314. [CrossRef]
36. Jeffrey, A.; Zwillinger, D. *Table of Integrals, Series, and Products*, 7th ed.; Elsevier Science: Amsterdam, The Netherlands, 2007.

Sample Availability: Matlab codes to reproduce the figures are accessible via the Code Oceans®, https://doi.org/10.24433/CO.742c3ef5-4861-4d48-92a2-b2c45c669d3d; Simulation models to verify the modelling results are available via FigShare®, https://figshare.com/s/1116824ed2b8477319d2

© 2018 by the author. Licensee MDPI, Basel, Switzerland. This article is an open access article distributed under the terms and conditions of the Creative Commons Attribution (CC BY) license (http://creativecommons.org/licenses/by/4.0/).

![electronics logo] *electronics*

MDPI

Article

A 2.5-GHz 1-V High Efficiency CMOS Power Amplifier IC with a Dual-Switching Transistor and Third Harmonic Tuning Technique

Taufiq Alif Kurniawan [1,*] and Toshihiko Yoshimasu [2]

[1]   Department of Electrical Engineering, Universitas Indonesia, Depok 16424, Indonesia
[2]   Graduate School of Information, Production and Systems, Waseda University, Kitakyushu-shi 808-0135, Japan; yoshimasu@waseda.jp
*   Correspondence: taufiq.alif@ui.ac.id; Tel.: +62-21-7270078

Received: 29 October 2018; Accepted: 2 January 2019; Published: 8 January 2019

Abstract: This paper presents a 2.5-GHz low-voltage, high-efficiency CMOS power amplifier (PA) IC in 0.18-μm CMOS technology. The combination of a dual-switching transistor (DST) and a third harmonic tuning technique is proposed. The DST effectively improves the gain at the saturation power region when the additional gain extension of the secondary switching transistor compensates for the gain compression of the primary one. To achieve high-efficiency performance, the third harmonic tuning circuit is connected in parallel to the output load. Therefore, the flattened drain current and voltage waveforms are generated, which in turn reduce the overlapping and the dc power consumption significantly. In addition, a 0.5-V back-gate voltage is applied to the primary switching transistor to realize the low-voltage operation. At 1 V of supply voltage, the proposed PA has achieved a power added efficiency (PAE) of 34.5% and a saturated output power of 10.1 dBm.

Keywords: dual-switching transistor; third harmonic tuning; low voltage; high efficiency; CMOS power amplifier IC

1. Introduction

For modern communication systems, such as short-range wireless applications, a high-efficiency power amplifier plays an important role in maintaining the battery life. To increase the efficiency, switching-mode amplifiers, such as class-E and class-F, are widely used [1–4]. By minimizing the overlap between the drain current and voltage waveforms, the dc power dissipation of the amplifiers can be diminished. The output power of switching-mode amplifiers is also comparable to current-mode ones for the same device peak voltage and current [5].

Unfortunately, efficiency and supply voltage represent a trade-off in switching-mode power amplifiers (PAs). Several techniques have been proposed for improving efficiency at low supply voltages [6–11]. The fully integrated PAs with a power combiner, such as multiple LC baluns [6], a transformer [7], and a distributed active transformer [8,9], were proposed to boost the power added efficiency (PAE) at low supply voltages. PAE is the ratio of the produced signal power (the difference between the output and input power) and the dc power consumption. However, large combiners lead to high insertion losses and enlarge the chip size. Another approach to increase the efficiency of a low supply voltage PA is an injection-locking technique [10,11]. Although this technique provides high gain and high efficiency, the circuit is complicated.

The harmonic manipulation techniques, such as class-J [12] and a tuned amplifier [13–15], are also attractive to improve efficiency. However, class-J employs only second harmonic tuning at the output port, while the tuned amplifier in [15] utilizes second and third harmonic tuning at the input and output ports. Both of the techniques increase the fundamental output power with high efficiency. The amplifier using a second harmonic short in [16] has a higher output power capability of 6.6% than the conventional class-E amplifier. However, the peak drain voltage increases significantly.

In this work, a high-efficiency CMOS PA IC operating at a low supply voltage is proposed using 0.18-μm CMOS technology. To boost the PAE, a dual-switching transistor (DST) was adopted in combination with a third harmonic tuning technique [17]. The class-E PA topology was employed as a basic structure, and 0.5-V positive, back-gate voltage was injected for low-voltage operation. A detailed theoretical and circuit analysis was performed, and the optimum circuit parameters were derived.

This paper is organized as follows. In Section 2, the circuit analysis of the proposed configuration is described, and the optimum circuit parameters are derived. Section 3 shows the simulation results of the proposed PA. Section 4 discusses the measurement results and compares them with recently reported PAs. Our conclusions are presented in Section 5.

2. Circuit Analysis

2.1. Dual-Switching Transistor (DST)

Figure 1a,b show the operation principle and the input/output characteristics of the DST structure, respectively. The structure consists of two switching transistors: (1) the primary switching transistor M_1 biased at class-AB and (2) the secondary switching transistor M_2 biased at class-B, which are connected in parallel. Only single input and output matching circuits were employed, thus providing less complexity for the single chip implementation. To realize the class-AB operation of M_1, the positive back-gate voltage V_{bg} was injected. Hence, the threshold voltage decreased so that the overdrive voltage of M_1 would be sufficient for the class-AB operation.

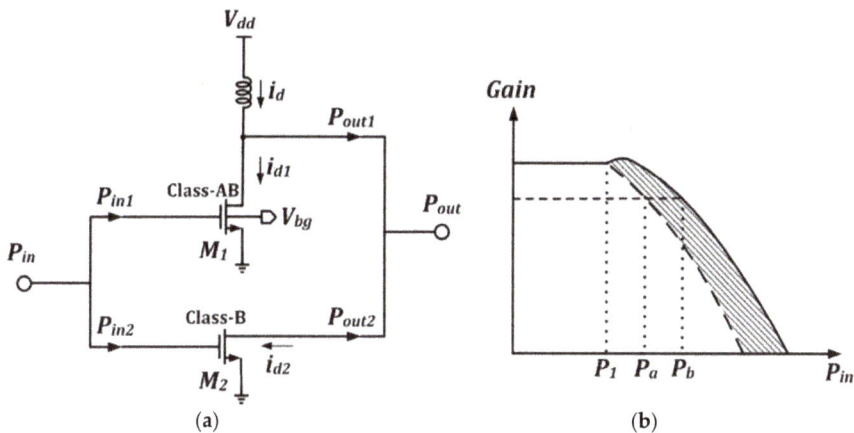

Figure 1. (a) Operation principle and (b) Input/output characteristics of the dual-switching transistor (DST).

At small input power ($P_{in} < P_1$), the total output power is mainly contributed by M_1 ($P_{out} = P_{out1}$), because M_1 is in active operation. In this condition, the total dissipation power P_{DC} of the amplifier is equal to $V_{dd}I_{d1}$.

At large input power ($P_{in} > P_1$), both of the switching transistors are active. M_1 operates near the saturation power level and corresponds to its output power ($P_{out1} = P_{sat1}$), while M_2 delivers the output power of P_{out2}. Therefore, the total output power of the amplifier P_{out} is the sum of the two transistors ($P_{sat1} + P_{out2}$). To maximize the PAE, the size of M_2 is set to be sufficiently smaller than M_1, hence the $i_{d2} \ll i_{d1}$ and P_{DC} is expected to be slightly higher than $V_{dd}I_{d1}$. The overlapping between the drain voltage and current waveforms of the DST should be minimized by adjusting the conduction angle of M_2. Consequently, the PAE at the saturation power level is approximately represented as follows:

$$PAE = \frac{(P_{sat1} + P_{out2}) - P_{in}}{V_{dd}I_{d1}} * 100\%. \tag{1}$$

In addition, the additional gain expansion generated at the saturation power level by the DST (shaded area in Figure 1b) improves the linearity of the amplifier, $P_b > P_a$, where P_b is the input P1dB of the proposed amplifier and P_a is the input P1dB of the conventional class-AB amplifier.

2.2. Third Harmonic Tuning Technique

The third harmonic tuning technique shapes the voltage and current waveforms of the switching transistor at the drain node to minimize the dc power dissipation [18]. The technique is realized by connecting a series C_3–L_3 resonated at $3f_o$ to the output node of the conventional class-E amplifier, as shown in Figure 2. Due to the resonator C_3–L_3, under specified conditions, the third harmonic component of the drain voltage and current waveforms can be controlled.

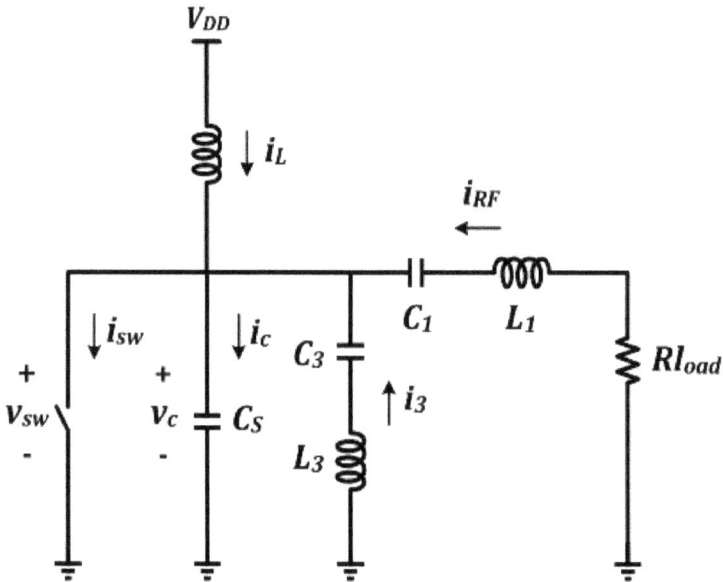

Figure 2. Basic configuration of the third harmonic tuning circuit.

Based on a zero-voltage switching (ZVS) condition, a series of mathematical analyses was conducted to describe the operation principle of the amplifier with the third harmonic tuning circuit. The following assumptions were considered: (1) the transistor is an ideal switch, and (2) all passive elements are lossless and linear. For the operation, the transistor was driven at frequency f_o and at 50% of the duty cycle (on at $0 \leq \omega t < \pi$ and off at $\pi \leq \omega t < 2\pi$).

The loaded Q-factor of C_1–L_1 and C_3–L_3 were assumed to be infinite. The fundamental output current I_{RF} and the third harmonic current I_3 are represented as follows:

$$i_{RF}(\omega t) = I_{RF}\sin(\omega t + \theta). \tag{2}$$

$$i_3(\omega t) = I_3\sin(3\omega t). \tag{3}$$

2.2.1. On-State Condition, $0 \leq \omega t < \pi$

When the switch is turned on, the current flowing to the capacitor i_c is equal to 0. Therefore, the capacitor voltage v_c is equal to 0. At this interval, the current that flows to the switch i_{sw} is given by

$$i_{SW}(\omega t) = I_L + I_{RF}\sin(\omega t + \theta) + I_3\sin(3\omega t) \tag{4}$$

where I_L and θ denote the dc current and the initial phase angle, respectively.

Due to the characteristics of the shunt capacitor C_s, the initial current of i_{sw} during the on/off transition is zero. Hence, the following is true:

$$\frac{I_L}{I_{RF}} = -\sin\theta. \tag{5}$$

2.2.2. Off-State Condition, $\pi \leq \omega t < 2\pi$

When the switch is turned off, i_{sw} is equal to 0. Therefore, i_c can be defined as follows,

$$i_C(\omega t) = I_L + I_{RF}\sin(\omega t + \theta) + I_3\sin(3\omega t) \tag{6}$$

generating the voltage across the capacitor v_c as,

$$
\begin{aligned}
v_c(\omega t) &= \frac{1}{\omega C_s}\int_\pi^{\omega t} I_C(\omega t)d\omega t \\
&= \frac{1}{\omega C_s}\left\{ \begin{array}{c} I_L(\omega t - \pi) - \frac{1}{3}I_3(\cos 3\omega t + 1) \\ -I_R[\cos(\theta + \omega t) + \cos\theta] \end{array} \right\}.
\end{aligned} \tag{7}
$$

To solve the above equations, it is assumed that the real part of the third harmonic output impedance Z_3 is equal to zero. The Fourier series analyses for $v_c(\omega t)$ and $i_o(\omega t)$ are derived to obtain the third harmonic component. Note that $i_o(\omega t)$ is the sum of $i_{RF}(\omega t)$ and $i_3(\omega t)$, i.e.,

$$i_o(\omega t) = I_{RF}\sin(\omega t + \theta) + I_3\sin(3\omega t). \tag{8}$$

The third harmonic coefficient of v_c is given by

$$V_{c,3} = \frac{1}{2\pi}\int_\pi^{\omega t} v_c(\omega t)e^{-j3\omega t}d\omega t. \tag{9}$$

The third harmonic coefficient of i_o is expressed as

$$I_{o,3} = \frac{1}{2\pi}\int_\pi^{\omega t} i_o(\omega t)e^{-j3\omega t}d\omega t. \tag{10}$$

Substituting Equations (7)–(10), Z_3 can be obtained as

$$
\begin{aligned}
Z_3 &= \frac{V_{c,3}}{I_{o,3}} \\
&= \frac{j(-4jI_L + 3\pi jI_3 + I_L3\pi - 2I_3 - 6I_{RF}\cos\theta)}{9I_3\pi\omega C}
\end{aligned} \tag{11}
$$

Setting the real part of Z_3 to zero and substituting Equation (5) for Equation (11) yields the following:

$$\frac{I_L}{I_3} = \frac{2}{3\pi + 6\cot\theta}. \tag{12}$$

The initial phase angle (θ) can be obtained by deriving the third harmonic coefficient of i_{sw}, as follows:

$$I_{sw,3} = \frac{1}{2\pi} \int_0^{2\pi} i_{sw}(\omega t) e^{-j3\omega t} d\omega t$$
$$= -j\left(\frac{I_L}{3\pi} + \frac{I_3}{4}\right) \tag{13}$$

At the initial condition $i_{sw}(0) = 0$, we have the following:

$$\frac{I_3}{I_L} = \frac{4}{3\pi}. \tag{14}$$

Substituting Equation (12) with Equation (14) generates the initial phase angle θ of $-35°$. Thus, the ratio I_L/I_{RF} of 0.57 and I_L/I_3 of 2.34 are obtained.

Considering that 100% efficiency is realized when the total output power is equal to the dc power consumption, the following is true:

$$P_{DC} = P_{out}$$
$$I_L V_{DD} = \frac{1}{2} I_{RF}{}^2 R_{load} \tag{15}$$

where V_{DD} can be defined by applying the Fourier series expansion to Equation (7), i.e.,

$$V_{DD} = \frac{1}{2\pi} \int_0^{2\pi} v_{sw}(\omega t) d\omega t$$
$$= -\frac{1}{\omega C_s}\left(\frac{I_3}{3} + I_{RF}\cos\theta\right) \tag{16}$$

To realize a good switching condition, the optimum shunt capacitor C_s can be determined by substituting Equation (16) with Equation (15):

$$C_s = \frac{\pi \sin^2\theta}{\omega R_{load}}. \tag{17}$$

In the circuit implementation, C_s is set as the total external capacitance and output capacitance of the switching transistor.

2.2.3. Switching Waveforms

The normalized switching waveforms of drain current $i_{sw}(\omega t)$ for $0 \le \omega t < \pi$ and drain voltage $v_c(\omega t)$ for $\pi \le \omega t < 2\pi$ are given by using Equations (4), (7), and (16) and the ratio of I_L/I_{RF} and I_L/I_3 as follows:

$$\frac{i_{sw}(\omega t)}{I_L} = 1 - \sin\omega t \cot\theta - \cos\omega t + \frac{4}{3\pi}\sin 3\omega t \tag{18}$$

$$\frac{v_c(\omega t)}{V_{DD}} = \frac{(\pi - \omega t) + \frac{4}{9\pi}(\cos 3\omega t + 1) - \cot\theta(\cos\omega t + 1) + \sin\omega t}{\frac{4}{9\pi} - \cot\theta}. \tag{19}$$

Figure 3 illustrates the normalized drain voltage (dotted line) and current waveforms (solid line) during a time period T. Because the transistor is turned on at $0 \le T < T/2$, there is no voltage across the switch and the current flowing to the switch consists of dc, fundamental, and the third harmonic components. Because the transistor is turned off at $T/2 \le T < T$, all the currents flow to the shunt capacitor C_s. As shown in Figure 3, the third harmonic tuning technique flattens the waveforms and reduces the overlapping. Therefore, it is expected to achieve low dc power dissipation and high PAE. Because the technique only controls a single harmonic component at the output node, it offers a more suitable structure for single chip PA solutions than class F or class F^{-1}.

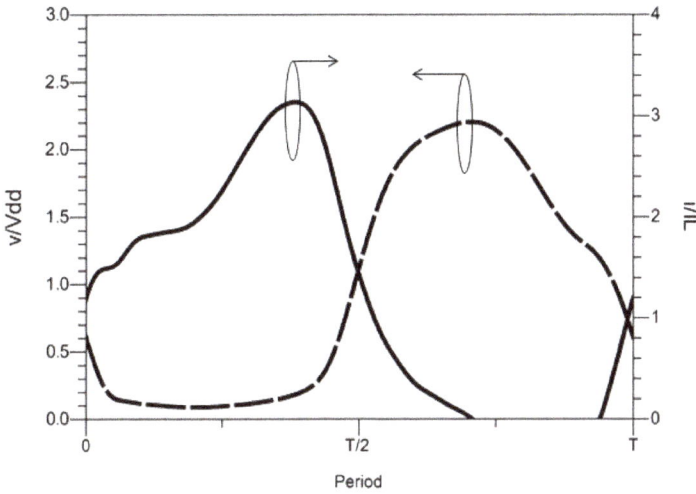

Figure 3. Simulated normalized switching waveforms.

Compared with the conventional class-E PA [19] and the second harmonic tuning PA [16], the PA with third harmonic tuning technique has a lower peak voltage waveform. The voltage stress of the switching transistor is reduced, and it is expected to get higher output power capability. For the CMOS process, the output power capability P_c calculates the maximum achievable output power for a given voltage stress V_{peak} and rms drain current I_{rms} and is expressed as follows:

$$P_C = \frac{P_{out}}{V_{peak}I_{rms}}. \tag{20}$$

The resonator C_3–L_3 induces a parasitic capacitance at the fundamental frequency and reduces the impedance at the output port. As a result, the loaded Q-factor of C_1–L_1, which is resonated at f_o, is decreased. The third harmonic resonator is designed by choosing an inductor with a high Q-factor at $3f_o$ as L_3. To compromise with the overall chip area and insertion loss, the layout dimension of the inductor is optimized. Then, the series capacitance C_3 is defined by the equation $C_3 = 1/9(\pi f_o)^2 L_3$. The parasitic capacitance of the resonator should be minimized to maintain the optimum shunt capacitance C_s of the amplifier. Consequently, by selecting the Q-factor of 10, the third harmonic resonator is optimized to $L_3 = 2.5$ nH and $C_3 =180$ fF.

2.3. Back-Gate Bias Technique

The back-gate bias technique is a method to modify the threshold voltage level and the on-resistance (R_{on}) of the MOSFET by connecting the body terminal to the positive or negative voltage. In this paper, a positive back-gate voltage was applied at the body of M_1 so that the threshold voltage and the R_{on} decreased. The technique led to the p-well of the body and n+ of the source being connected in a forward bias condition. Therefore, the positive back-gate bias voltage selection is very important to minimize an excessive dc leakage current from the body to the source [20]. A solid back-gate body potential should have an ideal connection to the ground, hence a large by-pass capacitance is placed between the body and the ground. Figure 4 shows the comparison between the positive back-gate voltage versus the threshold voltage at 0.5-V bias voltage.

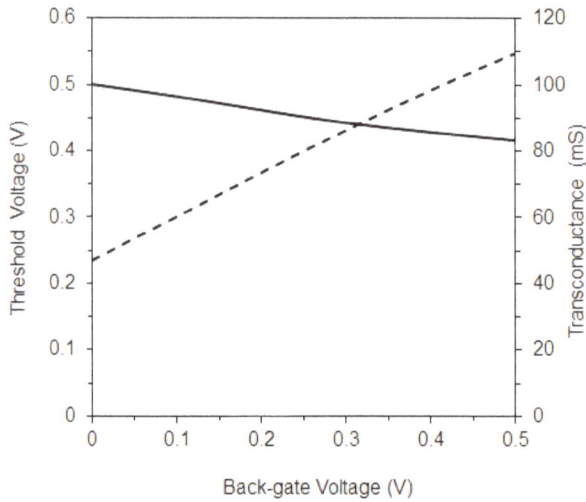

Figure 4. Simulated threshold voltage and transconductance versus back-gate voltage variations of the M_1 at 0.5-V bias voltage.

2.4. Circuit Configuration of the Proposed PA

Figure 5a shows the circuit schematic of the proposed CMOS PA IC. To achieve high PAE at a low supply voltage, a combination of the DST and the third harmonic tuning circuit (C_3–L_3) is proposed. The supply voltage V_{dd} of 1 V was applied with the bias voltage V_{bias} of 0.5 V. In addition, a 0.5-V positive body bias was injected into the body terminal (B) of M_1 by the external biasing terminal (V_{bg}) to decrease the threshold voltage level of 75 mV, which was optimum for the class-AB operation when there was no input power. In addition, the body of M_2 was connected to the ground for the class-B operation. The RC feedback network was employed to modify the input and the output resistance for the impedance matching requirement, as well as to increase the stability of the device; hence, the PA was always unconditionally stable.

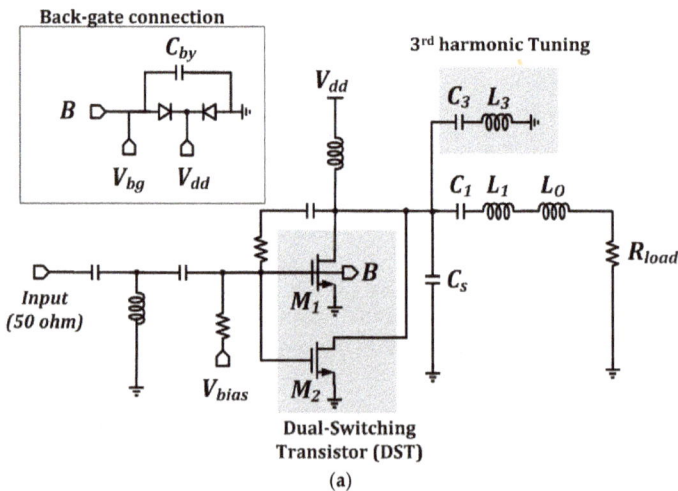

Figure 5. *Cont.*

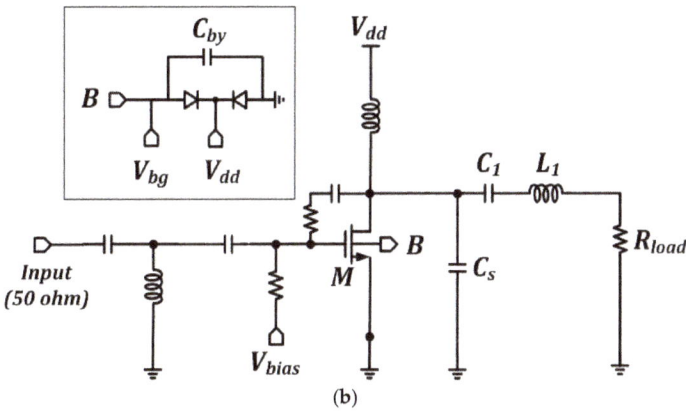

Figure 5. Circuit schematics of (**a**) the proposed power amplifier (PA) IC, and (**b**) the conventional class-E PA IC.

To minimize the drain loss, a large transistor size was selected. The large size, however, decreased the gain and increased the input power. In this work, the optimized total gate width of 464-µm with a gate length of 0.18-µm was utilized to obtain the power gain higher than 10 dB at 2.5-GHz. Because a large transistor size leads to high output capacitance, the shunt capacitor C_S was selected by considering the total output capacitance of the DST. Furthermore, the small inductor L_O was inserted at the output for impedance matching.

To verify the effectiveness of the proposed configuration, a 2.5-GHz conventional class-E PA was designed as illustrated in Figure 5b. The size of the switching transistor M was set equal to the total size of the switching transistors in the proposed PA. The 0.5-V back-gate voltage V_{bg} was injected, and the switching transistor was biased at class-AB with 1-V supply voltage

3. Simulation Results

The prototype of the proposed circuits was designed and fabricated using six metal layer (1P6M) 0.18-µm CMOS technology by TSMC. This CMOS process offered two ultra-thick top metal layers, 4-µm or 2-µm thick, for inductor implementation. High-density MIM capacitors of 1fF/µm^2 and 2fF/µm^2 were also provided. For noise isolation from the P-substrate, deep n-wells were available.

The small-signal and large-signal responses of the proposed PA IC were simulated on a wafer using ADS 2011 by Keysight [21] and Virtuoso ADE IC 6.1.5 by Cadence [22]. Input and output matching circuits were simulated and optimized using Momentum EM simulation in ADS 2011, while layout and verification were performed by Virtuoso ADE IC 6.1.5, respectively.

Figure 6 illustrates the small-signal input–output response of the proposed PA IC. The S_{11} and the S_{22} are −13.9 dB and −12.5 dB at 2.5 GHz, respectively. The maximum S_{21} is 11.2 dB with a 3-dB bandwidth from 1.65 GHz to 3.8 GHz.

Because the third harmonic tuning technique was very effective to decrease the dc power consumption, it is expected that the proposed CMOS PA has high efficiency. Figure 7 shows the simulated dc drain current of the proposed PA IC (solid line) and the conventional PA IC (dotted line) versus the input power. The proposed CMOS PA exhibited a lower dc current than the conventional one under small-signal conditions.

Figure 8 shows the simulated input–output response of the proposed and conventional class-E PAs with sweeping the input power. It confirmed that the proposed PA IC achieves better gain linearity with better PAE than the conventional class-E PA IC. At an input power of lower than -10 dBm, M_2 was turned off and the input signal was mainly amplified by M_1. When the input power increased, the output power of M_2 became higher to compensate for the gain compression of M_1, achieving a higher linear performance. From the simulation, it was shown that the proposed PA had a power gain of 11.5 dB, an output P1dB of 8.1 dBm, and a peak PAE of 38.5%.

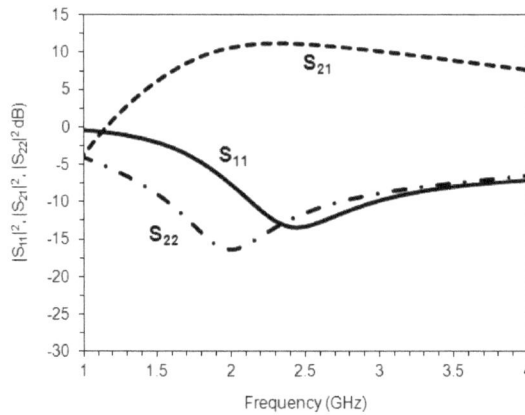

Figure 6. Simulation results of S_{11}, S_{21}, and S_{22} of the proposed PA IC.

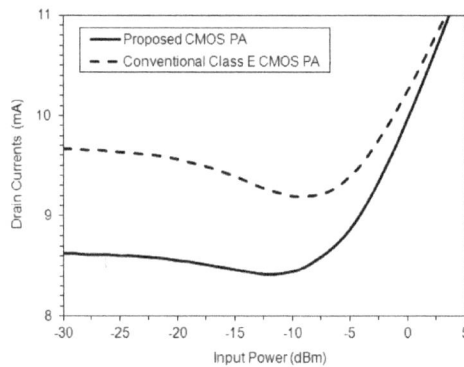

Figure 7. Simulation results of the drain currents versus the input power of the proposed PA IC (solid line) and the conventional PA IC (dotted line).

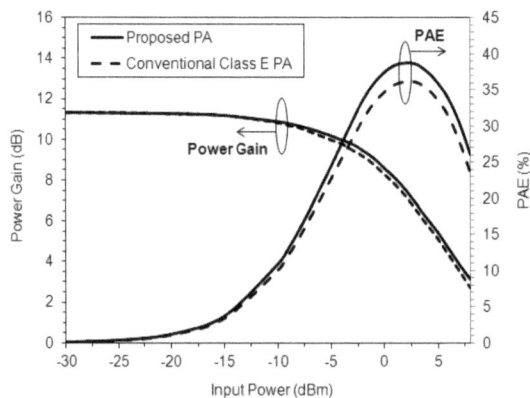

Figure 8. Simulated input–output response of the proposed PA (solid line) and the conventional class-E PA (dotted line) PA.

4. Measurement Results

The chip photograph of the proposed PA IC is depicted in Figure 9. The proposed PA was designed in 0.18-μm CMOS technology and measured on a wafer. The chip size is 0.9-mm by 1.1-mm. Figure 10 shows the measurement setup and probing situation of the proposed CMOS PA IC. The chip was probed using Summit 11201B Cascade Microtech with a single-ended GSG probe at the input and output RF signals. The RF input was generated using an Agilent E8267D vector signal generator, while the dc source was provided by using Yokogawa GS200.

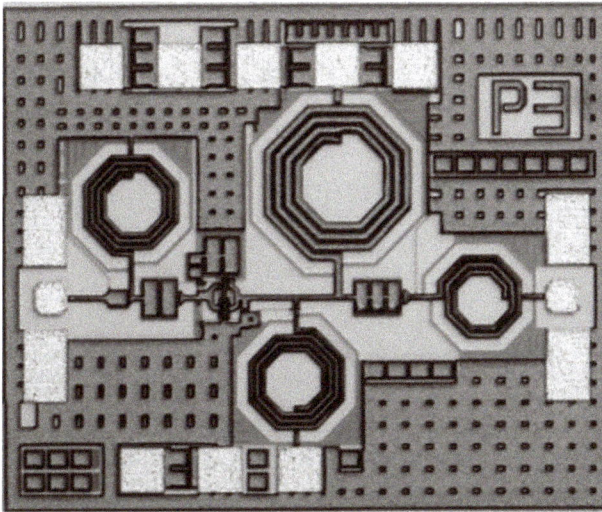

Figure 9. Photograph of the chip of the proposed circuits.

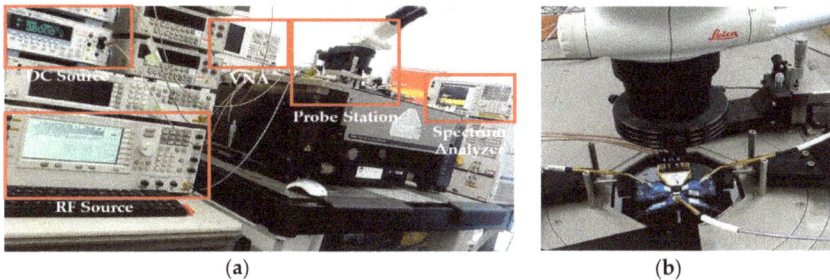

(a)　　　　　　　　　　　　　　　　　　(b)

Figure 10. (**a**) Measurement setup; and (**b**) probing situation of the proposed CMOS PA IC.

In the measurement process, 5 chips were measured for whole data variations, i.e., dc, small-signal, and large-signal performances. The average of the standard deviation of the I_{DS} measurement was 0.03 mA, and the averages of the standard deviations of S_{11}, S_{21}, and S_{22} were 0.05 dB, 0.06 dB, and 0.1 dB, respectively. To demonstrate the uniformity performance of the PA chips, the I_{DS} characteristics of chip 00 and chip 20 over the supply voltage (V_{dd}), bias voltage (V_{bias}), and back-gate voltage (V_{bg}) variations are illustrated in Figure 11.

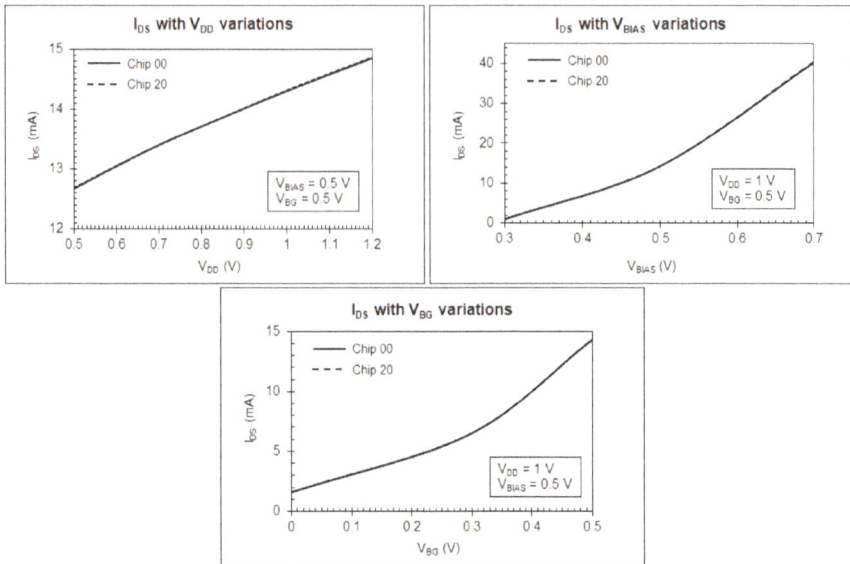

Figure 11. The uniformity dc performance measurement results of the proposed CMOS PA IC.

In order to demonstrate the small-signal performance based on S-parameters, the output RF was measured using the Agilent E8361A vector network analyzer. Figure 12 illustrates the comparison between the measured small-signal S-parameters (solid line) and the simulated small-signal S-parameters (dashed line) of the proposed circuits at 1 V of supply voltage. The quiescent current was 14.31 mA at 0.5 V of gate bias voltage. As expected, the measured S_{11} and S_{22} at 2.5 GHz were lower than -10 dB. The maximum small signal gain was 11.0 dB at 2.5 GHz with a 3-dB bandwidth from 1.7 GHz to 4.1 GHz. Figure 12 shows that measurement results agree well with the simulation results.

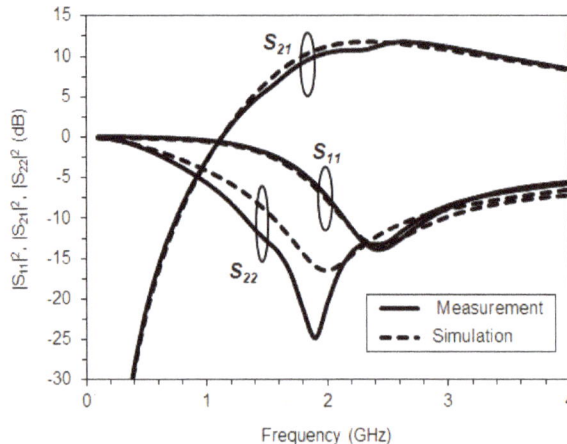

Figure 12. The measurement results (solid line) versus the simulation results (dashed line) of S_{11}, S_{21}, and S_{22} of the proposed PA IC at 1 V of supply voltage.

The measurement results of the power gain and the output power of the proposed CMOS PA IC are shown in Figure 13. To measure the output power, an Agilent E4448A spectrum analyzer was utilized. At a supply voltage of 1 V, the proposed PA achieved a saturated output power of 10.1 dBm and an output P1dB of 8.0 dBm. Figure 14 illustrates the measurement results of the PAE of the proposed PA (solid line) and the conventional class-E PA (dotted line) at 1 V of supply voltage. The proposed PA achieved a higher PAE than the conventional PA, with a peak PAE of 34.5%.

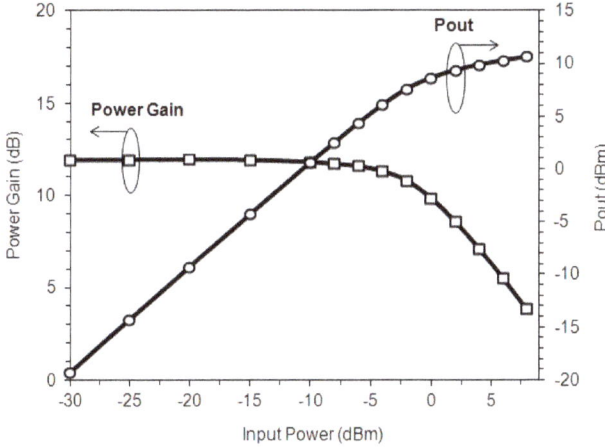

Figure 13. Measurement results of the power gain and pout of the proposed PA IC versus the input power.

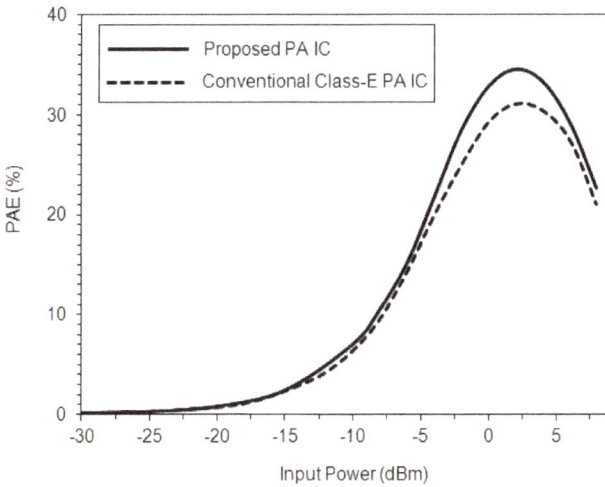

Figure 14. PAE of the proposed PA (solid line) and the conventional class-E PA (dotted line) versus the input power at a supply voltage of 1 V.

Figure 15 shows the measurement results of the output power and the P1dB of the proposed PA (solid line) and the conventional class-E PA (dotted line) versus the supply voltage variations from 0.5 V to 1.2 V. The proposed PA achieved larger saturated output power and P1dB than did the conventional PA. The dependency of the efficiency performances (drain efficiency (DE) and PAE) on the supply voltage is illustrated in Figure 16. The proposed PA achieved higher efficiency than the conventional PA at low-voltage operation. Because the load line was adjusted to obtain high efficiency at a supply voltage of 1 V, the efficiency at 1.2 V was slightly lower than that at 1 V.

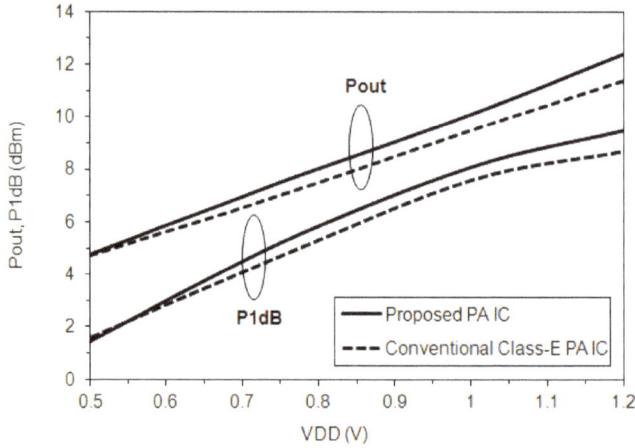

Figure 15. The pout and P1dB of the proposed PA (solid line) and the conventional class-E PA (dotted line) versus supply the voltage variations.

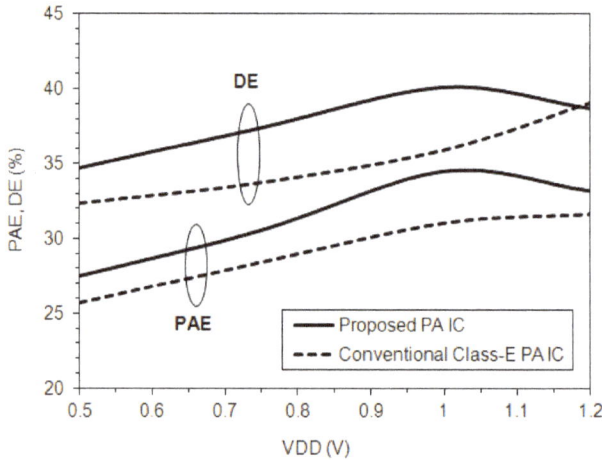

Figure 16. The drain efficiency (DE) and PAE of the proposed PA (solid line) and the conventional class-E PA (dotted line) versus the supply voltage.

Table 1 shows the performance of the proposed PA IC with several published works of low-voltage operation PAs. The proposed PA IC has achieved excellent PAE with sufficient output power at low-voltage operation compared with the previously published works.

Table 1. Summary of the proposed PA IC with several published works.

Reference	Technology	Frequency	Supply Voltage	PAE	Pout	Chip Size
[11]	0.18 μm	2.4 GHz	1.5 V	36%	7.6 dBm	0.6 mm^2
[23]	90 nm	2.4 GHz	1.2 V	32%	1.2 dBm	N/A
[24]	0.18 μm	2.4 GHz	1.8 V	21%	6.4 dBm	1.8 mm^2
[25]	90 nm	2.4 GHz	1.2 V	30%	9 dBm	1 mm^2
[26]	0.13 μm	1.9 GHz	1.2 V	26%	4.1 dBm	N/A
This work	180 nm	2.5 GHz	1.0 V	34.5%	10.1 dBm	0.99 mm^2

5. Conclusions

A 2.5-GHz fully integrated CMOS PA IC has been designed and fully evaluated in 0.18-μm CMOS technology for low-voltage operation. As the input power increased, the proposed dual-switching transistor (DST) effectively improved the gain linearity with high efficiency. In addition, the third harmonic termination technique compressed the voltage waveform and modified the current waveform to realize low dc power dissipation.

At 1 V of supply voltage, the proposed CMOS PA IC exhibited a power gain of 11.0 dB, an output P1dB of 8.0 dBm, and a peak PAE of 34.5%. The measured P1dB and PAE of the proposed PA were higher than those of recently reported works. Therefore, the proposed PA IC is applicable for low-voltage operation.

Author Contributions: T.A.K. developed the idea, performed the circuit design and experiments, and wrote the initial draft. T.Y. supervised the process of this work and reviewed and revised the draft.

Funding: This work is funded by Japan Society for the Promotion of Science (JSPS) KAKENHI Grant-in-Aid for Scientific Research (B) Number 23360162. The publication process received no external funding.

Acknowledgments: This work is supported by VLSI Design and Education Center (VDEC), the University of Tokyo in collaboration with Cadence Design Systems Inc., Mentor Graphics Inc., and Keysight Technologies Japan Ltd.

Conflicts of Interest: The authors declare no conflicts of interest.

References

1. Chen, P.; Yang, K.; Zhang, T. Analysis of a Class-E Power Amplifier with Shunt Filter for Any Duty Ratio. *IEEE Trans. Circuits Syst. II Express Briefs* **2017**, *64*, 857–861. [CrossRef]
2. Abbasian, S.; Johnson, T. Power-Efficiency Characteristics of Class-F and Inverse Class-F Synchronous Rectifiers. *IEEE Trans. Microw. Theory Tech.* **2016**, *64*, 4740–4751. [CrossRef]
3. Liu, C.; Cheng, Q.F. A Novel Compensation Circuit of High-Efficiency Concurrent Dual-Band Class-E Power Amplifiers. *IEEE Microw. Wirel. Compon. Lett.* **2018**, *28*, 720–722. [CrossRef]
4. Dong, Y.; Mao, L.; Xie, S. Extended Continuous Inverse Class-F Power Amplifiers with Class-AB Bias Conditions. *IEEE Microw. Wirel. Compon. Lett.* **2017**, *27*, 368–370. [CrossRef]
5. Kee, S.D.; Aoki, I.; Hajimiri, A.; Rutledge, D. The Class-E/F Family of ZVS Switching Amplifiers. *IEEE Trans. Microw. Theory Tech.* **2003**, *51*, 1677–1690. [CrossRef]
6. Reynaert, P.; Steyart, M.S.J. A 2.45-GHz 0.13-um CMOS PA With Parallel Amplification. *IEEE J. Solid-State Circuits* **2007**, *42*, 551–562. [CrossRef]
7. Haldi, P.; Chowdhury, D.; Reynaert, P.; Liu, G.; Niknejad, A.M. A 5.8 GHz 1 V Linier Power Amplifier Using a Novel On-Chip Transformer Power Combiner in Standard 90 nm CMOS. *IEEE J. Solid-State Circuits* **2008**, *43*, 1054–1062. [CrossRef]
8. Aoki, I.; Kee, S.D.; Rutledge, D.B.; Hijimiri, A. Fully Integrated CMOS Power Amplifier Design Using the Distributed Active-Transformer Architecture. *IEEE J. Solid-State Circuits* **2002**, *37*, 371–383. [CrossRef]
9. Khan, H.R.; Sajid, U.; Kanwal, S.; Zafar, F.; Wahab, Q. A Fully Integrated Distributed Active Transformer Based Power Amplifier in 0.13 um CMOS Technology. In Proceedings of the 2013 Saudi Electronics Communications and Photonics Conference (SIECPC), Fira, Greece, 27–30 April 2013. [CrossRef]
10. Oh, H.S.; Song, T.; Yoon, E.; Kim, C.K. A Power-Efficient Injection-Locked Class-E Power Amplifier for Wireless Sensor Network. *IEEE Microw. Wirel. Compon. Lett.* **2006**, *16*, 173–175. [CrossRef]
11. El-Desouki, M.; Deen, M.; Haddara, Y.; Marinov, O. A Fully Integrated CMOS Power Amplifier using Superharmonic Injection-Locking for Short Range Applications. *IEEE Sens. J.* **2011**, *11*, 2149–2158. [CrossRef]
12. Wright, P.; Lees, J.; Benedikt, J.; Tasker, P.J.; Cripps, S.C. A Methodology for Realizing High Efficiency Class-J in a Linear and Broadband PA. *IEEE Trans. Microw. Theory Tech.* **2013**, *57*, 3196–3204. [CrossRef]
13. Kusunoki, S.; Hatsugai, T. Harmonic-Injected Power Amplifier with 2nd Harmonic Short Circuit for Cellular Phones. *IEICE Trans. Electron.* **2005**, *E88-C*, 729–738. [CrossRef]
14. Bae, K.-T.; Lee, I.-J.; Kang, B.; Sim, S.; Jeon, L.; Kim, D.-W. X-Band GaN Power Amplifier MMIC with a Third Harmonic-Tuned Circuit. *Electronics* **2017**, *6*, 103. [CrossRef]

15. Colantonio, P.; Giannini, F.; Leuzzi, G.; Limiti, E. Multiharmonic Manipulation for Highly Efficient Microwave Power Amplifier. *Int. J. RF Microw. Comput. Aided Eng.* **2001**, *11*, 366–384. [CrossRef]

16. You, F.; He, S.; Tang, X.; Deng, X. High-Efficiency Single-Ended Class-E/F2 Power Amplifier with Finite DC Feed Inductor. *IEEE Trans. Microw. Theory Tech.* **2010**, *58*, 32–40. [CrossRef]

17. Kurniawan, T.A.; Yang, X.; Xu, X.; Yoshimasu, T. A 2.5-GHz Band Low Voltage High Efficiency CMOS Power Amplifier IC Using Parallel Switching Transistor for Short Range Wireless Applications. In Proceedings of the 45th European Microwave Conference 2015, Paris, France, 6–11 September 2015. [CrossRef]

18. Kurniawan, T.A.; Yang, X.; Xu, X.; Itoh, N.; Yoshimasu, T. A 2.5-GHz Band, 0.75-V High Efficiency CMOS Power Amplifier IC With Third Harmonic Termination Technique in 0.18-um CMOS. In Proceedings of the 2015 IEEE Wireless and Microwave Technology Conference (WAMICON), Cocoa Beach, FL, USA, 12–15 April 2015. [CrossRef]

19. Grebenikov, A.; Sokal, N.O. *Switch Mode RF Power Amplifiers*; Newnes: Newton, MA, USA, 2007.

20. Kurniawan, T.A.; Yang, X.; Xu, X.; Sun, Z.; Yoshimasu, T. A 2.5-GHz band low-voltage high efficiency class-E power amplifier IC with body effect. In Proceedings of the 2014 International Symposium on Integrated Circuits (ISIC), Singapore, 10–12 Desember 2014. [CrossRef]

21. Agilent Technologies: Advanced Design System 2011. Available online: http://edadownload.software. keysight.com/eedl/ads/2011/pdf/adstour.pdf (accessed on 1 January 2019).

22. Virtuoso ADE Product Suites. Available online: https://www.cadence.com/content/cadence-www/global/ en_US/home/tools/custom-ic-analog-rf-design/circuit-design/virtuoso-ade-product-suite.html (accessed on 1 January 2019).

23. Huang, X.; Harpe, P.; Wang, X.; Dolmans, G.; Groot, H. A 0 dBm 10 Mbps 2.4 GHz Ultra-Low Power ASK/OOK Transmitter with Digital Pulse-Shaping. In Proceedings of the 2010 IEEE Radio Frequency Integrated Circuits Symposium, Anaheim, CA, USA, 23–25 May 2010; pp. 263–266. [CrossRef]

24. Haridas, K.; Teo, T.H.; Yuan, X. A 2.4-GHz CMOS Power Amplifier for Low Power Wireless Sensor Network. In Proceedings of the 2009 IEEE International Symposium on Radio-Frequency Integration Technology (RFIT), Singapore, 9 January–11 December 2009; pp. 299–302. [CrossRef]

25. Chironi, V.; Debaillie, B.; D'Amico, S.; Baschirotto, A.; Craninckx, J.; Ingels, M. A Digitally Modulated Class-E Polar Amplifier in 90-nm CMOS. *IEEE Trans. Circuits Syst. I Reg. Paper* **2013**, *60*, 918–925. [CrossRef]

26. Chee, Y.; Rabaey, J.; Niknejad, A. A Class A/B Low Power Amplifier for Wireless Sensor Networks. In Proceedings of the 2004 IEEE International Symposium on Circuits and Systems (IEEE Cat. No.04CH37512), Vancouver, BC, Canada, 23–26 May 2004; pp. 409–412. [CrossRef]

© 2019 by the authors. Licensee MDPI, Basel, Switzerland. This article is an open access article distributed under the terms and conditions of the Creative Commons Attribution (CC BY) license (http://creativecommons.org/licenses/by/4.0/).

electronics

MDPI

Article

A Hierarchical Vision-Based UAV Localization for an Open Landing

Haiwen Yuan [1,2,*], Changshi Xiao [1,3,4,*], Supu Xiu [1], Wenqiang Zhan [1], Zhenyi Ye [2], Fan Zhang [1,3,4], Chunhui Zhou [1,3,4], Yuanqiao Wen [1,3,4] and Qiliang Li [2,*]

[1] School of Navigation, Wuhan University of Technology, Wuhan 430063, China; sp_xiu@whut.edu.cn (S.X.); zwq626197298@whut.edu.cn (W.Z.); michael_zf@whut.edu.cn (F.Z.); church_zhou@whut.edu.cn (C.Z.); yqwen@whut.edu.cn (Y.W.)
[2] Department of Electrical and Computer Engineering, George Mason University, Fairfax, VA 22030, USA; zye@gmu.edu
[3] Hubei Key Laboratory of Inland Shipping Technology, Wuhan University of Technology, Wuhan 430063, China
[4] National Engineering Research Center for Water Transport Safety, Wuhan University of Technology, Wuhan 430063, China
* Correspondence: hw_yuan@whut.edu.cn (H.Y.); cs_xiao@hotmail.com (C.X.); qli6@gmu.edu (Q.L.); Tel.: +86-155-2727-2422 (H.Y.); +86-136-6727-3296 (C.X.); +1-703-993-1596 (Q.L.)

Received: 11 April 2018; Accepted: 9 May 2018; Published: 11 May 2018

Abstract: The localization of unmanned aerial vehicles (UAVs) for autonomous landing is challenging because the relative positions of the landing objects are almost inaccessible and the objects have nearly no transmission with UAVs. In this paper, a hierarchical vision-based localization framework for rotor UAVs is proposed for an open landing. In such a hierarchical framework, the landing is defined into three phases: "Approaching", "Adjustment", and "Touchdown". Object features at different scales can be extracted from a designed Robust and Quick Response Landing Pattern (RQRLP) and the corresponding detection and localization methods are introduced for the three phases. Then a federated Extended Kalman Filter (EKF) structure is costumed and utilizes the solutions of the three phases as independent measurements to estimate the pose of the vehicle. The framework can be used to integrate the vision solutions and enables the estimation to be smooth and robust. In the end, several typical field experiments have been carried out to verify the proposed hierarchical vision framework. It can be seen that a wider localization range can be extended by the proposed framework while the precision is ensured.

Keywords: UAV; vision localization; hierarchical; landing; information integration

1. Introduction

Unmanned aerial vehicles (UAVs) are popular among civil and military situations that are hazardous to human operators. Automated localization is therefore highly desirable while the vehicles are required to land on stationary or moving platforms. Therefore, a real-time relative localization is desirable, which refers to the ability to localize themselves relying on onboard sensors, such as Global Positioning System (GPS), Inertial Measurement Unit (IMU), vision, lidar, etc. Currently the GPS, IMU or their combination is the most common method used to determine the pose of a UAV. However, these require the transmission of information between the air vehicle and the landing platform. The use of vision sensors for localization has many advantages. As one low-cost sensor vision is mostly passive and does not rely on an external signal. It is worth noting that vision can have millimeter-level accuracy and can determine not only the distance but also the relative orientation between two objects. This paper describes a vision-based localization framework and the key enabling technologies for an open landing.

In recent years, there has been a wealth of research and various vision-based methods available for UAV landing. These include both feature-based methods and direct methods. Some of the approaches require prior knowledge of the targets and others extract information from the surroundings in real time. These vision-based methods work with a good localization precision but are limited by the detection range. Especially, UAVs can hardly extract constant pose features from landing objects as the relative distance increases or decreases. To solve this problem, a hierarchical vision-based localization strategy is designed to extract reliable visual features at different scales in this paper.

It is noted that open landing refers to a complete decline process from a high altitude to touchdown, which normally requires a wide localization range. For this purpose, this paper describes a hierarchical vision-based UAV localization demonstration in which the pose (position and orientation) can be estimated by using the onboard camera. A Robust and Quick Response Landing Pattern (RQRLP) is designed for the hierarchical vision detection. The RQRLP is able to provide various scaled visual features for UAV localization. In detail, for an open landing, three phases—"Approaching", "Adjustment", and "Touchdown"—are defined in the hierarchical framework. First, in the "Approaching" phase the UAV is relatively far from the vessel and the contour of the RQRLP is detected and used as the main visual feature. Second, as the UAV approaches the landing object, detailed location markers can be extracted from the RQRLP. This phase is called "Adjustment" and the aerial vehicle can calculate the current pose with respect to the RQRLP with a high precision and adjusts its pose for the touchdown. In the final "Touchdown" phase, the UAV is so close to the vessel that the location markers are almost out of the field of view (FOV). As one alternative solution, an optic-flow based tracker is employed to calculate the current pose by tracking the previous one until the touchdown. To obtain a robust localization estimation, the three phases work in parallel as nodes. A federated filter based on the Extended Kalman Filter (EKF) is costumed to integrate these vision solutions. In the end, the proposed framework is tested and verified by several field experiments and results, which illustrate its performance.

The remainder of the paper is organized as follows. In Section 2, some related work is introduced. In Section 3, the design of the RQRLP is described as the landing object in the hierarchical vision localization framework and provides visual information for the UAV pose calculation. Section 4 introduces a hierarchical vision-based localization framework, which enables the three phases and integrates the UAV pose. Section 5 presents the experiments and results to verify and illustrate our proposed hierarchical vision-based framework. Finally, the conclusions are presented in Section 6.

2. Previous Work

Currently, vision-based localization is one of the most-adopted ways to actively study UAV autonomous landing. In general, for a spot landing, such as landing on a moving vehicle, UAVs with onboard cameras are able to calculate the pose by recognizing a referenced object [1]. In these related works, it is assumed that the image pattern and size of the referenced object are known in advance. The relative localization can be acquired by analyzing the projection image. For example, depending on the inertia moments of the image, the landing object could be distinguished from the background [2]. The UAV orientation is calculated by matching real-time images with a stored dataset of labeled images that have been calibrated offline. Due to image blurring, the cooperative feature points cannot be accurately extracted from the images. As one solution, a special pattern consisting of several concentric white rings on a black background was designed as a landing object [3]. Each of the white rings is recognized by a unique ratio of its inner to outer border radius. However, only the height with respect to the landing platform is provided by this method. Based on the feature lines of the cooperative object, a pose estimation algorithm was reported in [4]. In the algorithm, feature lines and vanishing lines were extracted to reduce the influence of image blurring. An initial 5 Degree-Of-Freedom (DOF) pose with respect to a landing pad was obtained by calculating the quadratic equation of the projected ellipse [5]. The IMU data was integrated to eliminate the remaining geometric ambiguity. The remaining one DOF of the camera pose, the yaw angle, was calculated by fitting an ellipse to the projected contour of the letter "H". The homography between the image frame and the object reference plane was also used to estimate the UAV initial pose [6]. With four correspondences between the world plane and the image plane, a minimal solution of the homography was estimated. A similar work was also reported in [7], where the homography between current and previous frames was decomposed and accumulated for ego-motion estimation. Moreover, the relative pose between current and previous frames could be estimated by tracking a structured-unknown object [8–10]. Similar to the dead reckoning of an inertial navigation system (INS), these type methods would suffer from signal drift as time elapsed.

In another case, it is assumed that the reference object for UAV landing and localization is unknown. Optical flow is the typical method and it is used to track or stabilize the UAV pose [11–16]. A biological guidance system was reported in [17], where some cues from the natural environment were detected and analyzed, such as the horizon profile and sky compass. An image Coordinates Extrapolation (ICE) algorithm [18] calculated the pixel-wise difference between the current view (panoramic image) and a snapshot taken at a reference location to estimate the real-time UAV 3D position and velocity. An optic flow-based vision system is reported in [19,20], where the optic flow was calculated and used for autonomous localization and scene mapping. Relevant control strategies using vision information are also discussed in detail. The combination of the vision and IMU data reported in [21,22] assumed that the IMU had the ability to provide a good roll and pitch attitude estimation, and four infrared spots on the target or the landing spot could be detected using the vision system. In addition, stereo vision using triangulation has been applied during a UAV autonomous landing [23,24].

Following the works described above, it is therefore expected that feature detection and recognition would be a key issue with regard to localization precision. In addition, for an open landing, such as in the wilderness or maritime environment, the UAV must have the ability to process the detection problem in a wide and consecutive working range. For this purpose, a hierarchical detection and localization framework is proposed and studied to detect and extract various scale features from the landing object. In one of our preliminary works, a UAV autonomous visual navigation system was reported in [25].

3. Feature Recognition and Pose Recovery

The UAV localization for automatic landing is a complex but solvable problem that can be achieved by the means of vision. In this section, the RQRLP as a reference object is designed for UAV vision and consists of a set of friendly artificial location markers, shown in Figure 1. By detecting and recognizing the RQRLP, the UAV can estimate its pose at different heights. The corresponding detection and pose recovery algorithm, based on homography decomposition, is also introduced in this section.

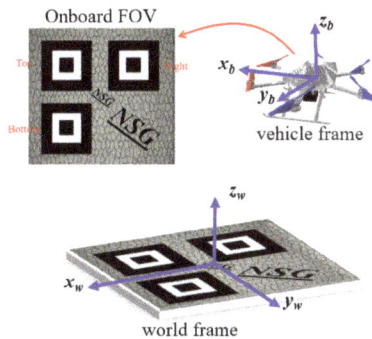

Figure 1. The landing object: Robust and Quick Response Landing Pattern (RQRLP).

3.1. The RQRLP as Landing Object

A vision-based localization is any one that makes use of visual information. The visual information can be used for navigation, vehicle stabilization, vehicle guidance, obstacle avoidance, or target tracking.

The visual feature at several scales can be provided by the designed landing object RQRLP. A series of structured and non-structured graphs are set in the RQRLP. The structured graphs comprised of several nested rectangles are used to provide the scale information for pose calculation, while the non-structured ones are good feature points for pose tracking. Considering the QR (Quick Response) code popularly applied in the field of current information recognition, each set of nested rectangles is regarded as the location markers Top, Right, and Bottom, respectively. These location markers can be detected and recognized robustly by contour extraction and statistics. First, since each location marker has a constant contour number, they can be extracted from the background using contour detection and statistics. Second, the "Top" marker is distinguished by calculating the straight-line distances between any two markers and is the one that is not on the longest line. Third, the "Right" and "Bottom" location markers are also recognized by calculating the slope of the longest line and the distance from the "Top" marker to the longest line. So far, these location markers are recognized uniquely. Assuming that the size of these markers is known, enough corners of the markers can be obtained as the corresponding information between the RQRLP and its image plane. The corresponding points are used for recovering the 6-DOF pose of the UAV. Except for the structured markers, random texture is designed as the background and consists of rich traceable feature points. The use of such a RQRLP mode can reduce algorithm complexity and run-time, and allows relative poses to be measured when the onboard camera system has been correctly calibrated.

3.2. Pose Recovery Based on Image Homography

The 6-DOF pose of a UAV (position and orientation) can be recovered by homography decomposition. Here, homography is a non-singular 3×3 matrix H that defines the projection between the RQRLP and its image plane, and can be calculated using the acquired corresponding points. Assuming that the 3D coordinate system is built on the RQRLP plane, the Z-axis of all the extracted points are zeros. As a result, the 3D coordinates of all points on the RQRLP are defined to be $\begin{bmatrix} X_i & Y_i & 0 \end{bmatrix}^T$. And the corresponding image points are $\begin{bmatrix} u_i & v_i \end{bmatrix}^T$, the homography relation can be described as follows,

$$\begin{bmatrix} u_i \\ v_i \\ 1 \end{bmatrix} = H \begin{bmatrix} X_i \\ Y_i \\ 1 \end{bmatrix}, \text{ with } H = K_{3\times3} \begin{bmatrix} r_1 & r_2 & t \end{bmatrix} \tag{1}$$

Using the extracted corresponding points, one rough solution about the matrix H can be obtained by Singular Value Decomposition (SVD) [26] or Gaussian Eliminate (GE) [27]. Then, using the Random Sample Consensus (RANSAC) method, the matrix H can be optimized to remove the errors from the mismatched points. The goal is achieved after iteratively selecting a random subset of the original data points by testing it to obtain the model and evaluating the model consensus, which is the total number of original data points that best fit the model.

As shown in Figure 2, the matrix H can be decomposed to require the onboard camera pose with respect to the RQRLP, since the homography contains the information of the camera intrinsic and extrinsic parameters. As shown in Equation (1), assuming that the camera parameter matrix $K_{3\times3}$ is known, the 3×3 rotation matrix R and the 3×1 translation vector t are involved in the remaining part and can be calculated based on the camera projection model [26],

$$\begin{cases} r_1 = \lambda K^{-1} h_1 \\ r_2 = \lambda K^{-1} h_2 \\ t = \lambda K^{-1} h_3 \end{cases}, \text{ with } \lambda = \frac{1}{\|K^{-1} h_1\|} = \frac{1}{\|K^{-1} h_2\|} \tag{2}$$

where the 3×1 vector h_i is the i-th column of H and the 3×1 vector r_i is the i-th column of R. Since all the columns of the rotation matrix are orthonormal to each other, r_3 can be determined from $r_1 \times r_2$. However, the data noise causes the resulting matrix to not satisfy the orthonormality condition, and SVD is used to form a new optimal rotation matrix that is fully orthonormal.

With this, $-R^{-1}t$, R^{-1} represents the position and the orientation of the onboard camera in the 3D coordinate system of the RQRLP. As a result, the UAV's pose can also be determined since the camera is fixed on the body.

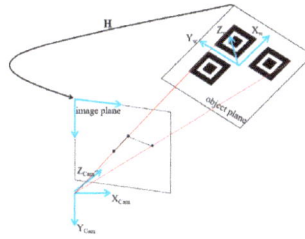

Figure 2. Vision based 3D coordinate systems and the object-to-image homography.

4. A Hierarchical Vision-Based Localization Framework

Except for the algorithm of object detection and pose recovery, as the limited image resolution and the fixed focal-length the employed vision system would have an effect on localization precision specially when the UAV is at different height. To solve the problem, a hierarchical vision-based localization framework is proposed, which can extract different scaled features for the corresponding detection phases, as shown in Figure 3. In this section, how to achieve a vision solution for the three phases and how to achieve the UAV pose by integrating the pose solutions is described.

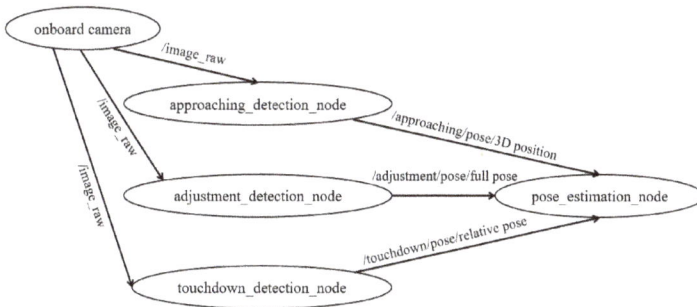

Figure 3. The hierarchical vision-based localization framework.

4.1. Hierarchical Localization

Considering an open landing, there are three phases—"Approaching", "Adjustment", and "Touchdown"—defined in the hierarchical framework. A different vision solution is employed in each phase.

At the beginning, the UAV is remote from the landing object RQRLP so that it cannot see the details of the RQRLP clearly. At this point, the outline and dimensions of the RQRLP are detected by the onboard vision system. Hence, a simple matching-based method is used to find the landing object from the scene. By tracking the four corners from the rectangular landing object, it is possible to calculate the relative pose of the UAV. But since the localization is rough, only the 3D position remains for guiding the vehicle towards the landing. By detecting and tracking the visual information, it is thought that the UAV can approach the RQRLP and more vision details can be acquired. This phase is therefore called "Approaching". In detail, the image dimensions and coordinates of the RQRLP are used to provide the relative 3D position for the UAV movement. It is noted that the image projection of an object implies the relative distance between current view and the object when the camera parameters are fixed.

For the next "Adjustment" phase of the landing, it is assumed that the flying vehicle is sufficiently close to the RQRLP so that the detail of the RQRLP can be detected as the visual information for localization. Using image corners from the location markers of the RQRLP, the relative position and orientation of the UAV can be calculated exactly by the pose recovery method presented in the previous section. As a result, a 6-DOF pose of the UAV can be acquired in the "Adjustment" phase. The obtained real-time pose is used to adjust the UAV to an appropriate state for landing. In particular, the movement of a landing object (surface or ground vehicle) can be also observed when the UAV is hovering over the object.

When the UAV is near the end of a landing, the view of the onboard vision system is limited and the image of the RQRLP can only be captured in part. This phase is called "Touchdown" in our work. Either one of the two visual features for the last two phases is out of work in this phase, and an optical flow-based pose tracker is designed to infer the current pose by calculating the optic-flow between current and previous image frames. Rich textures distributed in the RQRLP can provide vast traceable feature points. In detail, these points from the planer RQRLP are matched successfully by a nearest neighbor method and the matched point-pairs are used to calculate the homography H_i^{i+1} between current frame $i + 1$ and last frame i. Then, the UAV 6-DOF pose at the current time can be obtained by Equation (3). Such a pose tracking method is feasible as certain good corners in the RQRLP enable convenient tracking and the process is sufficiently short that accumulative error is negligible.

$$H_w^{i+1} = H_w^0 H_0^1 H_1^2 \cdots H_i^{i+1} \tag{3}$$

The visual information at different scales are detected in three defined landing phases for UAV localization with respect to the RQRLP. All of the vision-based feature detection and pose calculations constitute a hierarchical localization framework. The framework is practical and can guarantee a consecutive pose solution for a UAV relative landing, such as landing on a maritime vehicle.

4.2. Pose Integration

It should be noted that vision-based solutions for the three landing phases are not strictly separated and no less than one solution can be acquired during overlapping. To obtain an optimal localization by integrating these solutions, a federated filter that involves only three local filters is customized. The federated filter enables the final estimated localization to be consecutive and smooth.

The total structure of the customized federated filter is described in Figure 4. Three local filters are customized for the three vision solutions, Z_1, Z_2 and Z_3, respectively. Each local filter is a typical extended Kalman filter that involves prediction and update modules, and takes the localization solution from the vision nodes as the measurement input. The integration part is in charge of calculating the optimal pose solution and the allocation coefficient β_i. Moreover, an Inertial Measurement Unit (IMU) is used as the reference system of the federated filter and provides real-time angular velocities and accelerations for the UAV state prediction. In detail, the estimated state of each local filter X_i is a 7-dimensional vector, which involves the UAV position and orientation, as in Equation (4). In such a local filter, the state X_i is predicted by the IMU, and then is updated by the visual measurement. As is known from the above, the measurements Z_i ($i = 1, 2, 3$) from the three phases are absolute or relative pose, as shown in Equation (5),

$$X_i = [x\ y\ z\ q_1\ q_2\ q_3\ q_4]^T,\ i = 1,2,3 \tag{4}$$

$$Z_1 = Z_2 = [x\ y\ z\ q_1\ q_2\ q_3\ q_4]^T;\ Z_3 = [\Delta x\ \Delta y\ \Delta z\ \Delta q_1\ \Delta q_2\ \Delta q_3\ \Delta q_4]^T \tag{5}$$

Acquired from these local filters, these estimated state $\hat{X}_i$ with the corresponding covariance P_i that is a 7×7 matrix, are passed to the integration module. It is noted that the covariance P_i can imply the performance of the filter i, which means the current detection or measurement precision for the vision node i can be reflected by P_i. By summing all available $\hat{X}_i$ weighted with the corresponding covariances P_i from the local filters, the global process noise $\hat{Q}_g$, state variance $\hat{P}_g$ and state $\hat{X}_g$ are calculated as Equations (6)–(8),

$$\hat{Q}_g = (\sum_{i=1}^{N} Q_i^{-1})^{-1} \tag{6}$$

$$\hat{P}_g = (\sum_{i=1}^{N} P_i^{-1})^{-1} \tag{7}$$

$$\hat{X}_g = \hat{P}_g \times \sum_{i=1}^{N} P_i^{-1} \hat{X}_i \tag{8}$$

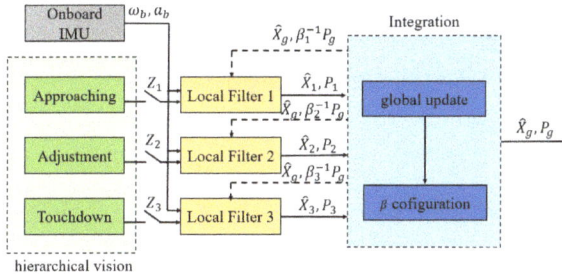

Figure 4. The customized federated filter for the hierarchical vision-based localization.

In addition, it is necessary for a federated filter to introduce a coefficient β_i ($\sum_{i=1}^{N} \beta_i = 1$), which can be used to allocate the prediction noise Q_i and the state covariance P_i for the local filters at next time. In general, β_i is fixed for a typical Carlson federated filter. However, the detection precision of the vision measurements is changing for different landing phases. Thus the allocation coefficient β_i should also be set to dynamical. As a result, β_i is defined to be related with P_i, and the calculation is shown in Equation (9). It can be thought that β_i is inversely proportional to P_i. By such a dynamical allocation, some disastrous influence could be reduced to output on other filters while one measurement of the vision localizations is out of work or unreliable. At the same time, the recovery capability for the failed filter can be also enhanced.

$$\beta_{i,k} = \frac{trace\left(P_{i,k}^{-1}\right)}{trace\left(\sum_{i=1}^{N} P_{i,k}^{-1}\right)} = \frac{trace\left(P_{i,k}^{-1}\right)}{trace\left(P_{g,k}^{-1}\right)}, \, i = 1, 2, 3 \tag{9}$$

5. Field Experiments and Results

The results of our hierarchical vision-based localization field experiments are presented in this section. In these experiments, a motion camera GOPRO4 with a resolution of 1080 p is installed to look downwards relative to the UAV. The camera is assumed to be calibrated correctly in advance, and the intrinsic parameters are known. The employed UAV is a six-rotor aircraft with an arm length of 1.6 m, which is armed by a Global Positioning System (GPS) and an IMU. The UAV rotors can have a manual or autonomous flight depending on the GPS & IMU system. A XSENS product (MTi-G-700) [28] is used as the IMU module, which can output high-precision angular and accelerated velocities at 100~400 Hz and has a low latency (<2 ms). The details of the employed UAV and onboard sensors are shown in

Table 1. In addition, a base station is set on the ground and Differential GPS is employed to provide a centimeter level accuracy for UAV position. Figure 5 shows the employed aircraft with onboard sensors and the landing object. All calculations are programed as nodes and the flight data is recorded from onboard sensors and considered as the ground truth for comparisons. The designed RQRLP is placed on the ground and the relative height is approximately zero m. The experiments begin when the landing object can be detected by the flying vehicle. The position and size of the three location markers in the RQRLP object are assumed to be known in advance.

Figure 5. Left: Landing object RQRLP and the employed six-rotors aircraft system. Right: Onboard camera and the Inertial Measurement Unit (IMU) module.

Table 1. Specification of the devices in the experiment.

Device	Specification
UAV frame	six rotors, arm length 1.6 m, total weight 3.2 kg
GOPRO camera	1080 p, 50 fps, focal length 1280 pixels, principal point: (948.9, 543.2), distortion factor $(-0.00908, -0.03128, 0.00109, -0.00198, 0)$
Xsens IMU	100~400 Hz, Gyroscope: full range 450 deg/s, noise density 0.01 deg/s/$\sqrt{\text{Hz}}$; Acceleration: full range 50 m/s2, noise density 80 µg/$\sqrt{\text{Hz}}$.

5.1. RQRLP-Based Localization

In the first experiment, the UAV is required to perform a series of typical movements, such as forward, backward, left, right, up and down, and several 360° spins. These movements involve all possibilities of a general UAV flight, and could be recovered by the onboard vision. The localization result and the ground truth from the onboard inertial sensors are shown with time in Figure 6. The corresponding errors have been also calculated: there is a small error with a RMSE (Root Mean Square Error) of 0.0239 m in the 3D position, while the RMSE in the orientation is 0.0818 rad. The results show good performance for the proposed vision-based pose recovery method with our designed RQRLP.

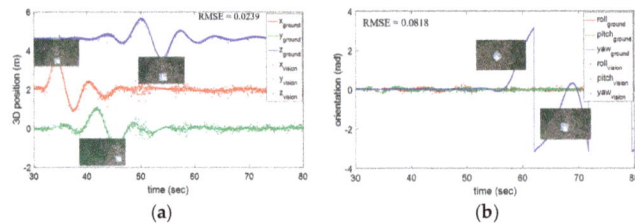

Figure 6. Vision-based pose recovery by the RQRLP landing object. (**a**) 3D position with respect to the RQRLP, in meters (m); (**b**) Orientation angles involving the roll, pitch and yaw, in radians (rad).

5.2. Hierarchical Localization for an Open Landing

To test the presented hierarchical vision-based framework in the previous section, another flight experiment has been carried out. The employed aircraft starts to decline at a of height 20 m. At the beginning, the object RQRLP is so small in the field-of-view of the UAV that the detailed detection for the RQRLP is inaccurate and the recovered pose based on the vision "Adjustment" node has a large error. Alternatively, a rough outline of the RQRLP could be segmented from the background, and is used in the "Approaching" node to provide the relative position and orientation information, as shown in Figure 7a. As the UAV is declining below the height of 10 m, as shown in Figure 7b, the location markers in the RQRLP can be recognized and the calculated pose solution from the "Adjustment" node tends to stabilize gradually. In the end of the landing, the UAV is so close with the landing object that the RQRLP is almost out of the view of the onboard vision, as shown in Figure 7c. At this moment, the optic-flow tracker in the "Touchdown" node is able to calculate the relative pose continuously by tracking the feature points between image frames and to ensure the final landing pose. The corresponding detection process of these three vision nodes are also shown in Figure 7d–f, respectively. The localization result from the hierarchical vision framework and the independent solutions from the three vision nodes have been displayed in Figure 8. To enable a smooth visual localization, all available solutions are used as the measurements of the federated EKF framework and contribute to the final estimation. As a result, an optimal estimation could be acquired by the proposed hierarchical vision-based framework. In addition, the optimal estimation is compared with the 3D flight trajectory based on DGPS (Differential Global Positioning System) in Figure 9.

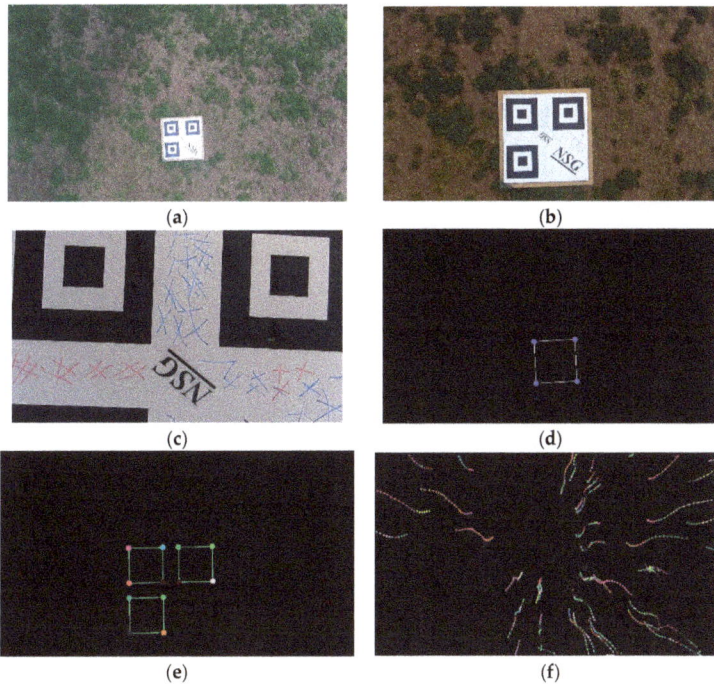

Figure 7. Feature extraction in different landing phases by the hierarchical vision-based framework. (**a–c**) Real images captured in the "Approaching", "Adjustment", and "Touchdown" phases, respectively; (**d–f**) the corresponding extracted features.

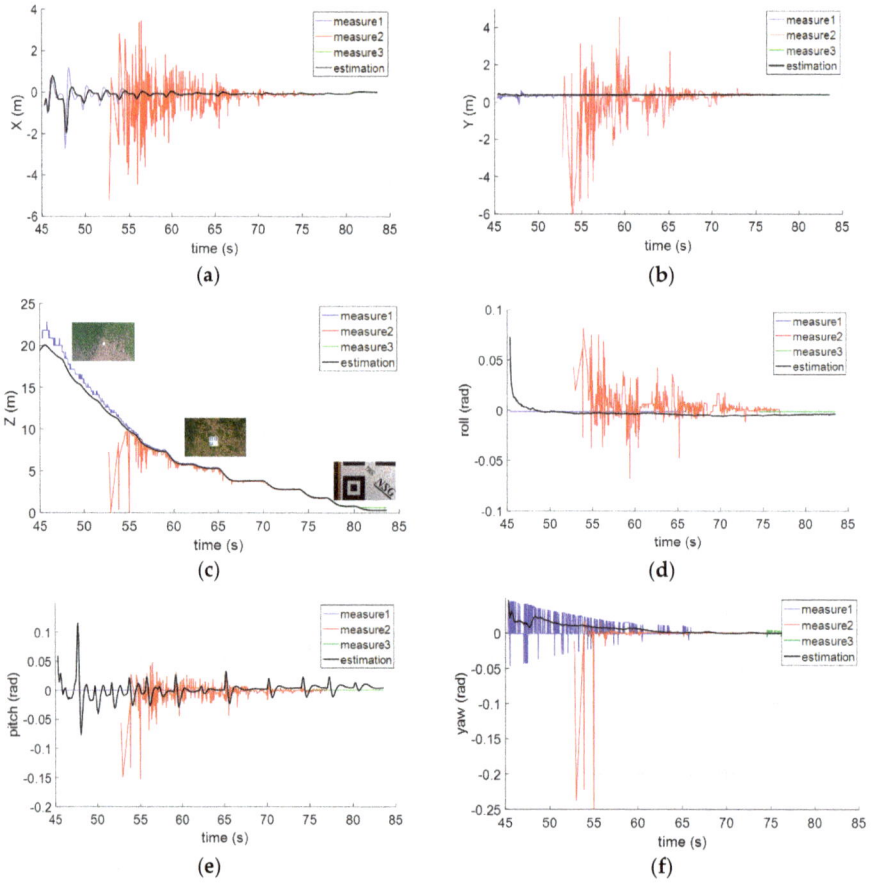

Figure 8. The hierarchical vision-based localizations and estimation for an open landing. (**a–c**) 3D position with respect to the RQRLP in meters (m); (**d–f**) orientation angles involving the roll, pitch, and yaw in radians (rad).

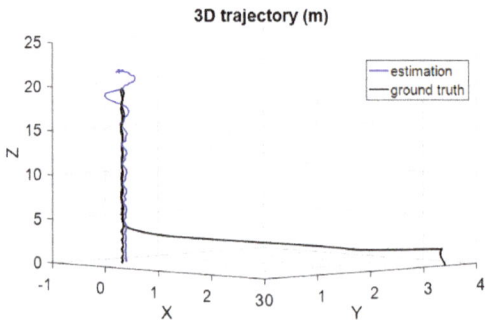

Figure 9. 3D trajectory estimation using the hierarchical vision framework and ground truth (DGPS).

5.3. Performance Analysis and Comparison

It can be noted that detection or localization precision would be affected when the object feature is almost out of view. During such a landing process, from a height of 20 m, the onboard camera kept detecting the landing object RQRLP to calculate the relative pose by using the three vision nodes in the proposed hierarchical framework. The height measurements with timestamp were acquired and have been shown in Figure 10a. It is found to be true that the localization is unstable or failed when the feature is too small to detect in the view. In other words, the detection or localization precision is dynamically changing as the detecting range. As shown in Figure 6b, the absolute errors on three measurements have been calculated and can illustrate this point. In our work, it is of great importance to fuse no less than two measurements in the overlaps in Figure 6a. For example, in the region of Overlap 1, the allocation parameter $\beta_i (i = 1, 2)$ is updated according to the posterior covariance P_i of each local filter , calculated in Figure 10c, and these can be used to trade off either measurement or local filter, whichever is be more credible. Depending on the real-time acquired β_i, the final height estimation is available in Figure 10d. By the comparison with the ground truth, the height RMSE of local or global estimations have been calculated to be 0.065 m, 0.0835 m, and 0.037 m, respectively. It can be observed that localization is improved by such a federated fusing strategy.

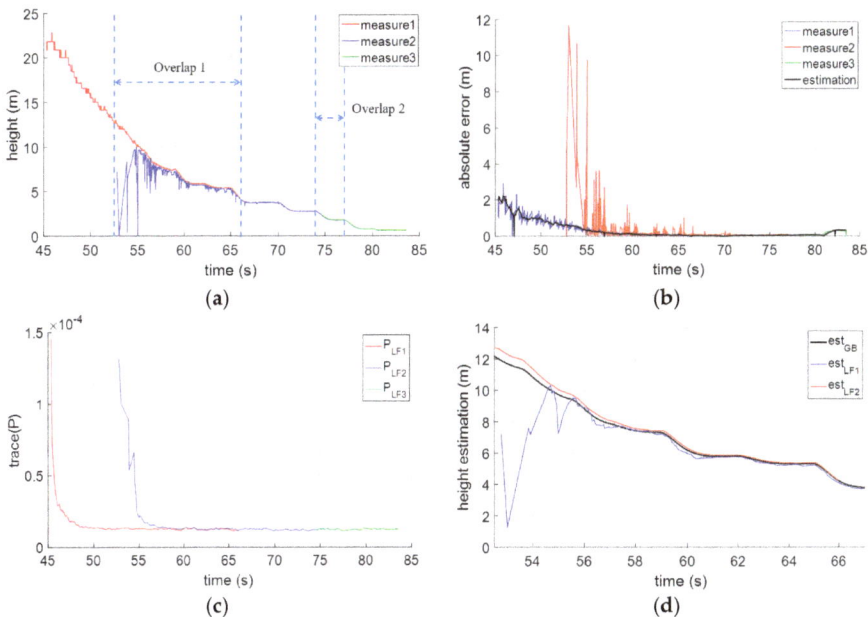

Figure 10. The landing height estimation by the hierarchical vision localization framework. (**a**) The height measurements by three vision nodes "Approach", "Adjustment" and "Touchdown"; (**b**) The absolute errors of visual measurements and estimation; (**c**) Traces of the covariance (P) of the three local filters (LF1~3); (**d**) height estimations from LF1, LF2 and GloBal update (GB).

In addition, the performance of the proposed hierarchical localization framework is illustrated by comparison with other typical methods. First, localization precision and range are considered as two main factors for such a comparison. As mentioned above, the RMSE is used to evaluate the localization precision. Range refers to relative distance, and these vision-based methods have been studied in the literature. Some methods could provide a 6-DOF or less pose for the UAV. Moreover, either the employed camera parameters or the landing object's size would affect the performance of these vision methods. Image resolution and the field-of-view are considered as camera parameters. Hence, the reference information, involving the landing object size, the full- or semi-orientation, the employed camera resolution and field-of-view, have been also collected for a scientific comparison. These detailed characteristics of four vision-based methods and our method are given in Table 2. The four methods were selected since almost all the information of interest has been provided in their work. It can be seen that the smaller or narrower the range is from the onboard camera to the landing object, the better precision is ensured. While a centimeter-level precision has been obtained by the other referenced methods, our hierarchical vision localization is able to calculate the 6-DOF pose between a larger range of 0~20 m. This means that, based on such a hierarchical framework, the UAV pose guidance would be enough and practical for an open landing.

Table 2. Comparisons between other methods from References (Ref) and our method.

Methods	Ref [6]	Ref [29]	Ref [5]	Ref [30]	Ours
Position RMSE (m)	0.2467	0.015	0.0392	<0.05	≤0.0639
Orientation RMSE (rad)	0.0653	0.0209	0.0436	<0.0872	≤0.0818
Range (m)	3~10	0.677~1.741	1	1~1.1	0~20
Pose (DOF)	4-DOF	6-DOF	6-DOF	6-DOF	6-DOF
Object size (m)	0.91 × 1.19	0.01 (diameter)	0.18 (diameter)	unknown	0.85 × 0.85
Vision Resolution (pixels)	640 × 480	752 × 480	640 × 480	320 × 240	1920 × 1080
field of view (deg)	unknown	90	90	unknown	90

6. Conclusions

In this paper, a hierarchical vision-based localization framework has been presented for UAV landing. In the hierarchical framework, the landing was defined in three phases: "Approaching", "Adjustment", and "Touchdown". For the three phases, object features at different scales are able to be extracted for UAV pose recovery. And the landing object RQRLP has also been designed for hierarchical vision detection. These original localizations are then integrated as independent measurements into a costumed federated EKF framework, which enables the final localization to be smooth and robust. Several typical experiments have been performed in the field and the results were analyzed to display the performance of our hierarchical vision localization framework. Such a hierarchical vision-based framework with a centimeter-level precision has a longer localization range and is thought to be significant for an open landing.

Author Contributions: H.Y. and C.X. conceived and designed the experiments; H.Y., W.Z. and S.X. performed the experiments; Y.W., C.Z. and F.Z. contributed the quadrotor platform and the experimental materials; H.Y. and Z.Y. analyzed the data; H.Y. and Q.L. wrote the paper.

Acknowledgments: This work reported in this paper is the product of several research stages at George Mason University and Wuhan University of Technology and has been sponsored in part by the Natural Science Foundations of China (51579204 and 51679180), Double First-rate Project of WUT (472-20163042). Qiliang Li would like to acknowledge the support of the Virginia Microelectronics Consortium (VMEC) research grant.

Conflicts of Interest: The authors declare no conflict of interest.

References

1. Kendoul, F. A survey of advances in guidance, navigation and control of unmanned rotorcraft systems. *J. Field Robot.* **2012**, *29*, 315–378. [CrossRef]
2. Saripalli, S.; Montgomery, J.F.; Sukhatme, G.S. Visually Guided Landing of an Unmanned Aerial Vehicle. *IEEE Trans. Robot Auton.* **2003**, *19*, 371–380. [CrossRef]
3. Lange, S.; Sunderhauf, N.; Protzel, P. A Vision Based Onboard Approach for Landing and Position Control of an Autonomous Multirotor UAV in GPS-Denied Environments. In Proceedings of the International Conference on Advanced Robotics (ICAR), Munich, Germany, 22–26 June 2009; pp. 22–26.
4. Xu, G.; Zeng, X.; Tian, Q.; Guo, Y.; Wang, R.; Wang, B. Use of Land's Cooperative Object to Estimate UAV's Pose for Autonomous Landing. *Chin. J. Aeronaut.* **2013**, *26*, 1498–1505. [CrossRef]

5. Yang, S.; Scherer, S.A.; Schauwecker, K.; Zell, A. Autonomous Landing of MAVs on an Arbitrarily Textured Landing Site Using Onboard Monocular Vision. *J. Intell. Robot. Syst.* **2014**, *74*, 27–43. [CrossRef]

6. Mondragon, I.F.; Campoy, P.; Martinez, C.; Olivares-Méndez, M.A. 3D pose estimation based on planar object tracking for UAVs control. In Proceedings of the 2010 IEEE International Conference on Robotics and Automation, Anchorage, AK, USA, 3–7 May 2010; pp. 35–41.

7. Martinez, C.; Mondragon, I.; Olivares-Mendez, M.A.; Campoy, P. On-board and Ground Visual Pose Estimation Techniques for UAV Control. *J. Intell. Robot. Syst.* **2011**, *61*, 301–320. [CrossRef]

8. Brockers, R.; Bouffard, P.; Ma, J.; Matthies, L.; Tomlin, C. Autonomous landing and ingress of micro-air-vehicles in urban environments based on monocular vision. In Proceedings of the Micro- and Nanotechnology Sensors, Systems, and Applications III, Orlando, FL, USA, 25–29 April 2011; p. 803111.

9. Sanchez-Lopez, J.L.; Pestana, J.; Saripalli, S.; Campoy, P. An Approach Toward Visual Autonomous Ship Board Landing of a VTOL UAV. *J. Intell. Robot. Syst.* **2014**, *74*, 113–127. [CrossRef]

10. Lin, S.; Garratt, M.A.; Lambert, A.J. Monocular vision-based real-time target recognition and tracking for autonomously landing an UAV in a cluttered shipboard environment. *Auton. Robots* **2016**, *41*, 881–901. [CrossRef]

11. Li, A.Q.; Coskun, A.; Doherty, S.M.; Ghasemlou, S.; Jagtap, A.S.; Modasshir, M.; Rahman, S.; Singh, A.; Xanthidis, M.; O'Kane, J.M.; et al. Experimental Comparison of open source Vision based State Estimation Algorithms. *Int. Symp. Exp. Robot.* **2016**, 775–786. [CrossRef]

12. Srinivasan, M.V. Honeybees as a model for the study of visually guided flight navigation and biologically inspired robotics. *Physiol. Rev.* **2011**, *91*, 413–460. [CrossRef] [PubMed]

13. Chahl, J.S.; Srinivasan, M.V.; Zhang, S.W. Landing strategies in honey bees and applications to uninhabited airborne vehicles. *Int. J. Robot. Res.* **2004**, *23*, 101–110. [CrossRef]

14. Strydom, R.; Thurrowgood, S.; Srinivasan, M.V. Visual Odometry: Autonomous UAV Navigation using Optic Flow and Stereo. In Proceedings of the Australasian Conference on Robotics and Automation, Melbourne, Australia, 2–4 December 2014.

15. Shen, S.; Mulgaonkar, Y.; Michael, N.; Kumar, V. Vision-based state estimation for autonomous rotorcraft mavs in complex environments. In Proceedings of the 2013 IEEE International Conference on Robotics and Automation, Karlsruhe, Germany, 6–10 May 2013; pp. 1758–1764.

16. Forster, C.; Pizzoli, M.; Scaramuzza, D. SVO: Fast semi-direct monocular visual odometry. In Proceedings of the 2014 IEEE International Conference on Robotics and Automation (ICRA), Hong Kong, China, 31 May–7 June 2014; pp. 15–22.

17. Thurrowgood, S.; Moore, R.J.D.; Soccol, D.; Knight, M.; Srinivasan, M.V. A Biologically Inspired, Vision-based Guidance System for Automatic Landing of a Fixed-wing Aircraft. *J. Field Robot.* **2014**, *31*, 699–727. [CrossRef]

18. Denuelle, A.; Thurrowgood, S.; Strydom, R.; Kendoul, F.; Srinivasan, M.V. Biologically-inspired visual stabilization of a rotorcraft UAV in unknown outdoor environments. In Proceedings of the 2015 International Conference on Unmanned Aircraft Systems (ICUAS), Denver, CO, USA, 9–12 June 2015; pp. 1084–1093.

19. Kendoul, F.; Fantoni, I.; Nonami, K. Optic flow-based vision system for autonomous 3D localization and control of small aerial vehicles. *Robot. Auton. Syst.* **2009**, *57*, 591–602. [CrossRef]

20. Herisse, B.; Hamel, T.; Mahony, R.; Russotto, F.X. Landing a VTOL Unmanned Aerial Vehicle on a Moving Platform Using Optical Flow. *IEEE Trans. Robot.* **2012**, *28*, 77–89. [CrossRef]

21. Wenzel, K.E.; Rosset, P.; Zell, A. Low-Cost Visual Tracking of a Landing Place and Hovering Flight Control with a Microcontroller. *J. Intell. Robot. Syst.* **2010**, *57*, 297–311. [CrossRef]

22. Li, P.; Garratt, M.; Lambert, A. Monocular Snapshot-based Sensing and Control of Hover Takeoff and Landing for a Low-cost Quadrotor: Monocular Snapshot-based Sensing and Control. *J. Field Robot.* **2015**, *32*, 984–1003. [CrossRef]

23. Kong, W.; Hu, T.; Zhang, D.; Shen, L.; Zhang, J. Localization Framework for Real-Time UAV Autonomous Landing: An On-Ground Deployed Visual Approach. *Sensor* **2017**, *17*, 1437. [CrossRef] [PubMed]

24. Ma, Z.; Hu, T.; Shen, L. Stereo vision guiding for the autonomous landing of fixed-wing UAVs: A saliency-inspired approach. *Int. J. Adv. Robot. Syst.* **2016**, *13*, 43. [CrossRef]

25. Yuan, H.; Xiao, C.; Xiu, S.; Wen, Y.; Zhou, C.; Li, Q. A new combined vision technique for micro aerial vehicle pose estimation. *Robotics* **2017**, *6*, 6. [CrossRef]

26. Zhang, Z. A flexible new technique for camera calibration. *IEEE Trans. Pattern Anal. Mach. Intell.* **2000**, *22*, 1330–1334. [CrossRef]

27. Bazargani, H.; Bilaniuk, O.; Laganière, R. A fast and robust homography scheme for real-time planar target detection. *J. Real-Time Image Proc.* **2015**, 1–20. [CrossRef]

28. XSENS MTi-G-700. Available online: https://www.xsens.com/products/mti-g-700/ (accessed on 30 March 2018).

29. Breitenmoser, A.; Kneip, L.; Siegwart, R. A Monocular Vision based System for 6D Relative Robot Localization. In Proceedings of the IEEE/RSJ International Conference on Intelligent Robots and Systems (IROS), San Francisco, CA, USA, 25–30 September 2011.

30. Sharp, C.S.; Shakernia, O.; Sastry, S.S. A vision system for landing an unmanned aerial vehicle. In Proceedings of the IEEE International Conference on Robotics and Automation (Cat. No.01CH37164), Seoul, Korea, 21–26 May 2001; pp. 1720–1727.

© 2018 by the authors. Licensee MDPI, Basel, Switzerland. This article is an open access article distributed under the terms and conditions of the Creative Commons Attribution (CC BY) license (http://creativecommons.org/licenses/by/4.0/).

MDPI

St. Alban-Anlage 66

4052 Basel

Switzerland

Tel. +41 61 683 77 34

Fax +41 61 302 89 18

www.mdpi.com

Electronics Editorial Office

E-mail: electronics@mdpi.com

www.mdpi.com/journal/electronics

www.ingramcontent.com/pod-product-compliance
Lightning Source LLC
Chambersburg PA
CBHW051730210326
41597CB00032B/5667